ATOMISTIC SPIN DYNAMICS

Atomistic Spin Dynamics

Foundations and Applications

Olle Eriksson

Department of Physics and Astronomy, Uppsala University, Box 516, SE-751 20 Uppsala, Sweden

Anders Bergman

Department of Physics and Astronomy, Uppsala University, Box 516, SE-751 20 Uppsala, Sweden

Lars Bergqvist

Department of Materials and Nano Physics, KTH Royal Institute of Technology, Electrum 229, SE-164 40 Kista, Sweden

Johan Hellsvik

Department of Materials and Nano Physics, KTH Royal Institute of Technology, Electrum 229, SE-164 40 Kista, Sweden

OXFORD
UNIVERSITY PRESS

OXFORD

UNIVERSITY PRESS

Great Clarendon Street, Oxford, OX2 6DP,
United Kingdom

Oxford University Press is a department of the University of Oxford.
It furthers the University's objective of excellence in research, scholarship,
and education by publishing worldwide. Oxford is a registered trade mark of
Oxford University Press in the UK and in certain other countries

First Edition published in 2017
Impression: 1

Published in the United States of America by Oxford University Press
198 Madison Avenue, New York, NY 10016, United States of America

British Library Cataloguing in Publication Data
Data available

Library of Congress Control Number: 2016938429

ISBN 978–0–19–878866–9

Printed and bound by
CPI Group (UK) Ltd, Croydon, CR0 4YY

Preface

The purpose of this book is to provide a theoretical foundation and an understanding of atomistic spin dynamics and to give examples of where the atomistic Landau–Lifshitz–Gilbert equation can and should be used. The choice of an atomistic description of magnetization dynamics, instead of using a continuous vector field, as is done in micromagnetic simulations, has several advantages. First, as argued in this book, the building block of materials is the atom, and hence a description of magnetism in an atomistic way is very natural and allows for an interpretation of experimental results in a deeper and clearer way. This description also allows for calculations, from first principles, of all parameters needed to perform the spin dynamics simulations, without using experimental results as input to the simulations. In addition, atomistic spin dynamics simulations experience no problems with treating anti ferromagnets or ferri magnets. The contents presented in this book involve a description of density functional theory both from a fundamental viewpoint as well as a practical one, with several examples of how this theory can be used for the evaluation of ground-state properties like spin and orbital moments, magnetic form factors, magnetic anisotropy, Heisenberg exchange parameters, and the Gilbert damping parameter. This book also outlines how interatomic exchange interactions are relevant for the effective field used in the temporal evolution of atomistic spins. The equation of motion for atomistic spin dynamics is derived, starting from the quantum mechanical equation of motion of the spin operator. It is shown that this leads to the atomistic Landau–Lifshitz–Gilbert equation, provided a Born–Oppenheimer-like approximation is made, where the motion of atomic spins is considered slower than that of the electrons. It is also described how finite temperature effects may enter the theory of atomistic spin dynamics, via Langevin dynamics. Details of the practical implementation of the resulting stochastic differential equation are provided, and several examples illustrating the accuracy and importance of this method are given. Examples are given of how atomistic spin dynamics reproduce experimental data for magnon dispersion of bulk and thin film systems, the damping parameter, the formation of skyrmionic states, all-thermal switching motion, and ultrafast magnetization measurements.

On a historical note, early and seminal steps were taken by Antropov et al. (1995, 1996) and by Stocks et al. (1998). There were also independent developments by Nowak et al. (2005) and Kazantseva et al. (2008), as reviewed by Evans et al. (2014) and Skubic et al. (2008). The results presented in this book reliy on the initial methodology, and subsequent improvements, of the UppASD method presented by Skubic et al. (2008). Among more recently developed methods for solving the atomistic spin dynamics problem, one may note the method developed by Thonig (2013). It is also noteworthy that other textbooks covering modern magnetization dynamics have been published (see e.g. Stöhr and Siegmann, 2006; Bertotti et al., 2009).

Acknowledgements

We would like to acknowledge fruitful discussions and a great collaboration with the following colleagues (in alphabetical order):

I. A. Abrikosov, B. Alling, G. Andersson, Y. Andersson, V. Antropov, P. Bessarab, S. Blügel, N. Bondarenko, A. Burlamaquie-Klautau, J. Chico, R. Chimata, M. Costa, R. de Almeida, I. di Marco, P. H. Dederichs, E. Delczeg, A. Delin, H. Ebert, A. Edström, S. Engblom, C. Etz, N. Fejes, J. Fransson, L. Genovese, A. Grechnev, O. Grånäs, J. Henk, B. Hjörvarsson, D. Iuşan, A. Jacobsson, B. Johansson, K. Kádas, O. Karis, M. I. Katsnelson, A. Kimel, A. Kirilyuk, R. Knut, K. Koumpouras, G. Kreiss, J. Kudrnovský, Y. Kvashnin, A. I. Lichtenstein, J. Lorenzana, S. Mankovsky, D. Meier, J. H. Mentink, K. Modin, P. Nordblad, L. Nordström, T. Nystrand, P. Oppeneer, F. Pan, S. Panda, M. Pereiro, S. Picozzi, M. Poluektov, Th. Rasing, D. Rodriques, J. Rusz, B. Sanyal, B. Skubic, A. Stroppa, P. Svedlindh, A. Szilva, L. Szunyogh, A. Taroni, D. Thonig, P. Thunström, I. Turek, L. Udvardi, L. Vitos, J. Werpers, J. M. Wills, D. Yudin, and J. Åkerman

The authors also acknowledge support from the Swedish Research Council (VR), eSSENCE, the Swedish e-Science Research Centre (SeRC), the Knut and Alice Wallenberg (KAW) Foundation (grants 2012.0031 and 2013.0020), and the Göran Gustafsson Foundation (GGS).

Contents

Part 1

Density Functional Theory and its Applications to Magnetism

Density functional theory (DFT) has been an invaluable tool for understanding and analysing the magnetism of materials. In the three introductory chapters forming Part 1, we review the most central and important features of this theory and give examples of magnetic properties that are accessible with this theory. We start this description with the well-established Hartree–Fock theory, to demonstrate the principal mechanism of intra-atomic spin pairing. The master equation of DFT, the so-called Kohn–Sham equation, is derived, and we show how it can be used for spin polarized calculations of spin and orbital moments, for magnetic form factors, for magnetic anisotropy, and for inter atomic (Heisenberg) exchange parameters. We also describe special considerations of this equation when applied to the solid state. We cover both translational and rotational symmetries, as well as complications that emerge when spin–orbit coupling is included in the calculations. Also outlined in Part 1 of this book is how concepts from DFT can be used for a multiscale approach to atomistic spin dynamics simulations. Hence, Part 1 shows how DFT calculations of only a few atoms can be used to enable simulations of billions of atomic spins. Throughout these three chapters, theoretical results are compared to existing experimental data, and the level of agreement between theory and observation is discussed.

1

Density Functional Theory

Density functional theory has been established as a very practical platform for model-ling, from first principles, the electronic, optical, mechanical, cohesive, magnetic, and structural properties of materials. Starting from the Schrödinger equation for the many-body system of electrons and nuclei, an effective theory has been developed allowing for material-specific and parameter-free simulations of non-magnetic as well as magnetic materials. In this chapter, an introduction will be given to density functional theory, the Hohenberg–Kohn theorems, the Kohn–Sham equation, and the formalism for how to deal with spin polarization and non-collinear magnetism.

1.1 Background of the many-electron problem

The basic problem in calculating the electronic structure and related properties of a ma-terial concerns how to deal mathematically with the interactions of a very large number of particles. To be more detailed, the Hamiltonian, in the non-relativistic case, can be written in terms of coordinates of electrons and nuclei and their kinetic energy in the following way:

$$
\hat{H} = -\frac{\hbar^2}{2} \sum_I \frac{\nabla_I^2}{M_I} + \frac{1}{2} \sum_{I \neq \mathcal{J}} \frac{1}{4\pi\epsilon_0} \frac{Z_I Z_{\mathcal{J}} e^2}{|\mathbf{R}_I - \mathbf{R}_{\mathcal{J}}|} - \frac{\hbar^2}{2m} \sum_i \nabla_i^2
$$
$$
+ \frac{1}{2} \sum_{i \neq j} \frac{1}{4\pi\epsilon_0} \frac{e^2}{|\mathbf{r}_i - \mathbf{r}_j|} - \sum_{i,I} \frac{1}{4\pi\epsilon_0} \frac{Z_I e^2}{|\mathbf{r}_i - \mathbf{R}_I|}, \tag{1.1}
$$

where the indices i, j denote electrons; $I, \mathcal{J}$ are for atomic nuclei; and the masses are denoted M_I for nuclei, and m for electrons. Furthermore, $\mathbf{R}_I$ and $\mathbf{r}_i$ stand for nucleus and electron coordinates, respectively, whereas Z_I denotes atomic number. In the following, we will adopt Hartree atomic units, that is, $e = m = \hbar = 4\pi\epsilon_0 = 1$. Since nuclei are much heavier than electrons, one may adopt the Born–Oppenheimer approximation and assume that the nuclei are fixed, while the electrons are dynamic objects. This allows, one to deal with electron states separately from the atomic nuclei. Thus, we are left 'only'

Atomistic Spin Dynamics. Olle Eriksson, Anders Bergman, Lars Bergqvist, Johan Hellsvik. First Edition.
© Olle Eriksson, Anders Bergman, Lars Bergqvist, Johan Hellsvik 2017. First published in 2017 by Oxford University Press.

with the description of the electron system, and the Hamiltonian acting on the electrons is written as

$$\hat{H} = -\frac{1}{2}\sum_i \nabla_i^2 + \frac{1}{2}\sum_{i\neq j}\frac{1}{|\mathbf{r}_i - \mathbf{r}_j|} - \sum_{i,I}\frac{Z_I}{|\mathbf{r}_i - \mathbf{R}_I|} = \hat{T} + \hat{W} + \hat{V}_{\text{ext}}. \qquad (1.2)$$

Here, $\hat{T}$ is the kinetic energy operator of the electrons, $\hat{W}$ is the operator determining the Coulomb energy of electron–electron interactions, and $\hat{V}_{\text{ext}}$ is the external potential accounting for the Coulomb interactions between the electrons and the nuclei. The corresponding total energy E is the expectation value of $\hat{H}$, that is

$$E = \langle \Psi | \hat{H} | \Psi \rangle = T + W + \int d^3r\, V_{\text{ext}}(\mathbf{r})n(\mathbf{r}), \qquad (1.3)$$

with T and W denoting the expectation values of the kinetic energy and electron–electron interaction operators, respectively, and $n(\mathbf{r})$ denoting the electron charge density.

1.2   The Hartree–Fock theory

The Hartree–Fock theory, discussed in several textbooks, (e.g. in Atkins and Friedman, 2005), assumes that electrons interact in a mean field fashion; it follows from approximating the many-electron wave function Ψ with a single Slater determinant constructed from the single-particle wave function ψ in the following way:

$$\Psi(\mathbf{x}_1, \mathbf{x}_2, \ldots, \mathbf{x}_i, \ldots, \mathbf{x}_N) = \frac{1}{\sqrt{N}}\det[\psi_i(\mathbf{x}_j)], \qquad (1.4)$$

where $\mathbf{x}$ is a composite coordinate of space and spin of an electron, N is the number of electrons, and we have used a compact expression for the Slater determinant. Forming the expectation value of the electron Hamiltonian with this approximate many-electron wave function yields an expression of the total energy. Minimization of this energy expression with respect to the single-particle wave functions that compose the Slater determinant gives the Hartree–Fock equation:

$$\left(-\frac{1}{2}\nabla_i^2 + V_{\text{ext}}(\mathbf{r}) + \sum_{i\neq j}\int d\mathbf{r}'\frac{\psi_j^*(\mathbf{x}')\psi_j(\mathbf{x}')}{|\mathbf{r}_i - \mathbf{r}_j|}\right)\psi_i(\mathbf{x})$$

$$-\sum_{i\neq j}\int d\mathbf{r}'\frac{\psi_j^*(\mathbf{x}')\psi_i(\mathbf{x}')}{|\mathbf{r}_i - \mathbf{r}_j|}\psi_j(\mathbf{x})\delta_{s_i s_j} = \epsilon_i\psi_i(\mathbf{x}), \qquad (1.5)$$

where V_{ext} represents the electron–nucleus interaction. This equation is similar to the Schrödinger equation for electrons that move in a potential and, in fact, the first two terms that follow the kinetic energy in Eqn (1.5) represent the effective potential that an electron experiences because of its attractive interaction with the nucleus and its repulsive interaction with all other electrons in the system (this interaction is referred to as the Hartree term; Atkins and Friedman, 2005). The last term on the left-hand side of Eqn (1.5) is referred to as the exchange interaction; it is active only if the spins of electrons i and j are the same, and is hence a term that lowers the energy if as many spins as possible have the same direction. This spin pairing (Jorgensen, 1962) is responsible for the fact that most atoms of the Periodic Table have a ground-state electronic configuration with a maximized number of parallel spins, provided that, the Pauli exclusion principle is obeyed. This microscopic mechanism behind Hund's first rule follows, interestingly, from a Hamiltonian that has no spin-dependent terms in it, Eqn (1.2). It is entirely quantum mechanical in nature and follows from the requirement that the wave function Ψ should be antisymmetric with respect to permutation of the coordinates of any two electrons. Note that, in Eqn (1.5), we have explicitly kept electron i from

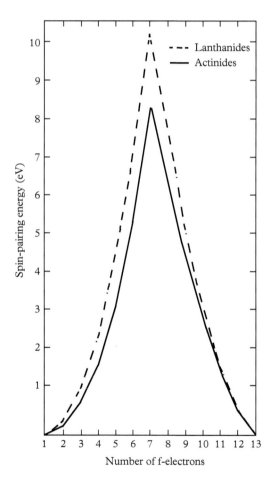

Figure 1.1 *Spin-pairing energy of actinides and lanthanides.*

interacting with itself, thus adopting a physically meaningful approach. If, however, we allow for the summations in Eqn (1.5) to also include terms where $i = j$, we would not introduce an error, since then the extra interaction in the Hartree term would be cancelled exactly by the extra interaction of the exchange, or Fock, term. This is easily seen by setting $i = j$ in the subscript of the wave functions of Eqn (1.5). This is normally described as the lack of self-interaction in the Hartree–Fock equation, and is relevant when discussing approximations to density functional theory, as will be discussed in Section 1.4. It should also be noted that the interactions in the Hartree–Fock equation are non-local, meaning that they involve interactions over distances described by $|\mathbf{r}_i - \mathbf{r}_j|$.

In order to illustrate the strength of the spin-pairing energy as defined by Jorgensen (1962), Fig. 1.1 shows the experimental values of the spin-pairing energy of lanthanides and actinides for the f-shell (Nugent, 1970). Note that the energies involved can be rather large, of the order of 10 eV. Hence, discussions of, for example, ultrafast demagnetization using intense laser pulses must take these strong intra-shell couplings into consideration. It is also clear from the figure that spin pairing is slightly larger for the lanthanide series than for the actinide series, because the $4f$ wave function is less extended in the lathanide series than in the actinide series and thus forces the electrons of the $4f$ shell to occupy a smaller volume than those in the $5f$ shell do. For this reason, the exchange energy is larger for the lanthanides than for the actinides, as may be seen from Eqn (1.5).

1.3 The Hohenberg–Kohn theorems

Using the Hartree–Fock theory is a rather time-consuming approach, especially when dealing with electronic structures of large systems and when dealing with solids. For this reason, other approaches have been explored, where maybe the most successful one is that provided by density functional theory (Hohenberg and Kohn, 1964; W. Kohn and Sham, 1965). This theory allows the use of the charge density, $n(\mathbf{r})$, and magnetization density, $m(\mathbf{r})$, of a system to be used as the key quantities that describe the ground-state properties. In this way, one avoids having to work with a many-electron wave function. Instead, one can resort to the single-electron theory of solids as is described in Sections 1.3 and 1.4. We give first, however, the foundation of density functional theory, which is based on the following two theorems.

Theorem 1.1 *The total energy of a system is a unique functional of the ground–state electron density.*

To demonstrate this, we consider the expectation value of Eqn (1.3). We evaluate the ground-state many-body wave function $\Psi_{gs}(\mathbf{r}_1, \mathbf{r}_2, \ldots, \mathbf{r}_N)$ for N electrons and their energy from the equation

$$\hat{H}\Psi_{gs} = E_{gs}\Psi_{gs}. \tag{1.6}$$

The electron density of this ground-state can be calculated from

$$n_{gs}(\mathbf{r}) = \sum_{i=1}^{N} \int d^3r_i \ | \ \Psi_{gs}(\mathbf{r}_1, \mathbf{r}_2, \ldots .\mathbf{r}_N) \ |^2 \ \delta(\mathbf{r} - \mathbf{r}_i). \tag{1.7}$$

We will now demonstrate that using two different external potentials in Eqn (1.3), for example, V_{ext} and V'_{ext}, gives rise to two different ground-state electron densities: n_{gs} and n'_{gs}, respectively. For simplicity, we consider first a non-spin polarized system, with $m = 0$. To show this, we note first that, for the system with external potential V'_{ext}, we have

$$\hat{H}' \Psi'_{gs} = E'_{gs} \Psi'_{gs}. \tag{1.8}$$

From the variational principles, it follows that

$$E_{gs} = \langle \Psi_{gs} | \hat{H} | \Psi_{gs} \rangle < \langle \Psi'_{gs} | \hat{H} | \Psi'_{gs} \rangle. \tag{1.9}$$

To estimate the energy of the rightmost term of Eqn (1.9), we add and subtract V'_{ext} to the expectation value so that

$$\langle \Psi'_{gs} | \hat{H} | \Psi'_{gs} \rangle = \langle \Psi'_{gs} | \hat{H} + V'_{ext} - V'_{ext} | \Psi'_{gs} \rangle = \langle \Psi'_{gs} | \hat{H}' + V_{ext} - V'_{ext} | \Psi'_{gs} \rangle. \tag{1.10}$$

This allows us to write

$$\langle \Psi'_{gs} | \hat{H} | \Psi'_{gs} \rangle = E'_{gs} + \int n'_{gs} (V_{ext} - V'_{ext}) \, d^3r. \tag{1.11}$$

Note that, in this expression, we have, for simplicity, omitted to write out the r-dependence of the external potential and density. Combining Eqns (1.9) and (1.11) yields the relationship

$$E_{gs} < E'_{gs} + \int n'_{gs} (V_{ext} - V'_{ext}) \, d^3r. \tag{1.12}$$

We could start all over from Eqn (1.9) and, going to Eqn (1.11), arrive at an expression similar to Eqn (1.12) but with all primed and unprimed symbols being interchanged, that is, the following expression would also follow from the variational principle:

$$E'_{gs} < E_{gs} + \int n_{gs} (V'_{ext} - V_{ext}) \, d^3r. \tag{1.13}$$

If we now assume that $n'_{gs} = n_{gs}$, an absurd relationship follows, since we can replace n'_{gs} in Eqn (1.12) with n_{gs} and then add Eqns (1.12) and (1.13) together to obtain

$$E_{gs} + E'_{gs} < E'_{gs} + E_{gs} + \int n_{gs} (V_{ext} - V'_{ext}) \, d^3r + \int n_{gs} (V'_{ext} - V_{ext}) \, d^3r, \tag{1.14}$$

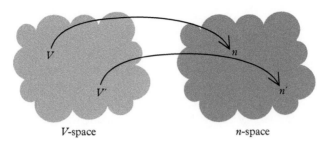

V-space   *n*-space

Figure 1.2 *Schematic figure showing the connection between external potential (V-space) and ground-state density (n-space).*

from which it would follow that $E_{gs} + E'_{gs} < E'_{gs} + E_{gs}$. This is clearly a result that is incorrect. Since deriving it follows from the rules of quantum mechanics, plus one assumption, we must conclude that this initial assumption was wrong, that is, that n'_{gs} and n_{gs} cannot be equal. Figure 1.2 shows a schematic of the relationship between external potential and electron density, following the above analysis of density functional theory. According to this figure, two different external potentials can never point to the same ground-state density; they exclusively will identify unique and separate ground-state densities. In addition, there are no points in *n*-space that cannot be reached from a point in *V*-space. Mathematically, such mappings are called bijective.

So far in this discussion, we have made the derivation following the direction of the arrows in Fig. 1.2, namely, starting from different potentials in *V*-space always results in unique positions in *n*-space. In principle, it is possible to follow these arrows backwards, starting from the ground state density and ending up in a unique spot of *V*-space, with this spot uniquely establishing the external potential where this density came from. Since the expressions for kinetic energy and electron–electron interaction are the same for any system, it is the form of the external potential which makes the Hamiltonian unique and hence specifies it. The arguments of density functional theory thus make the following claim: knowing n_{gs} implies we know which external potential was used in the Hamiltonian. This, in turn, specifies the full Hamiltonian and, from it, all states, even the excited ones. Thus we establish electron density as a property which describes a system, and we formally allow the use of electron density in place of a many-electron wave function. In particular, we are now able to express the following functional relationship between the ground-state energy of the Hamiltonian, and the ground-state electron density:

$$E[n(\mathbf{r})] = T[n(\mathbf{r})] + W[n(\mathbf{r})] + V_{ext}[n(\mathbf{r})]. \tag{1.15}$$

The three terms in Eqn (1.15) are the same as those defined in Eqn (1.3). Using similar, straightforward arguments, one can arrive at the second important theorem underlying density functional theory:

Theorem 1.2 *The exact ground-state density minimizes $E[n(r)]$ in Eqn (1.15).*

If we had an explicit form for $E[n(\mathbf{r})]$, we could go ahead and minimize it with respect to the electron density and in this way calculate the ground-state energy. The expression for $V_{ext}[n(\mathbf{r})]$ is straightforward, but those for $T[n(\mathbf{r})]$ and $W[n(\mathbf{r})]$ are more difficult. Attempts have been made at formulating such expressions, for example, Thomas–Fermi theory, but this theory comes short when compared to the topic of the next section, the Kohn–Sham approach, when results are compared to, for example, measured magnetic and cohesive properties.

1.4   The Kohn–Sham equation

The theorems described above are also valid for non-interacting systems where the part of the Hamiltonian that describes electron–electron interaction, W, is absent. In this case, electrons which move in the field of an external potential which we, for reasons that will be obvious, call V_{eff}, are solutions to a one-electron Schrödinger equation,

$$\left(-\frac{1}{2}\nabla^2 + V_{eff}(\mathbf{r})\right)\psi_i(\mathbf{r}) = \epsilon_i\psi_i(\mathbf{r}). \tag{1.16}$$

From this equation, which in the density functional theory community is also referred to as the Kohn–Sham equation, one can calculate an electron density from the occupied one-particle (op) states. Since no direct electron–electron interactions are considered when evaluating this density, it is referred to as a one-particle density:

$$n_{op}(\mathbf{r}) = \sum_{occ} |\psi_i(\mathbf{r})|^2 , \tag{1.17}$$

where the sum is over occupied states. In this case, the energy functional which describes the total energy may be written as

$$E_{op}[n_{op}(\mathbf{r})] = T_{op}[n_{op}(\mathbf{r})] + V_{eff}[n_{op}(\mathbf{r})] =$$
$$= \sum_{occ}\left\langle\psi_i(\mathbf{r})\left|-\frac{1}{2}\nabla^2\right|\psi_i(\mathbf{r})\right\rangle + \int n_{op}(\mathbf{r})V_{eff}(\mathbf{r})\,d^3r, \tag{1.18}$$

and the electron density which minimizes this functional is obtained from the requirement that the energy functional is stationary for small variations of the electron density around the ground-state density, that is $E_{op}[n_{op} + \delta n] - E_{op}[n_{op}] = 0$, which also can be written as

$$\delta T_{op}[n_{op}(\mathbf{r})] + \int \delta n(\mathbf{r})V_{eff}(\mathbf{r})\,d^3r = 0. \tag{1.19}$$

Carrying out this minimization, using Eqn (1.18) for the kinetic energy, leads to Eqn (1.16), and we have shown that non-interacting electrons which are the solutions to Eqn (1.16) result in an electron density which minimizes the total energy of this system via the functional in Eqn (1.18). The basic principle of the Kohn–Sham approach is the assumption that one can find an effective potential V_{eff} such that its density n_{op} is the same as the ground-state density of the fully interacting system, n_{gs}. The assumption is proven to hold for a homogenous electron gas and small deviations from it but no practical experience shows that the assumption holds in the general case. Nevertheless, since we know the coupling between potential and density, as described, for example, in Fig. 1.2, it seems like an efficient route to get the ground-state density of the interacting system, by careful selection of an effective potential, even if it is from a one-particle system.

The question now is, how do we determine V_{eff} so that n_{op} becomes equal to n_{gs}? In order to find a way to do this, we first recast the energy functional in Eqn (1.15) to the form

$$
E[n_{\mathrm{op}}(\mathbf{r})] = T_{\mathrm{op}}[n_{\mathrm{op}}(\mathbf{r})] + \int n_{\mathrm{op}}(\mathbf{r}) V_{\mathrm{ext}}(\mathbf{r}) \, d^3 r
$$

$$
+ \iint \frac{n_{\mathrm{op}}(\mathbf{r}) \cdot n_{\mathrm{op}}(\mathbf{r'})}{\mathbf{r} - \mathbf{r'}} \, d^3 r \, d^3 r' + E_{\mathrm{xc}}[n_{\mathrm{op}}(\mathbf{r})]. \tag{1.20}
$$

In Eqn (1.20), we have introduced the one-particle kinetic energy functional T_{op} instead of the true kinetic energy functional, and we have introduced the Hartree electrostatic interaction instead of the true electron–electron interaction. Hence in order to make Eqn (1.20) equal to Eqn (1.15) we must introduce a term that corrects for this, and this is what the exchange and correlation energy $E_{\mathrm{xc}}[n_{\mathrm{op}}(\mathbf{r})]$ does. Since the first three terms on the right-hand side of Eqn (1.20) can be calculated numerically, we have moved the complexity of the fully interacting system to finding the exchange and correlation functional. For a uniform electron gas, one can, however, calculate $E_{\mathrm{xc}}[n_{\mathrm{op}}(\mathbf{r})]$ for all values of the electron density, and parameterized forms of $E_{\mathrm{xc}}[n_{\mathrm{op}}(\mathbf{r})]$ as a function of $n_{\mathrm{op}}(\mathbf{r})$ are available. For uniform densities, it is hence possible to evaluate Eqn (1.20) with excellent accuracy, and obtain the total energy using the electron density as the decisive variable of the system.

The local density approximation (LDA; Hedin, 1965; Hedin and Lundqvist, 1971; Barth and Hedin, 1972; Ceperley and Alder, 1980) assumes that the parameterizations used for the uniform electron gas work even in cases where the electron gas is not uniform, and is applicable to molecules, solids, surfaces, and interfaces. This is done by assuming that locally, at a given point in space of, for example, a solid or molecule, one may consider the density as uniform and hence use the parameterized version from the uniform electron gas. This means that one uses the following expression for the exchange–correlation energy:

$$
E_{\mathrm{xc}}[n_{\mathrm{op}}(\mathbf{r})] = \int \epsilon_{\mathrm{xc}}[n_{\mathrm{op}}(\mathbf{r})] n_{\mathrm{op}}(\mathbf{r}) \, d^3 r, \tag{1.21}
$$

where $\epsilon_{\mathrm{xc}}[n_{\mathrm{op}}(\mathbf{r})]$ is the exchange–correlation energy density; in a parameterized form, its dependence on $n_{\mathrm{op}}(\mathbf{r})$ is relatively simple. We now have an expression for the energy functional that can be minimized with respect to the electron density, using an expression similar to Eqn (1.19) but now including electron–electron interaction via the Hartree term and the exchange–correction functional. This minimization results in a one-electron Schrödinger-like equation similar to Eqn (1.16). However, the minimization procedure leads to an explicit form of the effective potential V_{eff} from Eqn (1.16):

$$V_{\text{eff}}(\mathbf{r}) = V_{\text{ext}}(\mathbf{r}) + \int \frac{n_{\text{op}}(\mathbf{r}')}{|\mathbf{r} - \mathbf{r}'|}\, d^3 r' + \mu_{\text{xc}}[n_{\text{op}}(\mathbf{r})], \qquad (1.22)$$

where

$$\mu_{\text{xc}}[n_{\text{op}}(\mathbf{r})] = \frac{\partial\{\epsilon_{\text{xc}}[n_{\text{op}}(\mathbf{r})]n_{\text{op}}(\mathbf{r})\}}{\partial n_{\text{op}}(\mathbf{r})} = \epsilon_{\text{xc}}[n_{\text{op}}(\mathbf{r})] + n_{\text{op}}(\mathbf{r})\frac{\partial\{\epsilon_{\text{xc}}[n_{\text{op}}(\mathbf{r})]\}}{\partial n_{\text{op}}(\mathbf{r})}. \qquad (1.23)$$

We now can evaluate the total energy of a system, using electron density as the key variable that determines things. In practice, this means solving Eqn (1.16) with the effective potential specified by Eqn (1.22). Since the effective potential to be used in Eqn (1.22) depends on electron density, which is the property we want to calculate, one has to perform a self-consistent field calculation where an initial electron density is guessed and an effective potential is calculated from Eqn (1.22). This potential is then used to solve Eqn (1.16), and a new electron density is calculated from Eqn (1.17), which is then put back into Eqn (1.22). This procedure is repeated until convergence is obtained, that is, until the density does not change appreciably with successive iterations. Once a self-consistent electron density has been found, one can calculate the ground-state energy of the Kohn–Sham LDA energy functional, via Eqn (1.20). We comment here on the evaluation of the kinetic energy in Eqn (1.20). Since an accurate expression of it in terms of the electron density is missing, one evaluates it from $\sum_{\text{occ}}\langle\psi|-\frac{1}{2}\nabla^2|\psi\rangle$ by using Eqn (1.16), moving V_{eff} to the right-hand side of the equation and multiplying from the left with $\langle\psi|$ on both sides, yielding

$$T_{\text{op}} = \sum_{\text{occ}}\langle\psi_i| - \frac{1}{2}\nabla^2|\psi\rangle = \sum_{\text{occ}}\epsilon_i - \int d^3 r\, V_{\text{eff}}(\mathbf{r})n_{\text{op}}(\mathbf{r}). \qquad (1.24)$$

The last term on the right-hand side of this equation is sometimes referred to as the double-counting term, which is not to be confused with the double counting used in the (local density approximation + Hubbard parameter) LDA + U extension of density functional theory.

One of the effects that the exchange interaction Eqn (1.5) or the exchange–correlation interaction Eqn (1.21) has is that one electron can dig out a hole in the surrounding density provided by all the other electrons. This hole is normally referred to as the exchange–correlation hole (Gunnarsson and Lundqvist, 1976), in the density functional theory literature and, in fact, the exchange–correlation energy can be expressed as the energy due to interaction of an electron with its exchange–correlation hole.

It has been shown that only the spherical average of the exchange–correlation hole contributes to E_{xc} (Gunnarsson and Lundqvist, 1976). Hence, it may be argued that, even if the exact exchange–correlation hole may in general be strongly aspherical and non-local, it is not necessary for approximate functionals, for example, in LDA, to describe the non-spherical parts. Consequently, approximate functionals often do not experience major obstacles in evaluating materials properties with good accuracy, as will be described in Chapter 3.

Other differences between approximate functionals, like LDA, and exact functionals or even the Hartree–Fock theory are well known, in particular, that the self-interaction of the Hartree term cancels exactly the self-interaction of the Fock term. This means that one may include terms $i = j$ in the summations of Eqn (1.5) without causing an error. This could, for example, be done in order to make the Hartree–Fock theory more comparable to density functional theory, in which the contributions from all electrons are used to evaluate the density. Including all electrons in the terms of the Hartree–Fock theory therefore does not introduce an error, and one may say that the Hartree–Fock theory does not have a self-interaction error. However, most approximations of the exchange–correlation term of density functional theory do not generate a self-interaction-free functional and, in many cases, this deficiency has been known to cause unacceptable errors. Attempts to overcome this self-interaction error have been made; the most popular form is given by Perdew and Zunger (1981). It has been pointed out that this correction only partially removes the self-interaction (Lundin and Eriksson, 2001), and truly self-interaction-free functional forms of density functional theory have been suggested. However, the functionals suggested so far have not been shown to drastically improve the results of, for example, Perdew and Zunger (1981).

Parametrizations of $\mu_{xc}(n_{op}(\mathbf{r}))$ in Eqn (1.22), in terms of electron density, are available, where the exchange part is proportional to $n_{op}^{1/3}$. This was done by, for example, Hedin and Lundqvist (1971) and is discussed in several textbooks (e.g. in Ashcroft and Mermin, 1976; Marder, 2010). Extensions of this analysis of the exchange–correlation properties of the uniform electron gas to spin polarized situations allow majority spin-up (α) and minority spin-down (β) densities to not be the same, thus enabling calculations of finite magnetic moments. The exchange–correlation potential for electrons of a specific spin density n_{α} is shown from the work of Barth and Hedin (1972) to be

$$\mu_{xc}^{\alpha}[n_{\alpha}(\mathbf{r}), n_{\beta}(\mathbf{r})] = A[n_{op}(\mathbf{r})] \left(\frac{n_{\alpha}(\mathbf{r})}{n_{op}(\mathbf{r})} \right)^{\frac{1}{3}} + B[n_{op}(\mathbf{r})], \qquad (1.25)$$

with a corresponding expression for n_{β}, where $n_{op} = n_{\alpha} + n_{\beta}$. In Eqn (1.25), both $A(n_{op}(\mathbf{r}))$ and $B_{op}(n(\mathbf{r}))$ are negative for all densities (Barth and Hedin, 1972). Hence, majority spin-up and minority spin-down electrons travel through, for example, a crystal experiencing the separate effective potentials V_{eff}^{α} and V_{eff}^{β}; for magnetic materials, with spin polarization, one must treat the Kohn–Sham equation, Eqn (1.16), separately for the majority spin-up and minority spin-down states.

$$\left(-\frac{1}{2}\nabla^2 + V_{\text{eff}}^{\alpha}(\mathbf{r})\right)\psi_{i\alpha}(\mathbf{r}) = \epsilon_i\psi_{i\alpha}(\mathbf{r}),$$

$$\left(-\frac{1}{2}\nabla^2 + V_{\text{eff}}^{\beta}(\mathbf{r})\right)\psi_{i\beta}(\mathbf{r}) = \epsilon_i\psi_{i\beta}(\mathbf{r}), \tag{1.26}$$

respectively. The effective potential for the majority spin channel becomes a little deeper than that for the minority spin channel, which is why an imbalance between majority and minority spin states emerges in the first place. The difference between the effective potential of majority spin states and that of minority spin states often amounts to a constant shift between the electron states of the two spin channels. This shift is referred to as exchange splitting. It can be seen from Eqn (1.25) that the exchange contribution to the effective potential is proportional to the 'local density' of spin-up and spin-down states, and again a spin polarized calculation has to be done self-consistently, as described above, but now for each spin channel. Alternatively, one may express this as self-consistency which has to be achieved both for the charge density $n = n^{\alpha} + n^{\beta}$ and for the magnetization density $m = n^{\alpha} - n^{\beta}$. The latter description is to be preferred since, in the self-consistent cycle, a larger mixing of densities can be used for m than for n. Spin-polarized parameterizations of the LDA, normally referred to as the local spin density approximation (LSDA), have been established for some time (Barth and Hedin, 1972). The LSDA is known to give accurate total energies for many bulk systems. This theory is also known to reproduce with great accuracy the magnetic spin moments of most transition metals and their alloys. As an example, we show in Fig. 1.3 the measured and

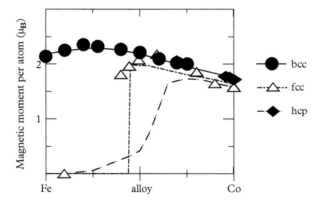

Figure 1.3 *Measured and calculated spin moments of FeCo alloys. Calculations represented by lines, and measurements by symbols. Note that data is shown for body-centred cubic (bcc), face-centred cubic (fcc), and hexagonal close-packed (hcp) phases in different symbols. Reprinted figure with permission from James, P., Eriksson, O., Johansson, B., and Abrikosov, I. A., Phys. Rev. B, 59, 419, 1999. Copyright 1999 by the American Physical Society.*

calculated (using LSDA) spin moments of Fe-Co alloys (James et al., 1999). Note that we have calculated spin moments by multiplying the integrated spin density with the electron g-factor. This corresponds to the standard formulation in quantum mechanics, where the size of the spin moment of one electron is calculated from $m_s = g_s s_z$, where $g_s = 2.002319$, to get moments in units of magnetic moment per atom (μ_B; used throughout this book). The phase diagram of the alloys in Fig. 1.3 shows that the crystal structure is body-centred cubic (bcc), for almost all concentrations. Only for the very Co-rich alloys does the hexagonal closed packed (hcp) crystal structure become stable. Also, for some concentrations, the face-centred cubic (fcc) phase has been stabilized, as precipitates in an fcc matrix of Cu. In Fig. 1.3 we see that, for the bcc phase, the measured and calculated magnetic moments (per 'average' atom of the alloys) agree with good accuracy. Theory is even found to reproduce the maximum value of the spin moment, the Slater–Pauling maximum, at ~25 % Co. For the fcc and hcp phases, theory also reproduces observation, for cases where a comparison can be made.

1.5   Non-collinear magnetism, and time-dependent density functional theory

In systems with a collinear magnetization and in which spin–orbit coupling is neglected, spin-up and spin-down states are orthogonal, and hence the density matrix is diagonal. This allows for the formulation of spin polarized density functional theory, with one Kohn–Sham equation for the majority (α) spin channel, and one for the minority (β) spin channel, as shown in Eqn (1.26). The majority and minority spin channels are connected only through the spin dependence of the exchange–correlation part of the effective one-particle potential. The more general situation is when the electronic magnetization density has a spin texture that is both non-collinear and time dependent, and the density matrix contains off-diagonal elements (Barth and Hedin, 1972; Nordström and Singh, 1996; Sandratskii, 1998; Kübler, 2009). Extending the spin polarized density functional theory formalism discussed, to the case of non-collinear magnetism and time dependence, the density matrix can be written as a 2×2 matrix:

$$\rho(\mathbf{r}, t) = \sum_{i \in occ} \begin{pmatrix} \psi_{i\alpha}(\mathbf{r}, t)\psi_{i\alpha}^*(\mathbf{r}, t) & \psi_{i\alpha}(\mathbf{r}, t)\psi_{i\beta}^*(\mathbf{r}, t) \\ \psi_{i\beta}(\mathbf{r}, t)\psi_{i\alpha}^*(\mathbf{r}, t) & \psi_{i\beta}(\mathbf{r}, t)\psi_{i\beta}^*(\mathbf{r}, t) \end{pmatrix}, \tag{1.27}$$

where the wave functions are two-component, time-dependent Pauli spinors; $\psi_i(\mathbf{r}, t) = [\psi_{i\alpha}^*(\mathbf{r}, t), \psi_{i\beta}^*(\mathbf{r}, t)]^T$; and the summation is over the occupied Kohn–Sham orbitals. The charge density and the magnetization density are calculated as traces of the density matrix:

$$n(\mathbf{r}, t) = \mathrm{Tr}[\rho(\mathbf{r}, t)],$$

$$\mathbf{m}(\mathbf{r}, t) = \mathrm{Tr}[\hat{\sigma}\rho(\mathbf{r}, t)], \tag{1.28}$$

where $\hat{\sigma} = (\hat{\sigma}_x, \hat{\sigma}_y, \hat{\sigma}_z)$ is a vector of Pauli matrices. Reciprocally the density matrix can be expressed in terms of $n(\mathbf{r}, t)$ and $\mathbf{m}(\mathbf{r}, t)$ as

$$\rho(\mathbf{r}, t) = \frac{1}{2}\left[n(\mathbf{r}, t)\mathbf{1} + \mathbf{m}(\mathbf{r}, t) \cdot \hat{\sigma}\right], \tag{1.29}$$

where $\mathbf{1}$ is the 2×2 identity matrix. Note that in relationship to Eqn (1.26), we have now made a generalization in that we consider non-collinear arrangements of magnetic moments, and hence the Kohn–Sham equation must be written in a somewhat more general form. However, before doing this, we note that Eqn (1.28) also contains the time dependence of the charge and magnetization density, and hence we should consider the time-dependent Kohn–Sham equation, normally written as a Pauli–Schrödinger-type equation,

$$i\frac{\partial \psi_{i\alpha}(\mathbf{r}, t)}{\partial t} = H_{\alpha\beta}\psi_{i\beta}(\mathbf{r}, t), \tag{1.30}$$

for single-particle orbitals $\psi_{i\alpha}(\mathbf{r}, t)$. Note that, in Eqn (1.30), we have used the Einstein convention of implicit summation over repeated indices. If an external magnetic field $\mathbf{B}_{\text{ext}}$ is included together with the spin–orbit coupling, the Hamiltonian for the electron system is written as

$$\hat{H}_{\alpha\beta} = -\frac{\nabla^2}{2}\delta_{\alpha\beta} + V_{\alpha\beta}^{\text{eff}}(\mathbf{r}, t) + \left\{\frac{1}{2c}\hat{\sigma} \cdot \mathbf{B}^{\text{eff}}(\mathbf{r}, t)\right\}_{\alpha\beta}$$

$$+ \left\{\frac{1}{4c^2}\hat{\sigma} \cdot \left(\nabla V^{\text{eff}}(\mathbf{r}, t) \times i\nabla\right)\right\}_{\alpha\beta}, \tag{1.31}$$

where c is the speed of light in vacuum. The effective non-magnetic potential

$$V_{\alpha\beta}^{\text{eff}}(\mathbf{r}, t) = V^{\text{ext}}(\mathbf{r}, t)\delta_{\alpha\beta} + V^{\text{H}}(\mathbf{r}, t)\delta_{\alpha\beta} + \mu_{\alpha\beta}^{\text{xc}}(\mathbf{r}, t)\delta_{\alpha\beta} \tag{1.32}$$

consists of the external potential $V^{\text{ext}}(\mathbf{r}, t)$, the Hartree potential $V^{\text{H}}(\mathbf{r}, t)$, and the non-magnetic part of the exchange–correlation potential $\mu_{\alpha\beta}^{\text{xc}}(\mathbf{r}, t)$. The effective magnetic field

$$\mathbf{B}^{\text{eff}}(\mathbf{r}, t) = \mathbf{B}^{\text{ext}}(\mathbf{r}, t) + \mathbf{B}^{\text{xc}}(\mathbf{r}, t) \tag{1.33}$$

can be decomposed into the weaker external magnetic field $\mathbf{B}^{\text{ext}}$ and the stronger exchange–correlation magnetic field $\mathbf{B}^{\text{xc}}(\mathbf{r}, t)$. The scalar and magnetic exchange–correlation potentials are calculated, following Eqn (1.23), as the functional derivatives of the exchange–correlation energy

$$\mu^{xc}(\mathbf{r}, t) = \frac{\delta E^{xc}[n, \mathbf{m}]}{\delta n(\mathbf{r})}, \tag{1.34}$$

$$\mathbf{B}^{xc}(\mathbf{r}, t) = -\frac{\delta E^{xc}[n, \mathbf{m}]}{\delta \mathbf{m}(\mathbf{r})}. \tag{1.35}$$

We have in this section provided the most general form of a practical scheme to calculate magnetic properties from density functional theory, in the sense that no restriction on the shape of the charge or on the magnetization direction is imposed and that time dependence is included. From this formalism, one can evaluate the 'direction' of the magnetization density, by diagonalization of the density matrix Eqn (1.27) (Nordström and Singh, 1996; Sandratskii, 1998; Kübler, 2000). This is described in Chapter 4, where we connect Eqn (1.31) to the concepts of atomistic spin dynamics. For most materials, the 'direction' of the magnetization density does not vary over regions where it is large; typically, this region is centred on each atomic nucleus, within a radius that is significantly smaller than interatomic distances. Examples of this are given in Chapter 2, and we note that this property implies that a collinear atomic description is meaningful in such a region (Sandratskii, 1998; Kübler, 2000), although examples of the opposite property have been discussed (Nordström and Singh, 1996). Before entering the discussion of how to derive from Eqn (1.31) the relevant equations of atomistic spin dynamics, we describe in Chapter 2 those aspects of a solid that are important when one solves the Kohn–Sham equation. We also give some examples of the magnetic properties of materials and how theory compares with observations (Chapter 3).

2

Aspects of the Solid State

Symmetries play an important role in the theory of the solid state. As will be developed in this chapter, density functional theory (DFT) calculations for crystalline materials are commonly performed for **k**-points in the irreducible part of the first Brillouin zone (BZ), an approach which relies on the use of translational and point-group symmetries. Two central properties that result from a calculation in reciprocal space are the wavevector-resolved energy spectra, that is, the so-called band structure or band dispersion, and the energy-resolved density of states. For magnetic materials, atomic magnetic moments can be defined and calculated, as well as effective interatomic exchange interactions. In this chapter, the essential aspects of such calculations are described.

2.1 Crystal systems and space groups

Normally, calculating of the electronic structure of materials is an application of DFT, that is, one tries to find a solution to the Kohn–Sham equation, wish was discussed in Chapter 1, to a crystalline environment. This means that one considers an infinite object with the effective potential described in Eqns (1.16) or (1.26). A crystal is characterized by its translational invariance, with a periodicity defined by the Bravais lattice vectors. The lattice points $\mathbf{R}$ for which the environment is identical are hence defined by $\mathbf{R} = n_1\mathbf{R}_1 + n_2\mathbf{R}_2 + n_3\mathbf{R}_3$, where n_1, n_2, and n_3 are integers, and $\mathbf{R}_j$, where j = 1, 2, 3, is the Bravais lattice vector. A crystal is made up of one or several atoms per unit cell. The symmetry properties of a crystal lattice are of great importance, since the computational cost may be reduced significantly when utilizing these symmetries. For a three-dimensional material, there are in general seven crystal systems (shown in Table 2.1) and 14 Bravais lattices. As an example, we mention that the group of cubic crystal systems contains three Bravais lattices; the body-centred cubic (bcc) lattice, the face-centred cubic (fcc) lattice and the simple cubic lattice (sc). The symmetry properties of a crystal are described by the space group, which in three dimensions is made from combinations of 32 crystallographic point groups combined with translations. The combination of all symmetry operations of a three-dimensional lattice results in a total of 230 space groups describing all possible crystal symmetries.

Atomistic Spin Dynamics. Olle Eriksson, Anders Bergman, Lars Bergqvist, Johan Hellsvik. First Edition.
© Olle Eriksson, Anders Bergman, Lars Bergqvist, Johan Hellsvik 2017. First published in 2017 by Oxford University Press.

Table 2.1 *Table showing the seven crystal systems.*

Crystal system	Point-group symmetry requirements
Cubic	4 threefold axes of rotation
Hexagonal	1 sixfold axis of rotation
Trigonal	1 threefold axis of rotation
Tetragonal	1 fourfold axis of rotation
Orthorhombic	Either 3 twofold axes of rotation, or 1 twofold axis of rotation and 2 mirror planes
Monoclinic	Either 1 twofold axis of rotation, or 1 mirror plane
Triclinic	All cases not satisfying the requirements of any other system

2.2   The Born–von Karman boundary condition, and Bloch waves

Due to the periodic symmetry of bulk materials, several simplifications evolve. A discussion of this can be found in standard textbooks on solid state physics (Ashcroft and Mermin, 1976; Marder, 2010). First of all, the effective potential which the electrons move in is periodic, a fact which enables the identification of a Bravais lattice vector. In the Kohn–Sham equation, this means that $V_{\text{eff}}(\mathbf{r}) = V_{\text{eff}}(\mathbf{r+R})$, where $\mathbf{R}$ is a Bravais lattice vector of the materials. The Born–von Karman boundary condition for the wave function of the electron states is then introduced:

$$\psi_{\mathbf{k}}(\mathbf{r} + N\mathbf{R}) = \psi_{\mathbf{k}}(\mathbf{r}), \tag{2.1}$$

where N is a (large) integer. It is then possible to show that electrons moving through an infinite, periodic crystal must obey Bloch's theorem (Ashcroft and Mermin, 1976; Marder, 2010), that is, that the one-electron wave function (e.g. the solution to the Kohn–Sham equation) must obey

$$\psi_{\mathbf{k}}(\mathbf{r} + \mathbf{R}) = e^{i\mathbf{k}\cdot\mathbf{R}}\psi_{\mathbf{k}}(\mathbf{r}). \tag{2.2}$$

A vector $\mathbf{k}$ has been introduced in this expression. This is a vector of reciprocal space.[1] In self-consistent DFT-based calculations one has only to consider $\mathbf{k}$-vectors which lie inside the first BZ for the calculation of charge and magnetization density. The BZ is defined as the Wigner–Seitz primitive cell of the reciprocal lattice and, as an example, we show in Fig. 2.1 the BZ for the bcc crystal structure. Note that a smaller region is marked that describes the so-called irreducible wedge of the BZ. Most crystal structures can undergo a rotation along one or several axis, such that the rotated geometry is identical to the starting geometry. These rotations define the point group of the crystal, and we give one example of a point group below. The point group operations imply, importantly,

---

[1] Reciprocal space is spanned by the vectors $\mathbf{G}_i$, defined as $\mathbf{G}_i \cdot \mathbf{R}_j = 2\pi\delta_{ij}$.

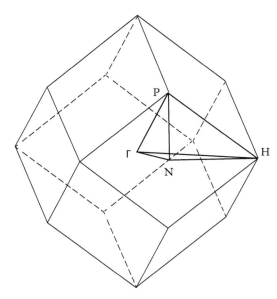

Figure 2.1 *Example of a Brillouin zone (BZ). The BZ is shown for a body-centred cubic crystal structure. In the figure, high-symmetry points and lines are marked and labeled with Γ, P, N, and H. The irreducible wedge is marked with thick lines.*

that any **k**-point of the full BZ can be generated from **k**-points inside the irreducible wedge of the BZ. Kohn-Sham eigenvalues of **k**-points that are connected by point-group operations are the same, and therefore one only need to consider **k**-vectors inside the irreducible wedge of the BZ when performing DFT calculations of crystalline materials. As this discussion has involved point-group operations, we consider as an example the different point-group operations of a cubic structure. We illustrate this structure as a cubic building block in Fig. 2.2. This building block is common for the sc, bcc, or fcc structure. The simplest symmetry operation which leaves the cube invariant is the identity operation. In the figure, the three arrows indicate the relevant axes for which we can perform rotations which leave this cubic building block invariant. Around the axis aligned along the 100 direction, it is possible to perform three 90° rotations, resulting in angles of 90°, 180°, and 270°, respectively. The fourth rotation, 360°, is identical to the identity. There are three axes of such rotations (100, 010, and 001) and hence there

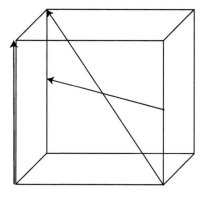

Figure 2.2 *Symmetry axes of a cubic building block.*

are nine rotations of this kind. Around the 111 axis, it is possible to make two 120°
rotations, since the third such rotation is identical to the identity. There are four such
axes, so eight rotations of this kind. Finally, the arrow along the 110 direction allows
for one 180° degree rotation, since a second such rotation is identical to the identity.
There are six such axes, and hence six rotations of this kind. Summing up the identity
and all possible rotations hence yields 24 symmetry operations. All these 24 operations
can, for the sc, fcc, or bcc crystal structure, be combined with a subsequent inversion of
the lattice, thus leading to an additional 24 allowed symmetry operations. Some of these
latter symmetry operations are equivalent to a reflection of the lattice through a plane,
but some of them cannot be described as such, and are referred to as roto-inversions or
improper rotations. Altogether, there are hence 48 point-group operations for a cubic
material with an sc, fcc, or bcc crystal structure. However, a Bravais lattice sc, fcc, or
bcc and which has more than one atom in the primitive cell (more than one atom in the
crystallographic basis) may have a lower number of point-group operations. An example
of this type of lattics is the zinc blende (diamond) structure, which has two atoms per
unit cell: one at (0, 0, 0), and one at (1/4, 1/4, 1/4). Si is an example of a material that
crystallizes in this structure. This material has no inversion symmetry, which means that
there are only 24 allowed point-group operations.

2.3   A variational procedure to obtain eigenvalues

In section 2.2, we showed that the Kohn–Sham equation must be solved for a number
of **k**-vectors which, for a given cycle in the self-consistent loop, may be treated as independent of each other. However, the toughest part of the problem is to actually find
a solution to the Kohn–Sham equation for any selected **k**-point. One approach is to
expand the unknown Kohn–Sham wave function in a set of known basis functions as
follows:

$$\psi_k(\mathbf{r}) = \sum_{l}^{l_{max}} c_{lk} \chi_{lk}(\mathbf{r}).$$

(2.3)

Note that l is here used as a compound index that involves the principal quantum
number, the orbital angular momentum quantum number, and the magnetic quantum
number. For compounds with several atom types, this index also has to incorporate atom
type. There are several choices of basis functions, which have given rise to different
electronic structure methods with different names, such as LCAO (linear combination
of atomic orbitals), LAPW (linear augmented plane waves), PAW (projector augmented
waves), and LMTO (linear muffin-tin orbitals). The sum in Eqn (2.3) is truncated after
sufficiently many basis functions $\chi_{lk}(\mathbf{r})$ have been included; the coefficients c_{lk} are, via
the Rayleigh–Ritz principle (Atkins and Friedman, 2005), determined from the equation
system

$$\sum_{l}^{l_{max}} H_{ll'} - \epsilon_k O_{ll'} c_{lk} = 0,$$

(2.4)

with the Hamiltonian matrix

$$H_{ll'} = \int_{\Omega_c} \chi_{l\mathbf{k}}^*(\mathbf{r}) \left(\frac{-\nabla^2}{2} + V_{\text{eff}} \right) \chi_{l'\mathbf{k}}(\mathbf{r}) \, d^3r \equiv \int_{\Omega_c} \chi_{l\mathbf{k}}^*(\mathbf{r}) \hat{h}_{\text{eff}} \chi_{l'\mathbf{k}}(\mathbf{r}) \, d^3r \qquad (2.5)$$

and the overlap matrix

$$O_{ll'} = \int_{\Omega_c} \chi_{l\mathbf{k}}^*(\mathbf{r}) \chi_{l'\mathbf{k}}(\mathbf{r}) \, d^3r, \qquad (2.6)$$

and the integral is over the unit cell (Ω_c). Once $H_{ll'}$ and $O_{ll'}$ have been evaluated, the eigenvalues $\epsilon_{\mathbf{k}}$ are determined from the secular equation (Atkins and Friedman, 2005)

$$\det |H_{ll'} - \epsilon_{\mathbf{k}} O_{ll'}| = 0, \qquad (2.7)$$

which is a standard numerical eigenvalue problem. In this stage of approximation, Eqn (2.7) can be solved separately for effective potentials of majority and minority spin states. Hence, eigenvalues and eigenstates are commonly written as $\epsilon_{\alpha\mathbf{k}}$ and $\psi_{\alpha\mathbf{k}}$, respectively with a corresponding expression for states with β spin projection. The one-electron density, discussed in Chapter 1, is then calculated as a sum over all possible $\mathbf{k}$-vectors and occupied eigenstates. This can be done for each spin channel separately:

$$n^\alpha(\mathbf{r}) = \sum_i^{occupied} \sum_{\mathbf{k}} \left| \psi_{i\mathbf{k}}^\alpha(\mathbf{r}) \right|^2, \qquad (2.8)$$

$$n^\beta(\mathbf{r}) = \sum_i^{occupied} \sum_{\mathbf{k}} \left| \psi_{i\mathbf{k}}^\beta(\mathbf{r}) \right|^2, \qquad (2.9)$$

from which the electron and the magnetization densities are obtained through

$$n(\mathbf{r}) = n^\alpha(\mathbf{r}) + n^\beta(\mathbf{r}), \qquad (2.10)$$

$$m(\mathbf{r}) = n^\alpha(\mathbf{r}) - n^\beta(\mathbf{r}), \qquad (2.11)$$

respectively. Note that the sums only have to be carried out over the $\mathbf{k}$-points of the irreducible wedge of the BZ and multiplied by the corresponding $\mathbf{k}$-point weight; for this reason, a higher point-group symmetry of the crystal requires a smaller part of the BZ to be sampled and hence a smaller computational effort. A popular way to display the result of an electronic structure calculation is to plot the eigenvalues of the Kohn–Sham equation as a function of the $\mathbf{k}$-point, preferably along high-symmetry lines and high-symmetry points in the BZ. An example of such an electronic structure, or band plot, is shown in Fig. 2.3 for bcc Fe. Here it may be seen that the eigenvalues of the Kohn–Sham equation depend more or less strongly on $\mathbf{k}$, a phenomenon that is referred to as band dispersion. In this figure, we show bands for spin-up states on the left-hand side, and spin-down states on the right-hand side. The figure shows that, due to the exchange splitting, as discussed in Chapter 1, the spin-down states are typically shifted up in energy, compared to spin-up states. This yields a higher number of occupied spin-up states, compared to the number of occupied spin-down states; a finite magnetic moment; and a difference in the spin-up and spin-down densities, with the latter leading, via Eqn (1.25), to a difference around the Fermi level effective potentials of the two spin projections. In Fig. 2.3, it is clear that, for both spin-up and spin-down electrons, the lowest band has a minimum at the Γ-point, around which the dispersion is essentially quadratic in **k**, similar to the result of the free-electron model. This is because the lowest band is essentially dominated by 4*s* orbitals. These states are quite extended in space and hence have a free-electron-like dispersion. However, the lowest eigenvalue at the H-point is dominated by 3*d* states, and the character of the lowest band when going from Γ to H hence changes gradually from being 4*s* dominated to being 3*d* dominated. Note from Fig. 2.3 that the highest occupied energy, the Fermi level, is marked as a horizontal dashed line. Electron states below this energy are hence occupied, and the states above are unoccupied. It is the occupied states that contribute to the charge and magnetization density in Eqns (2.8) and (2.9). A further inspection of Fig. 2.3 shows that the exchange splitting seems to be strongest around the Fermi level. This is due to that the bands around the Fermi level are dominated by the Fe 3*d* orbitals, which experience

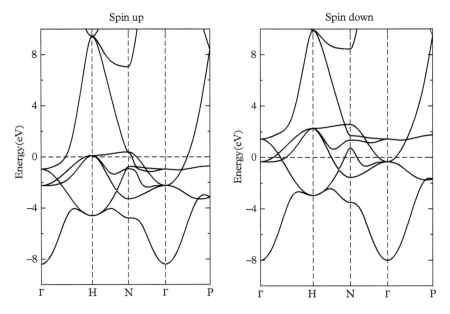

Figure 2.3 *Band structure of body-centred cubic Fe, with spin-up states to the left, and spin-down states to the right. The Fermi energy is marked as a dashed horizontal line. Picture kindly provided by Dr. Swarup Panda.*

the largest effect of the exchange interaction and hence have the largest exchange splitting. The data shown in Fig. 2.3 yield a magnetic spin moment that almost perfectly coincides with the experimentally observed magnetic moment, and it is found that more than 95 % of the calculated magnetic moment comes from the Fe $3d$ states. Figure 2.3 also shows that Kohn–Sham eigenvalues cut the Fermi level at specific points of **k**-space. If one moved in a circuit around the BZ via a path connecting all the **k**-points that had a Kohn–Sham eigenvalue equal to the Fermi level, one would trace out the shape of the Fermi surface. This has been done many times for transition metal elements and compounds. It is found that, for most of the calculations, the theoretical Fermi surface agrees with measurements, for example, as provided by the de Haas-van Alphen experiments or angular resolved photoemission spectroscopy (Ashcroft and Mermin, 1976; Marder, 2010), despite the fact that the eigenvalues in Kohn–Sham theory, strictly speaking, do not have any physical meaning. Eventual disagreements between observed and calculated Fermi surfaces are often ascribed to an inaccuracy of the effective potential used in Eqn (1.22) and have led to the development of more sophisticated methods of treating electron–electron repulsion in a solid. A description of these methods, is however, outside the scope of this book.

In Fig. 2.4, we show as an additional example of the output of an electronic structure calculation based on DFT; the magnetization density $m(\mathbf{r})$ of bcc Fe. In this cut of the crystal, there are atoms positioned at the corners of the plot. It can be seen that the magnetization density is largest around these atoms, as is natural, since most of the magnetic moment is due to exchange splitting of the $3d$ orbitals that are centred on these atoms. It can also be observed that the shape of the magnetization density, even when close to the atomic nuclei, deviates somewhat from spherical symmetry, as may be seen, for example, in the dark lobes in Fig. 2.4. However, this influence of the crystal geometry on the magnetization density is consistent with the point-group symmetry of the

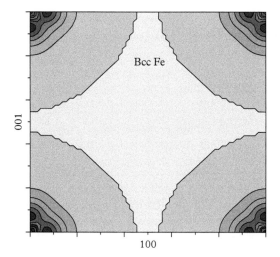

Figure 2.4 *Magnetization density of body-centred cubic (bcc) Fe in a plane spanned by the 100 and 001 vectors of the bcc crystal structure. Increase in density is indicated by an increase in the darkness of the colour.*

lattice, as is also clear from the figure. It can also be seen that the magnetization density is quite localized in space around the atomic nuclei. This property is not unique for bcc Fe but is found in most magnetic materials. It is, however, interesting that, although the magnetization density and magnetic moment seem rather localized in space, as illustrated in Fig. 2.4, the microscopic reason for the magnetism is due to exchange splitting of delocalized electron states with significant band dispersion, as shown in Fig. 2.3. Nevertheless, the pronounced localization in space of the magnetization density, and hence the magnetic moments, means that it is meaningful to treat magnetic excitations, for example, magnons, with an effective spin Hamiltonian. Such a Hamiltonian may be of the Heisenberg or extended Heisenberg type and will be discussed in more detail in Chapter 4.

2.4   Density of states

One of the more popular outputs of an electronic structure calculation, apart from total energy, energy bands, magnetic moments, and magnetization density, as discussed in Section 2.3, is the density of states of the electronic structure. This is a property which is very useful for analysing and understanding calculations. A derivation of the density of states may be found in most textbooks on solid state physics (Ashcroft and Mermin, 1976; Marder, 2010) and so is not repeated here. Instead, we use the result, that the density of states of spin-up (α) states can be calculated from

$$D^{\alpha}(E) = \sum_i \frac{1}{(2\pi)^3} \int_{BZ} \delta(E - \epsilon_{i\alpha\mathbf{k}}) \, d\mathbf{k}, \tag{2.12}$$

with a corresponding expression for spin-down (β) states. The Fermi level can now be calculated from the density of states via the expressions

$$N_{\text{val}}^{\alpha} = \int_{-\infty}^{E_F} D^{\alpha}(E) \, dE, \tag{2.13}$$

$$N_{\text{val}}^{\beta} = \int_{-\infty}^{E_F} D^{\beta}(E) \, dE, \tag{2.14}$$

$$N_{\text{val}} = N^{\alpha} + N^{\beta}, \tag{2.15}$$

where E_F stands for Fermi energy, and N_{val} is the number of valence electrons considered in the calculation. As an example, we show in Fig. 2.5 the density of states of the spin polarized electronic structure of bcc Fe. Here, the larger occupation of spin-up states, compared to that of spin-down states is clearly seen, as well as the fact that most of the features of the two spin channels are similar, apart from an exchange splitting primarily of $3d$ states and which is approximately 2 eV in size.

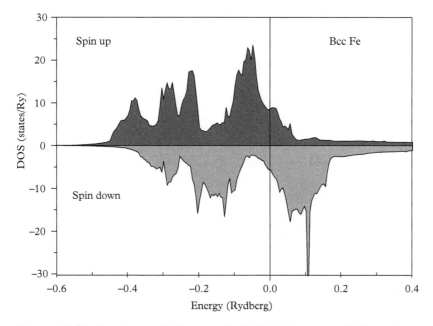

Figure 2.5 *Density of states of body-centred cubic Fe. Spin-up states (dark grey) are shown in the upper panel, and spin-down states (light grey) are reflected and shown in the lower panel. The Fermi level is at 0, and energies in Rydbergs (1 Ry equals 13.6058 eV)*

2.5   Relativistic effects

The symmetry properties of a spin polarized electronic structure calculation which includes spin–orbit coupling requires some discussion since, in this case, there is coupling between spin degrees of freedom, and orbital space. This can most easily be seen by noting that, in the simplest form, spin–orbit coupling enters as $\xi l \cdot s$ to the Kohn–Sham equation, Eqn (1.16), where s is an angular momentum operator of spin space, and l is an angular momentum operator of orbital space. The strength of the coupling between spin space and real space is given by $\xi \propto \frac{1}{r}\frac{dV_{\text{eff}}(r)}{dr}$, where V_{eff} is the effective potential of Eqn (1.16). For a $3d$ transition metal, the value of ξ for the d-orbital ranges between 10 and 100 meV whereas, for the $4d$ and $5d$ elements, this value is in the range of 20–200 meV and 150–600 meV, respectively (Christensen, 1984; Popescu et al., 2001). For the $4f$ orbitals of the lanthanides, ξ reaches values of 80-100 meV, whereas the $5f$ shell of the actinides experiences a spin–orbit coupling of the order of 200–300 meV (Schadler et al., 1986). The consequence of this coupling is that the spin and the orbital degrees of freedom become coupled and hence that the magnetization direction is locked to specific crystallographic orientations. This will be discussed in more detail in Chapter 3, Section 3.3. As regards symmetries, the influence of spin–orbit coupling is maybe best analysed using the example of a cubic crystal, described in Section 2.1. Here we discussed the 48 point-group operations which are allowed for a cubic crystal structure if magnetism is neglected. If we, for instance, consider a material where the magnetization (**M**) is pointing along the 001 axis (bcc Fe is an example of this, since the 001 axis is the easy magnetization axis of this element), a 180° rotation around the *x*- or *y*-axis is no longer allowed, since this rotates the magnetization direction 180°, and hence **M** → −**M**. In a similar fashion, a 120° rotation around the 111 axis rotates the magnetization from the 001 direction to the 010 direction, and again the original magnetic structure is not recovered. A further example of an operations which would be allowed if the coupling between spin space and real space were ignored but that otherwise would not be allowed is a 90° rotation around the *x*- or *y*-axis. In both of these cases, the magnetization would also be rotated away from the 001 direction. As a matter of fact, in the particular example of bcc Fe with (**M**) aligned in the 001 direction, the only operations that preserve **M** are the identity operation and three 90° rotations around the 001 axis, as well as the inversion operation following these four operations (since **M** is a pseudo-vector, it does not change sign after an inversion operation). There are hence eight point-group operations that preserve the bcc lattice with a magnetization along the 001 direction, corresponding to tetragonal symmetry. Fortunately, any operation which changes **M** → −**M** can be followed by the time reversal operator, which changes the sign of the magnetization, so we end up with an allowed compound symmetry operation. The inclusion of additional symmetry operations for magnetic materials, for example, compound operations containing both rotations and the time reversal operator, defines the so-called double group, which is described in detail, for example, by Bradley and Cracknell (1972). In practice, for our example of bcc Fe with **M** oriented along the 001 direction, the inversion leads to 8 additional operations that are allowed, and thus a total of 16 double-group operations. In general, a lattice with a magnetization direction oriented in a specific direction has a reduced symmetry, compared to the same lattice for a non-magnetic material or when spin–orbit coupling is neglected. Direct consequences of this reduction in symmetry are that the irreducible wedge of the BZ is larger than shown in Fig. 2.1 and that the evaluation of densities in Eqns (2.8) and (2.9) have to be done over a larger number of **k**-points because the irreducible BZ for the magnetic system is larger than the irreducible BZ for the non-magnetic system.

When spin–orbit coupling is added to the Kohn–Sham Hamiltonian, Eqn (1.16), it is no longer possible to obtain independent solutions for majority and minority spin states. The reason for this can be seen from the fact that $\xi l \cdot s$ can be written as ξ ($l_x s_x + l_y s_y + l_z s_z$)=$\xi$ [$\frac{1}{2}(l_+ s_- + L s_+) + l_z s_z$]. Since the terms involving s_+ and s_- mix spin states, one has to generalize the expression in Eqn (2.3) to the following:

$$\psi_{\mathbf{k}}(\mathbf{r}) = \sum_{l\sigma}^{l_{max}} c_{l\sigma\mathbf{k}} \chi_{l\sigma\mathbf{k}}(\mathbf{r}),$$

(2.16)

where the index σ indicates a sum over spin components, and the trial wave function χ explicitly has a component of a spinor function, leading to a doubling of the matrix size in

Eqn (2.4). In addition, the presence of that component means that the eigenstates of the Kohn–Sham equation are no longer pure spin states but admixtures of spin-up and spin-down components, similar to those in the non-collinear case, Eqn (1.27). For a system like bcc Fe, spin–orbit coupling is significantly weaker than exchange splitting, so one of the two spin-components still dominates each Kohn–Sham eigenvalue, a situation that is quite similar to the case in Fig. 2.3, where all states have pure spin quantum character.

An additional direct consequence of adding spin–orbit coupling to the Kohn–Sham equation is that an orbital moment now emerges as a property. Since both spin and orbital moments may be measured experimentally in magnetic materials, for example, from ferromagnetic resonance experiments (see Chapter 8), or from neutron scattering experiments, it is important to be able to theoretically analyse these separate contributions to magnetism. In most applications of DFT, the spin and orbital moments are evaluated from the expressions

$$m_s = \sum_i^{occupied} \sum_k \langle \psi_{ik} | \sigma_j | \psi_{ik} \rangle, \tag{2.17}$$

$$m_l = \sum_i^{occupied} \sum_k \langle \psi_{ik} | l_j | \psi_{ik} \rangle, \tag{2.18}$$

where $j = \{x, y, z\}$. Normally, one of the j-components dominates the expressions for spin and orbital magnetism in Eqn (2.17); consequently, spin moments and orbital moments are frequently reported only from this component, normally chosen as the z-component. There are, however, cases where the x- and y-contributions to the expectation values in Eqn (2.17) are also significant, sometimes even leading to there being a small angle between the calculated total spin moments and the calculated total orbital moments (Solovyev, 2005; Grånäs et al., 2014), that is, to their being non-collinear.

The way to include spin–orbit coupling, as discussed so far in this section, is to add it to the Kohn–Sham equation, Eqn (1.16). We have also discussed the consequences this has for the symmetry of the lattice; but, for valence states, this interaction is actually not the strongest relativistic correction. The most accurate Schrödinger-like (or Kohn–Sham-like) equation also includes relativistic terms labelled mass velocity and Darwin shift. These corrections follow from a transformation of the one-particle Dirac equation to an effective one-particle Pauli–Schrödinger equation (Foldy and Wouthuysen, 1950; Andersen, 1975; Koelling and Harmon, 1977). For Coulomb potentials, the Darwin term is finite only for $l = 0$, that is, s orbitals. The negative mass-velocity correction is a term which decreases in magnitude with increasing l. Adding these two terms to Eqn (1.16), while ignoring the spin–orbit interaction, is normally referred to as the scalar relativistic approximation.

Physically, the Darwin term is a result of the increase of the wave function at the nuclear site. A consequence of this enhanced localization in space, arguing from the view point of the Heisenberg uncertainty principle, is an increase of the electron momentum, and hence the kinetic energy. Since, in Schrödinger-like equations, only s states have a non-vanishing amplitude at $\mathbf{r} = 0$, the Darwin shift appears only for $l = 0$. The mass-velocity term is caused by the relativistic mass enhancement of fast particles, as this enhancement reduces the importance of the kinetic energy, hence leading to a contraction of the electron orbits; in addition, when the electrons move to regions where the potential is more attractive, the one-particle energy becomes lowered. An alternative to adding relativistic corrections to Eqn (1.16) is to consider the solution to the Dirac equation, in an effective potential given by Eqn (1.22). This is also an effective one-electron treatment, with the advantage that it is formally and practically the most precise description, since relativistic effects are naturally included. It is relevant to compare solutions of the Dirac equation to solutions given by Eqn (1.16) with relativistic corrections added, in order to estimate errors of the latter treatment. For this reason, we show in Fig. 2.6 the density of states of one of the heaviest elements of the Periodic Table, Pu, where relativistic effects are expected to be most important (Figure redrawn from O. Eriksson, 1989). In the upper panel, the Dirac equation has been solved, using an effective potential based on local density approximation whereas, in the lower panel,

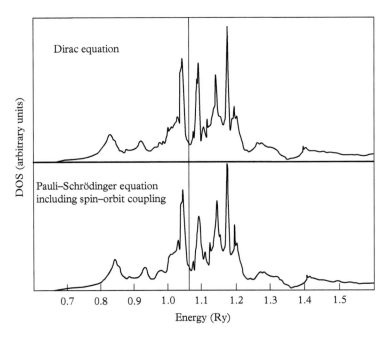

Figure 2.6 *Density of states of face-centred cubic Pu for a Wigner-Seitz radius of 3.10 a.u. The Fermi level is marked by a vertical solid line, and energies are given in Rydbergs (1 Ry equals 13.6058 eV). Figure redrawn after Eriksson, O., (1989), 'Electronic Structure, Magnetic and Cohesive Properties of Actinide, Lanthanide and Transition Metal Systems', PhD thesis, Uppsala University, Uppsala.*

the density of states from a Kohn–Sham equation of the same form as in Eqn (1.16), albeit with relativistic corrections, is shown. It may be seen that the two approaches result in extremely similar density of states functions, thus illustrating that adding relativistic corrections to Eqn (1.16) is indeed an excellent approach in most cases. Most modern computational software for solving the Kohn–Sham equation has the ability to include relativistic effects, in either of the two methods discussed above.

2.6 Green's function formalism, Heisenberg exchange, and a multiscale approach to spin dynamics

In Fig. 2.7, the result of a self-consistent calculation of magnetization density is shown for a typical itinerant ferromagnet, bcc Fe (the same result was also shown at a different scale in Fig. 2.4). It may be seen that the magnetization density is located primarily around the atomic nuclei and that the interstitial region contributes negligibly to the magnetic moment. This makes it possible to replace the quantum mechanical description, as provided by DFT, where the relevant information is provided by the difference of spin-up and spin-down densities, with an atomistic description of the magnetism, as shown in the upper right corner of Fig. 2.7 and represents a crucial step in a multiscale description of magnetization dynamics, where a property over some part of space, for example the magnetization density, is replaced by an averaged or integrated property, for example, the magnetic moment. It is desirable to take this step in such a way that dynamical properties are the same, or very similar, on the two length scales, thus making it possible to carry out simulations for larger systems, without loss of accuracy. In Chapters 4 and 5, such an approach will be derived, but we note here that one of the key properties needed in this process is the effective exchange interaction between any two atomic spins, for example, as shown in the upper right corner of Fig. 2.7. Before we continue, let us make a comment on time scales. The magnetization density in Fig. 2.7 is

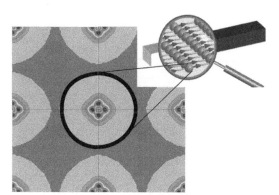

Figure 2.7 *Schematic illustration of multiscale spin dynamics. Darker regions inside spheres centred at atomic nuclei represent high magnetization density, and lighter regions represent low magnetization density. The homogenous grey region in the interstitial region between these spheres represents the lowest magnetization density (that in some cases even becomes negative). Integration of the magnetization density (coloured contours) over an 'atomic volume' can conceptually be replaced by atomic moments, as shown in the upper right corner of the figure, that then hold the relevant magnetic information of a material. Picture kindly provided by Prof. Olof Karis.*

generated by electrons moving through the material, where, for a spin polarized material, some regions of space contain more spin-up, or majority, electrons, than spin-down, or minority, electrons. If the time-dependent Kohn–Sham equation, Eqn (1.30), had been used to generate this, the fluctuations around the time-independent density shown in Fig. 2.7 would be apparent. One might ask with which frequency such fluctuations occur. This depends, of course, on the system at hand, and whether its magnetization is driven by an external stimulus, like a laser. Without an external driving force, the frequency of such fluctuations must come from the electrons as they travel through the lattice, spending some time at one atomic site before jumping over to the next. Since the magnetization density is generated by the electrons, the answer to the question about the typical frequencies of the oscillations of the magnetization density, Eqn (1.28), should be given by the frequencies with which electrons typically enter or leave an atomic site. In order to estimate this, we note that a typical distance between atoms in Fig. 2.7 is a few Ångströms, and the typical drift velocity is of the order of a percent of the speed of light. From these estimates of distance and velocity, one can conclude that, approximately once every tenth of a femtosecond, an electron will leave or enter an atomic site. If we now are interested in the magnetization dynamics of atomic spins, according to Fig. 2.7, over time scales larger than 10–100 fs, an atomistic description should be valid. In this case, we can ignore the smaller, faster fluctuations of the magnetization density, adopt a Born–Oppenheimer-like (adiabatic) approximation for the magnitude and direction of the atomic spins, and let these two quantities evolve in time. In Chapter 4, we return to details of how this is done. The dynamics of atomic spins is to a large extent governed by the interatomic exchange interaction, and hence the calculation of this interaction from DFT is a crucial step in establishing a method for atomistic spin dynamics simulations. The possibility of calculating the interatomic exchange interactions of an effective spin Hamiltonian, for example, of Heisenberg form, in a materials-specific way, via DFT theory, was shown by Liechtenstein, Katsnelson, and Gubanov (1984) and Liechtenstein, Katsnelson, Antropov and Gubanov (1987). They demonstrated that the parameters of an effective spin-Hamiltonian of the form

$$\mathcal{H}_{\text{Heis}} = -\frac{1}{2} \sum_{i \neq j} \mathcal{J}_{ij} \mathbf{m}_i \cdot \mathbf{m}_j, \tag{2.19}$$

can be evaluated from a multiple scattering formulation of DFT. In Eqn (2.19), $\mathbf{m}_i$ denotes an atomic magnetic moment of a particular site, i. The interatomic exchange parameter, denoted $\mathcal{J}_{ij}$, describes the energy change of the system when two moments at site i and j are rotated from their original orientation, for example, that of a collinear ferromagnetic alignment. As we will see in Chapter 4, an inter atomic exchange field is responsible for the time evolution of each atomic spin (see Eqn (4.1)). This exchange field can be obtained from an effective spin Hamiltonian, such as the one in Eqn (2.19). Hence, the ability to evaluate from DFT all terms of this effective spin Hamiltonian is crucial in the multiscale approach that forms the basis of modern atomistic spin dynamics simulations.

Since the arguments behind the derivation of interatomic exchange interactions and their connection to concepts from DFT are based on Green's functions of quantum systems (Economou, 1979), we briefly recapitulate some of the more important aspects of these objects. In Eqn (2.5), the Kohn–Sham equation is written as an effective operator, h_{eff}, acting on Kohn–Sham orbitals. For a Hermitian, time-independent operator that has a complete set of eigenstates, one can derive the relationship (Economou, 1979)

$$(z - \hat{h}_{\text{eff}}) \hat{G}(z) = 1, \tag{2.20}$$

where z is a complex energy variable whose real part is the energy E, as shown, for example, on the x-axis in the plots in Figs 2.5 and 2.6. Here, the Green's function $G(z)$ enters and is seen to basically be the inverse of the complex energy minus the Kohn–Sham operator. The real space representation of the Green's function is obtained via

$$G(\mathbf{r}, \mathbf{r}', z) = \langle \mathbf{r} | G(z) | \mathbf{r}' \rangle, \tag{2.21}$$

and, using this representation, it is possible to define a Green's function that connects two sites that are the centre of an atomic nucleus, at position $\mathbf{R}_i$ and $\mathbf{R}_j$ as $G(\mathbf{R}_i, \mathbf{R}_j, z)$ or $G_{ij}(z)$ for short. For spin polarized electronic structures, one obtains effective Hamiltonians for spin-up and spin-down states, Eqn (1.26), and corresponding Green's functions as

$$(z - \hat{h}_{\text{eff}}^{\alpha}) \hat{G}^{\alpha}(z) = 1,$$
$$(z - \hat{h}_{\text{eff}}^{\beta}) \hat{G}^{\beta}(z) = 1. \tag{2.22}$$

It was shown, after some algebra associated to multiple scattering theory, that the leading orders of the changes of the total energy, when any two atomic sites have the orientation of the magnetic moment modified with some small angle θ from the ground-state configuration, are proportional to θ^2 (Liechtenstein, Katsnelson, and Gubanov, 1984; Liechtenstein, Katsnelson, Antropov, et al., 1987). For small values of θ, Eqn (2.19) also results in energy excitations proportional to θ^2, and hence it was demonstrated that the interaction strength, $\mathcal{J}_{ij}$, between two magnetic moments, defined in Eqn (2.19), follows from multiple scattering theory, yielding the expression

$$\mathcal{J}_{ij} = \frac{\text{Im}}{4\pi} \int_{-\infty}^{E_F} \text{Tr} \left[\delta_i(E) G_{ij}^{\alpha}(E) \delta_j(E) G_{ji}^{\beta}(E) \right] dE, \tag{2.23}$$

where the trace is over orbital indices, and G_{ij}^{σ} is the Green's function that connects sites i and j for electrons with spin σ. The quantity $\delta_i(E)$ has units of energy and yields the local exchange splitting between spin-up and spin-down states, at site i. In transition metals, the exchange splitting δ_i is largest for the d states, and hence these states normally dominate the interatomic exchange interaction, according to Eqn (2.23), at least for short-range exchange. It should be noted that the derivation by (Liechtenstein, Katsnelson, and Gubanov, 1984; Liechtenstein, Katsnelson, Antropov, et al., 1987) was made for a non-relativistic treatment of a collinear magnetic system. Hence, the expression for interatomic exchange, Eqn (2.23), is in principle not relevant for a magnetic material at finite temperature, where the magnetic ordering typically is non-collinear. However, as we shall see in Chapters 7, 8, and 9, using exchange parameters from Eqn (2.23) for finite temperature simulations of the magnetization dynamics is often a very good approximation. We note, however, that generalizations of Eqn (2.23) to non-collinear situations (Szilva et al., 2013) and to relativistic cases (Udvardi et al., 2003) have been presented. The latter is important, since it leads to the Heisenberg exchange, as well as to the Dzyaloshinskii–Moriya interaction. We will show some examples of exchange parameters in Chapter 3, where we also will discuss their behaviour and microscopic origins.

We first end this chapter with a comparison of the forces between particles in molecular dynamics simulations, and the forces or torques between atomic moments in atomistic spin dynamics simulations, described in detail in Chapters 4 and 7. In molecular dynamics simulations, the forces are obtained by connecting information from the electronic structure, via the Hellman-Feynman force, which then allows for a dynamical treatment and time evolution of the nuclei (or atoms) that constitute the solid. In atomistic spin dynamics simulations, the corresponding information is provided by Eqn (2.23), which hence may be seen as the spin dynamics version of the Hellmann–Feynman theorem.

3

Applications of Density Functional Theory

In this chapter, we give examples of how density functional theorey (DFT) describes some of the most basic magnetic properties of a material. This involves spin and orbital moments, Heisenberg exchange parameters, and magnetic form factors. Relativistic effects couple spin and orbital space and make magnetic materials anisotropic, which means that the ground-state magnetization has a preferred orientation, which is most frequently parallel to a high symmetry-direction of the crystalline structure. We also illustrate how well DFT describes cohesive properties and how magnetism influences these properties. These examples serve to give a general picture of how well DFT, as described in Chapters 1 and 2, can reproduce relevant features of magnetic materials, as well as to illustrate how the onset of spin polarization can have a drastic influence on all the properties of a material.

3.1   Cohesive and structural properties

We start by discussing how magnetism influences chemical bonding and phase stability. In order to do this, it becomes relevant to compare the properties of magnetic materials with those of iso electronic non-magnetic materials. Hence, we show in Fig. 3.1 the nearest-neighbour distance of the $3d$, $4d$, and $5d$ transition metals (Janthon et al., 2014). Note that, for simplicity, the α-phase of Mn, with a basis of 29 atoms (D. Young, 1991) is not shown in this figure. This complex magnetic material was studied with the use of first-principles theory by Sliwko et al. (1994), who observed good agreement between measured and calculated moments; however, this study is not discussed further here. Figure 3.1 reflects the typical size of an atom and compares theoretical data with measurements. It may be seen that, overall, there is good agreement between theory and the measurements. This is the case for both non-spin polarized materials and spin polarized, magnetic materials. The trend of the nearest-neighbour distance, when a transition metal series is traversed, is that the curve follows a more or less well-defined parabolic trajectory. This reflects the filling of bonding orbitals in the early part of the series, and anti-bonding states for the heavier elements of a given series, a trend which may be

Atomistic Spin Dynamics. Olle Eriksson, Anders Bergman, Lars Bergqvist, Johan Hellsvik. First Edition.
© Olle Eriksson, Anders Bergman, Lars Bergqvist, Johan Hellsvik 2017. First published in 2017 by Oxford University Press.

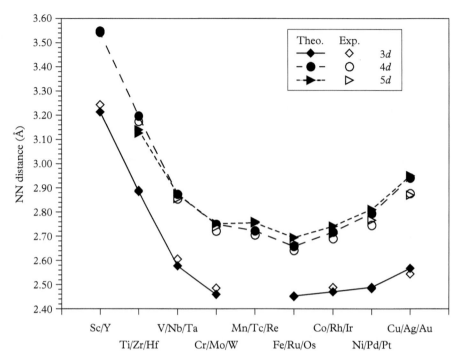

Figure 3.1 *Nearest-neighbour (NN) distances of 3d, 4d, and 5d transition metal elements. Theoretical (Theo.) data given as filled symbols, and experimental, (Exp.) data as open symbols. Picture kindly provided by Debora Carvalho de Melo Rodrigues.*

explained by the Friedel model (Harrison, 1989). Comparing different transition metal series, one can observe that the nearest-neighbour distances, or equivalently the atomic volumes, of the $3d$ elements in general are significantly smaller than the isoelectronic counterparts of the $4d$ and $5d$ series. This is due to the significantly smaller radial extent of the $3d$ wave function, as compared to the $4d$ and $5d$ wave functions. Since the $4d$ wave functions are required to be orthogonal to the $3d$ states, which are the filled core states for the $4d$ elements, they are forced to be more extended. Similarly, the $5d$ wave function is required to be orthogonal to the $3d$ and the $4d$ core levels and hence becomes even more extended. Consequently, the valence electron bands of the $5d$ elements are wider than those of the $4d$ elements, which in turn are wider than those of the $3d$ elements, although, in the latter case, the difference is not so dramatic. This also is part of the reason why the volumes of the $3d$ elements are smaller than those of isoelectronic $4d$ and $5d$ elements. Calculated electronic density of state (DOS) values can be structures and found in publically available databases.[1]

---

[1] e.g. the Electronic Structure Project (http://gurka.fysik.uu.se/ESP) and the Materials Project (http://materialsproject.org).

A similar comparison can be made between the lanthanide and actinide series, where the $4f$ wave function of the lanthanides has significantly smaller radial extent, compared to the $5f$ wave function of the actinides. The effect of magnetism is also visible for the equilibrium volumes of Fe, Co, and Ni. For the late $4d$ and $5d$ elements, the Friedel model results in a rather sharp upturn of the nearest-neighbour distances, since anti-bonding states start to become filled. This is seen from Fig. 3.1 to give a distinct minimum at Ru and Os. For the isoelectronic $3d$ element Fe, the minimum is much less conspicuous. In fact, the nearest-neighbour distances of Fe, Co, and Ni are rather similar to each other, giving a trend without a marked minimum, in contrast to the trend of the $4d$ and $5d$ series. This difference is caused by the onset of magnetism and spin-polarization of the electron states, which reduces the strength of the chemical bonding, leading to a magnetovolume effect, with increased nearest-neighbour distances. Had Fe, Co, and Ni been non-magnetic or non-spin polarized materials, they would have had stronger chemical binding with a resulting smaller nearest-neighbour distance than that shown in Fig. 3.1, and this would have led to a marked minimum for Fe. However, the large gain in exchange energy of body-centred cubic (bcc) Fe, hexagonal close-packed (hcp) Co, and face-centred cubic (fcc) Ni prevents this from happening and, instead, spin polarization sets in. For bcc Cr, which orders antiferromagnetically (to be precise, bcc Cr has a spin density wave, as described by Fawcett et al. (1994)) with relatively small exchange splitting in relation to the band width, the magnetovolume effect is less visible.

The total energy is, in density functional theory, composed of a kinetic part and the electrostatic electron–electron and electron–nucleus interactions, as well as the exchange energy, as discussed in Chapter 1. The presence or absence of magnetism in itinerant electron systems can to a good description be analysed as the balance between kinetic and exchange energy, which both depend on the spin polarization to a greater extent than the electron–electron and electron–nucleus interaction energies. In fact, these two interactions only depend on the electron density, and not the magnetization density, so it is possible to neglect them when analysing the possibility of the spin polarization of a material (Mohn, 2003). In order to illustrate this we consider a simplification of the DOS, namely the square shape, as shown in Fig. 3.2. The kinetic energy can be calculated from Eqn (1.24), where the sum of eigenvalues is evaluated from the integral $\int^{E_F} D \cdot E \, dE$ and, for the spin-degenerate case, as shown in the upper part of Fig. 3.2, we choose for simplicity a reference level of the energy so that the double-counting contribution to Eqn (1.24) becomes 0. In this way, we get for the spin-degenerate case a kinetic energy of $D(E_F^2 - W^2)$. For the spin polarized case, the integration of one-electron energies is straightforward, but since this corresponds to a case where an exchange splitting $\pm\Delta$ appears in the effective potential of states with spin down and spin up, respectively, one has to include this Δ in the double counting of Eqn (1.24). The resulting kinetic energy is in this case $D(E_F^2 - W^2) + D\Delta^2$. The exchange splitting of the spin polarized case results in a spin polarization, $m = 2D\Delta$. This spin polarization lowers the exchange and correlation energy, and in an approximate expression presented by M. Brooks and Johansson (1983), the energy is found to be lowered by $\frac{Im^2}{4}$, where I is the Stoner exchange parameter. Note that this mathematical form is consistent with

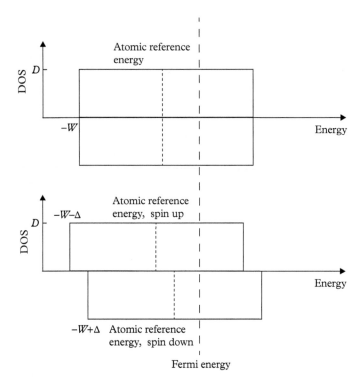

Figure 3.2 *Square-shaped density of states, of a spin-degenerate situation (upper panel) and a spin polarized situation (lower panel). The constant value of the DOS is D, and W is the bottom of the DOS for the spin-degenerate case. For the spin polarized case, the spin-up states are shifted down with Δ, and the spin-down states are shifted up with Δ, so that the Fermi energy is the same as in the spin-degenerate case. In the figure, the atomic reference energy, which separates bonding from anti-bonding states, is also drawn.*

the spin-pairing energies shown in Fig. 1.1. Balancing the increase of the kinetic energy with the reduction of exchange energy leads to the established Stoner criterion: $I \cdot D = 1$, signalling a situation where kinetic and exchange energy are equal, at least for small values of Δ. If the Stoner product, $I \cdot D$, is larger than 1, exchange interaction dominates the kinetic contribution, and ferromagnetism occurs. This happens for systems with narrow bands, where the value of D becomes sufficiently large and spin polarization sets in spontaneously to form a ferromagnetic state. Although based on simple arguments, the Stoner criterion is surprisingly accurate in predicting whether a material is ferromagnetic, simply by inspection of the DOS at the Fermi energy and with knowledge of the Stoner exchange integral. Accurate tables of the latter can be found in the work by M. Brooks and Johansson (1983). Among all the transition metal elements, it is only bcc Fe, hcp Co, and fcc Ni that have a Stoner product that exceeds 1, in agreement with their experimentally observed ferromagnetic states. An example of an element that has a Stoner product close to 1 is fcc Pd. For this reason, it is possible to induce large magnetization clouds in Pd by doping with a magnetic impurity, for example, Fe. As a matter of fact, huge induced moments have been observed in a Pd matrix doped with Fe, as every Fe impurity is associated with a magnetic spin moment of 10–$12\,\mu_B$ (Craig et al., 1962). This large moment is then induced in the Pd matrix, forming an extended spin cloud with a radius of ~ 10 Å and which involves significant moments several shells away from the Fe impurity. Stoner theory has been used in the past and is still used, with arguments such as the ones presented for Fig. 3.2, being generalized for the multi-orbital case (M. Brooks et al., 1988) as well as for a fully relativistic scenario (O. Eriksson et al., 1989).

The kinetic energy of a solid is connected to the dispersion of the electron states, that is, the degree to which electrons jump around from one lattice point to another, and thus provides a mechanism of stability for a material, since the energy is lower when the states are dispersive than when all the electrons stay put on an atomic site. This can be seen from Fig. 3.2 by comparing the energy of all electrons located at the atomic reference energy, that is, in the non-bonding case for an atom, to that obtained when the states from the bottom of the band to the Fermi energy are occupied, as the latter is clearly lower. The kinetic energy of the band situation is hence always lower than that of the non-binding case and provides a very strong mechanism for chemical bonding. However, the example in Fig. 3.2 shows that the lowering of the kinetic energy becomes less pronounced when spin polarization sets in. The simplest way to see this is to analyse how bonding and anti-bonding states are filled. In the upper part of Fig. 3.2, a situation is drawn where all the bonding states are filled while only some of the anti-bonding states are filled, for both spin-up and spin-down states. In the spin polarized case, shown in the lower panel of Fig. 3.2, a larger fraction of anti-bonding states are filled for the spin-up states than for the spin-down states. As analysed above, this increases the kinetic energy by the amount $D\Delta^2$, compared to the spin-degenerate situation. This is the microscopic explanation of the magnetovolume effect displayed in Fig. 3.1. Since Fe has a larger moment than Co or Ni, its value of Δ is larger, and hence its kinetic energy is influenced the most, with a correspondingly large magnetovolume effect. For Co the effect is slightly smaller and, for Ni, it is smaller yet. The balance between the magnetovolume effect and the expected parabolic trend of the nearest-neighbour distance of non-spin polarized states of $3d$ elements leads to Fe, Co, and Ni having very similar atomic volumes, as shown in Fig. 3.1.

If one subtracts the total energy of the atom from the total energy of the solid, one obtains the cohesive energy. This energy can also be compared directly to experimental data, as is done in Fig. 3.3 for the $3d$, $4d$, and $5d$ elements. It should be noted here that such calculations must be made with some care since the electronic configuration, that is, the numbers of s, p, and d electrons, of the atom may differ from that of the solid. As seen in Fig. 3.3, DFT reproduces experimental data with good accuracy, for all three transition metal series. As expected from the Friedel model, the trend of the cohesive energy is parabolic, with a maximum value in the middle of the series, since all bonding states are filled while anti-bonding states are empty. For the late $3d$ series, the influence of magnetism and spin polarization is clear, since the trend is drastically different compared

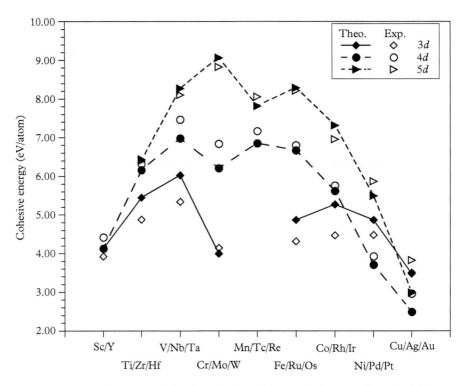

Figure 3.3 *Cohesive energy of 3d, 4d, and 5d transition metal elements. Theoretical (Theo.) data are given as filled symbols, and experiment al (Exp.) data as open symbols. Picture kindly provided by Debora Carvalho de Melo Rodrigues.*

to the isoelectronic 4d and 5d elements. Again, this is easily understood from the analysis of Fig. 3.2, where spin polarization is shown to diminish the influence of the kinetic energy.

We now turn our attention to the structural properties, since it is here that magnetism is very clearly influential. We start by commenting on the work of Skriver (1985), who showed that the crystal structures observed for all elements of the Periodic Table are, with very few exceptions, reproduced by DFT, that is, the approach described in Chapters 1 and 2. More importantly, it was pointed out that the distinct features of the DOS function of any crystal structure, for example, bcc, lead to an easy understanding of the crystal's structural stability. From this analysis, it was argued that isoelectronic elements, for example, from among the transition metals, should have identical crystal structures. This is, indeed, also the case for most elements that do not spontaneously develop magnetic ordering. For example, the trivalent elements Sc, Y, and Lu form in the hcp structure, as do tetravalent Ti, Zr, and Hf. The Group V elements V, Nb, and Ta all form in the bcc structure at low temperatures, as do the isoelectronic elements Cr, Mo, and W. For the late transition elements, the similarities of crystal structures among isovalent elements hold well for the 4d and 5d series. However, for the 3d series, magnetism disrupts this trend. For instance, Tc and Re have the same crystal structure, that is, hcp, and this structure is also shared by Ru and Os. For the heavier $4d$ and $5d$ transition metals, the trend of isoelectronic elements having similar structures continues, as elements in Groups IX–XI all have the same crystal structure: fcc. As Fe is isoelectronic with Ru, it would for this reason be expected to crystallize in the hcp phase, in contrast to the observed bcc structure. Co is isoelectronic with Rh and is hence expected to crystallize in the fcc structure, not in the observed hcp structure. It was pointed out by Söderlind et al. (1994) that, since Fe, Co, and Ni are ferromagnetic, their structures cannot be expected to be the same as those of their non-magnetic, isoelectronic counterparts, since the filling of the DOS function in ferromagnetic elements is different from that in non-magnetic elements. Taking bcc Fe as an example, we may see from Fig. 2.5 that the spin-up states are essentially filled, in such a way that both bonding and anti-bonding states are filled. A major chemical activity of this spin channel is hence not expected. However, the spin-down states are essentially half filled, so that the Fermi energy ends up in a pseudogap. This is a very favourable situation, since the eigenvalue sum used to calculate the kinetic energy, Eqn (1.24), is particularly low when much of the weight of the DOS function is located at lower energies, away from the Fermi energy. The band filling of the spin-down band of bcc Fe is in fact similar to the band filling of the degenerate spin-up and spin-down states of Nb, Mo, Ta, and W, all elements that form in the bcc structure. Hence, Fe should, from a structural point of view, not be compared to isoelectronic elements but rather to the Group V and VI elements. This is illustrated in Fig. 3.4. One may conclude that the electronic structure of spin polarized Fe dictates the stability of the bcc structure whereas, in the absence of spin polarization, the hcp phase should be stable. As a matter of fact, at high applied pressures, where the kinetic energy of band formation dominates over the spin-pairing energy, such that the Stoner criterion is not fulfilled, Fe is observed in the hcp structure (Saxena et al., 1994). In a similar fashion, the shape of the DOS and the band filling of spin polarized Co is similar to the situation in paramagnetic Tc, Ru, Re, and Os, and the observed hcp phase appears naturally. The connection of band filling of the $3d$ and $4d/5d$ elements is shown schematically in Fig. 3.4.

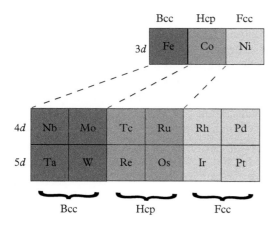

Figure 3.4 *The connection between band filling in spin polarized 3d elements and that in non-magnetic 4d/5d elements, and the coupling of band filling with crystal structure stability. Filling spin-down states in spin polarized body-centred cubic (bcc) Fe gives a band filling similar to that of non-magnetic Nb, Mo, Ta, and W. The common bcc crystal structure is hence natural. The same connection is made for hexagonal closed-packed (hcp) Co and Tc, Ru, Re, and Os; and for face-centred cubic (fcc) Ni and Rh, Pd, Ir, and Pt.*

3.2 Spin and orbital moments, and the magnetic form factor

In Fig. 1.3, we saw one example of how theoretically obtained magnetic spin moments compare to experimental ones. In this section, we broaden this discussion and make a similar comparison for several classes of materials. These results are shown in Table 3.1, where we have collected spin and orbital moments of some relevant materials that are discussed in detail below. We note first that, according to Fig. 2.4, the magnetization density is distributed over the entire unit cell and not just in the region close to the nuclei. This means that, particularly for compounds involving different atomic elements, it may not be straightforward to define a moment that is uniquely associated with a specific atom or atom type. However, as Fig. 2.4 shows for bcc Fe, most of the magnetization density is located within a sphere around the nuclei; hence, an 'atomic' description becomes relevant. For most materials, over 95 % of the magnetic spin moment is located in this sphere, with the rest representing a diffuse magnetism distributed over the interstitial region. The orbital moment often has an even more dominant contribution from a smaller region close to the nuclei, since it is a spin–orbit driven entity and, as discussed in Section 2.5, the spin–orbit coupling is dominated by contributions close to the nuclei. Hence, it is relevant to describe the magnetism of a material as being composed of a main atomic part and a smaller, diffuse part, as we will do below. It should be remembered, however, that the distribution of atomic and diffuse moments depends to some extent on the volume of the region used when integrating the magnetization density and that it is suitable to make this volume as large as possible. The total moment, that is, the sum of the atomic and the diffuse parts is, however, not sensitive to this choice.

In Table 3.1, we show calculated and experimentally derived spin moments and orbital moments for selected elements and compounds. For the elements, the total spin and orbital moments, including the atomic and diffuse parts are shown whereas, for the compounds, we only show the atomic contribution since, as mentioned, it is difficult to associate different diffuse contributions with different atomic types. For the magnetic transition metals bcc Fe, hcp Co, and fcc Ni (Stearns, 1984; Daalderop et al., 1991; Hjortstam et al., 1996) as well as for the rare-earth element hcp Gd (Moon et al., 1972; Roeland et al., 1975; M. Brooks et al., 1992), the agreement between theory and experiment is good for all elements concerning the spin moment. Gd is seen to have the largest spin moment, which primarily is due to the seven $4f$ electrons that are essentially atomic-like and form a spin moment of 7 μ_B/atom. These spin polarized $4f$ electrons couple their exchange field, via the local spin density approximation (LSDA) potential, to the itinerant s-, p-, and d-valence electrons, inducing a non-negligible theoretical moment of 0.65 μ_B/atom. This induced moment can be compared to a corresponding experimental value obtained as the difference between the measured total moment of 7.62 μ_B/atom, and the moment expected for the $4f$ shell, 7 μ_B/atom. It is clear that theory and experiment agree rather well for the induced magnetic moment of the valence states of Gd, and it may also be noted that this moment is essentially as large as the magnetic moment of fcc Ni. The level of agreement between theoretical and measured spin moments shown in Table 3.1 for elements also holds for the spin moments of ferromagnetic compounds

Table 3.1 *Calculated and experimentally derived atom-projected spin and orbital moments for selected transition metal elements and compounds, as well as the rare-earth element Gd. The local spin density approximation calculations for NiO are derived from work by Norman (1991) and Fernandez et al. (1998).*

Element/compound	μ_s	μ_l
Bcc Fe (theory-LSDA)	2.19	0.05
Bcc Fe (expt)	2.13	0.08
Hcp Co (theory-LSDA)	1.59	0.08
Hcp Co (expt)	1.52	0.14
Fcc Ni (theory-LSDA)	0.61	0.05
Fcc Ni (expt)	0.57	0.05
Hcp Gd (theory-LSDA)	7.65	—
Hcp Gd (expt)	7.62	—
Hex Fe_2P site I (theory-LSDA)	0.96	—
Hex Fe_2P site I (expt)	1.03	—
Hex Fe_2P site II (theory-LSDA)	2.04	—
Hex Fe_2P site II (expt)	1.91	—
$L1_0$ FePt (theory-LSDA)	Fe 2.93/Pt 0.36	Fe 0.067/Pt 0.45
$L1_0$ FePt (expt)	Fe 2.92/Pt 0.47	Fe 0.2
NiO (theory-LSDA) (1)	1.08	0.15
NiO (theory-LSDA) (2)	1.59	—
NiO (theory-LSDA+DMFT)	1.85	—
NiO (expt)	1.90	0.32
Bcc Cr (theory)	0.5 (LSDA) 0.9–1.4 (GGA)	—
Bcc Cr (expt)	0.6	—

Note: DMFT = dynamical mean field theory; expt = experimental; GGA = generalized gradient approxima-tion; LSDA = local spin density approximation; μ_l = orbital moment; μ_s = spin moment.
Source: Fernandez, V., Vettier, C., de Bergevin, F., Giles, C., and Neubeck, W. (1998). Observation of orbital moment in NiO. *Phys. Rev. B*, 57, 7870–6; Norman, M. R. (1991). Crystal-field polarization and the insulating gap in FeO, CoO, NiO, and La_2CuO_4. *Phys. Rev. B*, 44, 1364–7.

like Fe_2P (Fujii et al., 1979; Wäppling et al., 1975; Ishida et al., 1987; O. Eriksson et al., 1988) and FePt (Daalderop et al., 1991; Staunton et al., 2004; Burkert et al., 2005; C. Sun et al., 2006; Antoniak et al., 2009), compounds with relatively wide bands, where electron–electron interaction is expected to be reproduced well by LSDA. In the Fe_2P crystal structure, space group $P\bar{6}2m$ (Fujii et al., 1979), there are two Fe sites—one in the octahedral position, and one in the tetrahedral position—and experiments show that there is quite a drastic difference between the magnetic moments of these atoms. It is rewarding that theory, as Table 3.1 shows, can predict this large difference, as two atoms of the one and the same element experience a seemingly similar crystallographic environment. We note also that the P atom in Fe_2P only has a tiny moment which is due to exchange interaction with the magnetic Fe atoms and which is an induced moment that is not discussed further here. Another magnetic compound listed in Table 3.1 is FePt, space group $P4/mmm$, which is of technological interest due to its large magnetic anisotropy. It has a reported spin moment for both the Fe and the Pt sites, as Table 3.1 shows and, for both sites, the agreement between measured and observed values is good. The largest uncertainty of the measured moments of FePt is for the Pt site, and it is not clear if the difference between measurements and calculations are due to inaccuracies of the theory. The element specific probe used by Antoniak et al. (2009) is the so-called X-ray magnetic circular dichroism (XMCD) technique, which is known to sometimes give measured values that need calibration against values obtained with other techniques. It therefore becomes meaningful to compare the atomic resolved moments of FePt, listed in Table 3.1, to the total magnetic moment, which is obtained from magnetometry and which naturally incorporates all contributions to the magnetic moment, both atomic and interstitial. Measured values of the total moment of FePt, using magnetometry, have been reported to be ~ 3.4 μ_B/f.u. (Daalderop et al., 1991 and references therein). This moment coincides with the sum of XMCD values of the spin moments of Fe and Pt and is in good agreement with the calculated value from theory (3.3 μ_B/f.u.).

Many more examples of comparisons between measured and LSDA calculations of spin moments have been published, with similar good agreement between theory and observed spin moments, as shown in Table 3.1. However, exceptions exist to this excellent trend where theory and experiment normally agree with errors not exceeding 5–10 %. For instance, for antiferromagnetic NiO, the measured (Norman, 1991; Fernandez et al., 1998) and LSDA-calculated spin moments are in poor agreement, as Table 3.1 shows (Cheetham and Hope, 1983). The experimentally determined spin moment is significantly larger than the spin moment from an LSDA calculation. For NiO, the $3d$ levels form a narrow set of bands and, for such systems, mean field theories like LSDA do not capture well the energetics of electrons jumping around from lattice site to lattice site and causing fluctuations of the electron occupation of, for example, the Ni site, and involving large Coulomb energies of the order of 7–10 eV, which is the an estimated energy of repulsion between two electrons residing on the same Ni site in NiO. However, a treatment which includes correlations in a more direct way, like dynamical mean field theory (DMFT Metzner and Vollhardt, 1989; Anisimov et al., 1997; Kotliar et al., 2006), performs much better, as Table 3.1 shows (Kvashnin et al., 2015).

Next, we discuss a particularly difficult element, bcc Cr (Fawcett, 1988; Singh and Ashkenazi, 1992). As Table 3.1 shows, different parametrizations of the exchange and correlation energy result in drastically different magnetic moments, some of which agree with the measured values, and some that deviate very much. Although LSDA is formally expected to be a less accurate approximation than generalized gradient approximation (GGA), it is clear from Table 3.1 that LSDA performs better than any of the GGA functionals tested by Singh and Ashkenazi (1992) and which are quoted in Table 3.1. There is no clear reason why theory should describe bcc Cr so much worse than it does for Fe, Co, and Ni and, for this reason, bcc Cr is in some sense an unsolved mystery, at least from a DFT point of view. The $3d$ bands of Cr are not narrower than the $3d$

bands of Fe, Co, and Ni, and hence electron correlations are not expected to be more important for bcc Cr. It is, however, important to note that experimental studies show that bcc Cr is an antiferromagnetic spin density wave where nearest-neighbour magnetic moments are coupled antiferromagnetically: a magnetic structure that is superimposed with a long wavelength modulation. The theoretical results listed in Table 3.1 are for a normal antiferromagnetic phase without long wavelength modulation and hence do not correspond exactly to the experimental configuration. However, it has been argued (Singh and Ashkenazi, 1992) that the long wavelength modulation of the spin density wave of Cr cannot explain the large difference between calculated and measured moments for this intriguing element.

As concerns the orbital moment, it should first be noted that, for Fe, Co, and Ni, it is very small compared to the spin moment. This quenching of the orbital moment is explained by the crystal field effect, which arises in a crystal and results in electron states with equal admixtures of positive and negative magnetic quantum numbers. Spin–orbit coupling breaks this equal admixture and, as a result, a small orbital moment develops. As Table 3.1 shows, theory gives the right order of magnitude of the orbital moment and, for fcc Ni, it even gives the correct value. However, for bcc Fe and hcp Co, the calculated orbital moment is ∼40 % smaller than the measured values. Recent calculations based on DMFT by Chadov et al. (2008) give improved orbital moments for these elements: 0.08 for bcc Fe, 0.15 for hcp Co, and 0.06 for fcc Ni. Thus, using DMFT drastically improves the agreement between theory and observations; this result highlights the importance of using corrections to the regular exchange and correlation functional described the Chapter 1, in order to obtain accurate orbital moment for these elements, a fact which was also pointed out by O. Eriksson et al. (1990).

Another level of comparing experiments with theory is via the so-called magnetic form factor, which is available from polarized neutron scattering experiments and represents essentially the Fourier component of the magnetization density (Marshall and Lovesey, 1971), shown, for example, in Fig. 2.4. An example of a comparison between measured and theoretical form factors is shown in Fig. 3.5 for Gd. In this figure, the measured magnetic form factor (Moon et al., 1972) obtained for hcp Gd is compared to a form factor calculated by using a spin density from a Gd atom (Koelling and Harmon, 1977). Since the localized $4f$ density dominates the spin density, it is a valid approximation to consider the $4f$ states of an atom as representative of those of a solid and, as Fig. 3.5 shows, the agreement between calculation and measurement is good. This is evidence that the magnetization density has the correct shape or, in the case of a magnetization density obtained from an atomic calculation, the correct radial extension, since otherwise the agreement in Fig. 3.5 would not be observed. One may also note from the figure that, for low values of momentum transfer, the theoretical contribution to the form factor is lower than the experimental (fitted) values, a result which is to be expected since here the more diffuse contribution (non-$4f$) to the magnetization density becomes important and, in the calculations, this contribution had not been considered. In principle, it is possible to generate an experimental magnetization density that is similar to the theoretical one in Fig. 2.4, by measuring of the form factor for many momentum transfers and then performing an inverse Fourier transform to obtain the magnetization density in real space.

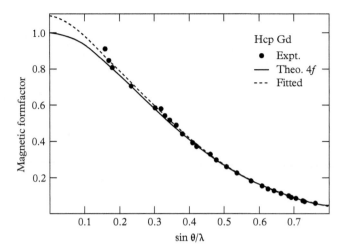

Figure 3.5 *Calculated and measured magnetic form factors of hexagsonal close-packed (hcp) Gd versus momentum transfer. The experimental (Exp.) curve is normalized to a measured total moment of 7.62 μ_B/atom (dashed line), while the theoretical (Theo.) 4f contribution (solid line) is normalized to 7.00 μ_B/atom. Figure redrawn after Koelling, D. D., and Harmon, B. N., (1977), A technique for relativistic spin-polarised calculations,* J. Phys. C, *10, 3107–14.*

3.3   Magnetic anisotropy energy

In Section 2.5, we discussed how the spin–orbit coupling locks the orientation of the magnetization to certain crystallographic directions. This has the effect that the energy of the magnetic crystal is lowest when the magnetization is pointing along a specific direction called the easy axis. As an example, we note that bcc Fe has the ⟨001⟩ direction as the easy axis. Each material has a unique easy or hard magnetization axis. For fcc Ni, the ⟨111⟩ direction is an easy axis whereas, for hcp Co, the easy axis is ⟨0001⟩. An applied field along any other direction can force the magnetization away from the easy axis into a hard-axis direction (Chikazumi, 1997), and because the energy of the crystal depends on the orientation of the magnetization, it is referred to as magnetic anisotropy energy. For the 3d elements, magnetic anisotropy energy causes the energy differences between easy and hard directions to be in the range of a few to ~100 μeV/f.u. As an example, we note that bcc Fe has an energy difference of the order of 1.4 μeV/f.u., which is equivalent to 0.05 MJ/m^3, between the ⟨001⟩ and the ⟨111⟩ axes. In contrast, in the most commonly used permanent magnet, $Nd_2Fe_{14}B$, where the energy difference between the easy and the hard magnetization directions is required to be high, the energy difference is 4.9 MJ/m^3. Typically, these numbers are absolutely tiny when compared to most other energies that are relevant of a material. For instance, the energy gained when the crystal is formed, the cohesive energy, is of the order of 2–10 eV/atom for the transition elements, as shown in Fig. 3.3. Despite being small compared to the cohesive energy, magnetic anisotropy energy has enormous relevance when the performance of a magnet is evaluated. Considerable effort is spent in designing materials with as large magnetic anisotropy as possible to produce so-called hard magnets, which are used in, for example, generators and electrical motors. 'Soft' magnets are also used, for example, for transformers; in this case, it is a low magnetic anisotropy that is desired.

The bcc phase of Fe allows for six different easy-axis orientations, whereas fcc Ni has eight different easy-axis orientations. Despite this difference, they are both examples of materials with cubic magnetic anisotropy, in contrast to hcp Co, which has uniaxial anisotropy, where the magnetization prefers to be either parallel or antiparallel to the c-axis of the hexagonal crystal structure. Mathematically the energy dependence of the direction of the magnetism is for uniaxial materials often expressed as

$$E_{\mathrm{MAE}} = K_1^u \sin^2 \theta + K_2^u \sin^4 \theta, \tag{3.1}$$

where MAE stands for magnetocrystalline anisotropy energy. Here, θ is the angle between the magnetization direction and the easy-axis direction, whereas the coefficients K_1^u and K_1^u are the (uniaxial) anisotropy constants. For materials with cubic crystal structure, the expression often used is

$$E_{\mathrm{MAE}} = K_1 (\alpha_1^2 \alpha_2^2 + \alpha_2^2 \alpha_3^2 + \alpha_3^2 \alpha_1^2) + K_2 (\alpha_1^2 \alpha_2^2 \alpha_3^2), \tag{3.2}$$

where, instead of the angle θ, we have used the direction cosines $\alpha_1, \alpha_2, \alpha_3$ of the magnetization with regard to the crystallographic axes and, as in Eqn (3.1), the coefficients K_1 and K_2 are the (cubic) anisotropy constants.

The microscopic mechanism behind the size and sign of the anisotropy constants in Eqns (3.1) and (3.2) can be found in the details of the electronic structure, as described in Chapter 2. The total energy difference relevant for the magnetic anisotropy energy is something which, in principle, DFT should describe well, since it involves differences of ground states with different symmetry. Early theories of magnetic anisotropy energy and how it couples to electronic structures can, for example, be found in the work by H. Brooks (1940); Fletcher (1954); Slonczewski (1962); Kondorsky and Straube (1972); Kondorsky (1974); Bruno (1989), and Daalderop et al. (1991). Typically, the small strength of the spin–orbit coupling, when compared to other terms in the Kohn–Sham Hamiltonian, allows for a treatment involving perturbation methods. As an example, we mention that, for uniaxial magnets, the leading term is of second order in the spin–orbit interaction. This has been analysed, for example, by Bruno (1989) and Andersson et al. (2007) in a rather general situation which describes the magnetic anisotropy energy of a compound built up of several elements. Using perturbation theory, the following expression of energy differences for a system with different directions ($\hat{n}_1$ and $\hat{n}_2$) of the magnetization was derived:

$$E_{\mathrm{MAE}} \propto \sum_{qss'} E^{ss'}_{\mathrm{soc},q}(\hat{\mathbf{n}}_1) - E^{ss'}_{\mathrm{soc},q}(\hat{\mathbf{n}}_2),$$

(3.3)

where

$$E^{ss'}_{\mathrm{soc},q}(\hat{\mathbf{n}}) = -\sum_{kij}\sum_{q'}\sum_{\{l\}} A_{kis,ql,q'l'} A_{kjs',q'l'',ql'''} \cdot \left[\frac{\langle qls|\xi_q l \cdot s|ql'''s'\rangle\langle q'l''s'|\xi_{q'}l \cdot s|q'l's\rangle}{\epsilon_{kj} - \epsilon_{ki}} \right].$$

(3.4)

In this expression, **k** is the wavevector, whereas i, j are indices of occupied and un-occupied Bloch states, respectively. The spin indices s and s' denote the spin of the Kohn–Sham states i and j. The index $\{l\}$ represents a set of quantum numbers of the wave function expansion in Eqn (2.16), whereas $l \cdot s$ is, as discussed in Section 2.5, the interaction between orbital and spin angular momentum operators. In addition, $\epsilon_{ki}, \epsilon_{kj}$ are Kohn–Sham energies of the unperturbed system, that is, ones without spin–orbit interaction. Finally, the coefficients $A_{kis,ql,q'l'} = c^*_{kis,ql} c_{kis,q'l'}$, where the c-coefficients are the expansion coefficients of the electron wave function Eqn (2.16), and the spin–orbit strength of atom q is denoted ξ_q in Eqn (3.4). Having Eqn (3.4) is useful, since it makes it possible to analyse the magnetic anisotropy of complex systems such as compounds with more than one atom type. An obvious observation is that states that are close in energy but on different sides of the Fermi energy contribute maximally to the magnetic anisotropy energy. In addition, since heavy elements have larger spin–orbit coupling than lighter ones do, it is possible that compounds composed of lighter elements at site q, such as a $3d$ element that provides a large exchange splitting and magnetic moment, and heavier elements at site q' may develop a large magnetic anisotropy. According to Eqn (3.4), a heavy element provides a large value of $\xi_{q'}$ in Eqn (3.4) and, if the hybridization of electron states is large, that is, if $A_{kis,ql,q'l'}$ is significant, a large magnetocrystalline anisotropy is possible. An example of such a material is FePt.

In order to illustrate the influence of the denominator in Eqn (3.4), we discuss the magnetic anisotropy energy of hcp Gd. In Fig. 3.6, we show the calculated and measured magnetic anisotropy energy of hcp Gd. It can be seen that experiment gives an easy axis which is at an angle approximately 30° away from the c-axis. The theory also gives an easy axis which is tilted away from the c-axis and which has a minimum at 30° from the c-axis and an energy which is ~2 μeV/atom lower compared to when the moment is parallel to the c-axis. It should be remembered here that the theory shown in Fig. 3.6 treats the $4f$ shell as part of the core, and all contributions to the theoretical magnetic anisotropy energy come from a relativistic treatment including spin–orbit interaction of the spin polarized, itinerant valence band, something called 'band contribution' in Fig. 3.6. When the classical dipole contribution is added to the 'band contribution' of hcp Gd, the calculated easy axis is at an angle of 20° from the c-axis (Colarieti-Tosti et al., 2003). Hence, theory reproduces this peculiar feature in Gd, where the easy axis is not aligned along a crystallographic direction but is pointing away from the c-axis. However, the angle is somewhat underestimated in the calculations, and this result illustrates that it is difficult to achieve accuracy at the level of microelectron volts per atom. The data in

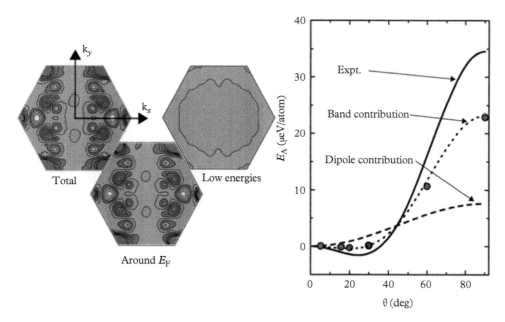

Figure 3.6 *Calculated magnetic anisotropy of hexagonal close-packed Gd. The reference energy (zero energy) is when the magnetization is parallel to the ⟨0001⟩ direction. The right side shows first-principles results, the classical dipole contribution to the magnetic anisotropy, as well as experimental data. The left side shows the contribution to the magnetic anisotropy from different energy regions as well as for different regions of the Brillouin zone, for a plane where $k_z = 0$. Darker colours indicate larger contributions; E_A, magnetic anisotropy energy; E_F, Fermi energy.*

Fig. 3.6 also show that additional terms in Eqn (3.1) are needed in the phenomenological expression for the magnetic anisotropy energy, for an accurate description of hcp Gd (Colarieti-Tosti et al., 2003). In order to analyse the importance of the denominator in Eqn (3.4), we show in the left part of Fig. 3.6 the contribution to the magnetic anisotropy energy, as it is broken up into different contributions, for example, from states which are close to the Fermi energy, and for states with energies lower in energy. This is shown for different regions in k-space, on the left side of Fig. 3.6. It may be seen in the figure that the majority of the contribution to the magnetic anisotropy energy comes from states close to the Fermi energy, as is expected from Eqn (3.4). Hence, in a sense, a calculation of the magnetic anisotropy energy of a material is, to a large extent, a matter of very accurately resolving the Fermi surface of that material.

3.4 Heisenberg exchange parameters

There are by now several implementations of Eqn (2.23), in different electronic structure software packages, that allow for accurate calculations of the Heisenberg exchange

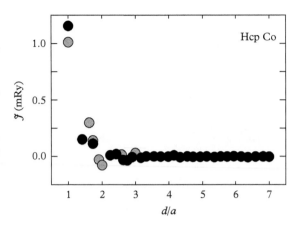

Figure 3.7 *Exchange interaction $\mathcal{J}$ of hexagonal close-packed Co as a function of the distance. The grey and black circles correspond to atoms lying in even and odd (0001) planes, respectively; d is the distance between two Mn atoms, and a is the lattice constant. Figure redrawn after Turek I., Kudrnovský, J., Drchal, V., and Bruno, P., (2006), Exchange interactions, spin waves, and transition temperatures in itinerant magnets,* Philos. Mag., **86**, *1713–52.*

parameters (e.g. in Liechtenstein, Katsnelson, and Gubanov, 1984; Liechtenstein, Katsnelson, Antropov and Gubanov, 1987; Frota-Pessôa et al., 2000; Turek et al., 2003; Udvardi et al., 2003; Belhadji et al., 2007; Szilva et al., 2013; Kvashnin et al., 2015). As an example of such calculations, we show in Fig. 3.7 the results for hcp Co (Turek et al., 2003). This is a typical example of exchange interactions of a metallic ferromagnet, where the interaction is dominated by near-range contributions but where, nevertheless, long-range interactions, up to a length of three lattice constants, are significant and may not be ignored. Note also that, since hcp Co does not have an ideal c/a ratio, there are, for each atom i, two sets of nearest-neighbour atoms j that are located almost, but not identically, at the same distance from i. These two sets of atoms are seen in Fig. 3.7 to have quite different strengths of interatomic exchange interaction: atoms in the same ⟨0001⟩ plane have a different interaction than those in different ⟨0001⟩ planes. The magnon curves and the spin wave stiffness constant calculated from exchange parameters obtained via first-principles electronic structure theory by using Eqn (2.23) are in general in good agreement with observations, for example, as reviewed by Etz et al. (2015) and as will be discussed in Chapter 9. In Fig. 3.8, exchange parameters for a dilute system, that is, Mn-doped fcc Cu, are shown. Only interactions between Mn atoms are shown, since these are the only atoms that carry significant magnetic moment in this system (Sato et al., 2010). Note that the figure contains results for various concentrations of Mn, including the extreme limit of two interacting impurities (in the inset). In this plot, the exchange interaction is multiplied by $(d/a)^3$ (d is the distance between two Mn atoms, and a is the lattice constant), since it is expected that the RKKY interaction should dominate the exchange interaction in this system. The RKKY interaction scales asymptotically as $\sim \frac{\cos(2k_F R)}{R^3}$, where R is the distance between the magnetic moments, and k_F is a so-called caliper vector (Pajda et al., 2001) that connects two different parts of the Fermi surface. If the material has many such vectors with equal length and direction, a situation referred to as nesting occurs, which leads to a distinct contribution to the cosine function of the RKKY interaction. Multiplying the exchange interaction with R^3 (or d^3) should then lead to a clear oscillatory behaviour. Figure 3.8 does indeed display

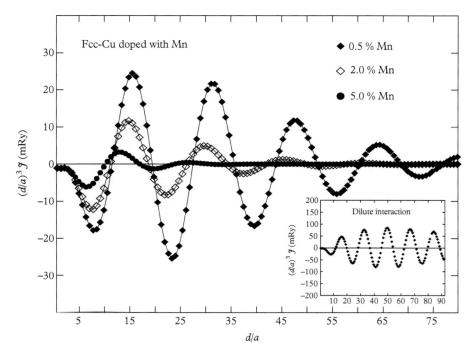

Figure 3.8 *Exchange interaction J along the dominating 110 direction in CuMn alloys for 0.5, 2.0, and 5.0 % Mn concentrations. Interactions are multiplied by the factor $(d/a)^3$, where d is the distance between two Mn atoms, and a is the lattice constant. Exchange interactions between two isolated impurities in a face-centred cubic (fcc) Cu host are shown in the inset. Figure redrawn after Turek, I., Kudrnovský, J., Drchal, V., Bruno, P., and Blügel, S., (2003), Ab initio theory of exchange interactions in itinerant magnets,* Phys. Status Solidi (b), *236, 318–24.*

such behaviour and provides a very good example of RKKY interaction, which, according to Fig. 3.8, seems to be present for all Mn concentrations and is particularly clear in the extremely dilute limit of two impurities interaction (shown in the inset). The period shown in Fig. 3.8 reflects very well the caliper vectors of the Fermi surface of fcc Cu.

3.5   Non-collinear magnets

So far, we have given examples of collinear magnets, which can be analysed theoretically using the Kohn–Sham equation, as described in Section 1.4. We end this chapter with one example of a non-collinear magnet, IrMnSi, space group *Pnma*. To analyse this material theoretically, one must use the non-collinear formulation of DFT, as described in Section 1.5. Before entering the discussion about the electronic structure and the magnetism of IrMnSi, we note that the theory of non-collinear magnetism dates back to the work of Herring (1966) and that the first practical theories were pioneered by

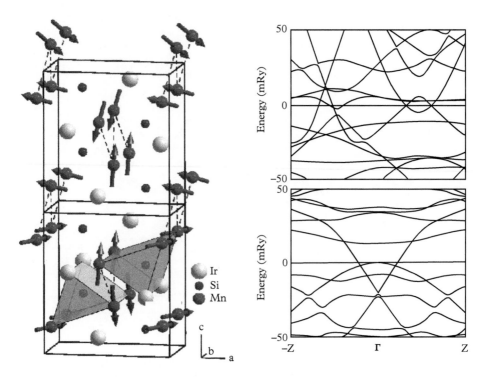

Figure 3.9 *(Left) The crystalline and magnetic structure of IrMnSi. The magnetic moment of each Mn atom is marked with an arrow. (Right) Band structure of collinear (upper panel) and non-collinear (lower panel) magnetic structures of IrMnSi. The Fermi level is marked with a horizontal line. Reprinted figure with permission from Eriksson, T., Bergqvist, L., Burkert, T., Felton, S., Tellgren, R., Nordblad, P., Eriksson, O., and Andersson, Y., Phys. Rev. B, 71, 174420, 2005. Copyright 2005 by the American Physical Society.*

Kübler and co-workers (Kubler et al., 1988; Kübler, 2009). Generalizations to these earlier works were published by Nordström and Singh (1996), and an extensive review on non-collinear magnetism was published by Sandratskii (1998).

IrMnSi has a cycloidal magnetic structure, that is, it has a non-collinear, magnetic structure, as was observed by T. Eriksson et al. (2005). The crystalline and the magnetic structures are displayed in Fig. 3.9. It may be seen from the figure that the Mn atoms carry a sizeable magnetic moment, while Ir and Si essentially are non-magnetic. Neutron scattering experiments have determined the size of the Mn moment to be 3.8 μ_B/atom, which can be compared to the 3.2 μ_B/atom obtained from DFT calculations (T. Eriksson et al., 2005). As the left side of Fig. 3.9 shows, the conventional chemical unit cell contains three sets of Mn atoms that are coupled in a non-collinear fashion, where the atoms closer to the edges of the cell point primarily along the *b*-axis while the Mn atoms in the centre of the unit cell are essentially antiparallel to each other and at an approximately 90° angle to the Mn atoms close to the cell edges. To complicate things further, in the next unit cell along the c-axis, the directions of all the Mn moments are more or less reversed with respect to those in the unit cell below. This results in a magnetic propagation vector along the $(0, 0, q_c)$ direction. If all the Mn moments in the top cell were reversed by 180° with respect to those in the bottom cell, the value of q_c would be 0.5. This was observed to almost be the case for IrMnSi: the measured value of q_c was 0.453 (T. Eriksson et al., 2005). This value of the propagation vector was reproduced well by DFT calculations, which gave a value of 0.43 (T. Eriksson et al., 2005). Hence, both experiment and theory indicate that IrMnSi is a material with complex non-collinear structure inside every conventional unit cell, a magnetic structure that is modulated along the c-axis in a way that is incommensurate with the underlying crystal structure.

One may ask why magnetic moments of any material would prefer such a complex orientation as that shown in Fig. 3.9. A possible explanation could lie in the analysis of the Heisenberg exchange parameters, which potentially may have competing ferromagnetic and antiferromagnetic interactions. This, of course, could shine some light into complex magnetic orderings, but this analysis immediately opens up questions concerning the microscopic background underlying why the Heisenberg interactions of a material like IrMnSi would be competing ferromagnetically and antiferromagnetically. A slightly different way to analyse complex magnetic orderings would be to investigate band dispersion. This can be done for IrMnSI, in a manner similar to what has been done in the past for several non-collinear magnetic materials (Kubler et al., 1988; Sandratskii, 1998; Lizarraga et al., 2004; Kübler, 2009). To illustrate this, we show the band dispersion of IrMnSi, for the ground-state magnetic configuration and for a hypothetical ferromagnetic configuration. The bands are shown in the right side of Fig. 3.9; the ferromagnetic bands are shown in the top, and the non-collinear bands are shown at the bottom. According to Chapter 1, Section 1.5, a non-collinear formulation of the Kohn–Sham equation results in energy bands that, in general, have admixtures of spin-up and spin-down character. This is in sharp contrast to the case with a collinear formulation, where, if spin–orbit coupling is neglected, pure spin states emanate from the DFT machinery. Hence, the band dispersion of a non-collinear magnet allows for hybridization between many more levels, compared to that of a collinear magnet, where spin-up and spin-down states are orthogonal and cannot mix. If hybridization of the non-collinear magnetic configuration pushes energy levels away from E_F to lower energies, the total energy of the material is lowered. When comparing the two panels on the right side of Fig. 3.9, we observe that this, indeed, is the case: the ground-state magnetic configuration of IrMnSi has a significantly reduced number of bands close to E_F, as compared to the ferromagnetic configuration. Hence, IrMnSi is an example of a magnetic material with a complex magnetic structure that is stabilized due to the band mechanism proposed by Kübler (2009); Kubler et al. (1988); Sandratskii (1998), and Lizarraga et al. (2004).

Part 2

Equation of Motion for Atomistic Spin Dynamics

The connection between the Kohn–Sham equation and the main equation for atomistic spin dynamics, the atomistic, stochastic Landau–Lifshitz (or Landau–Lifshitz–Gilbert) equation is established in Part 2 of this book. The formal aspects of a multiscale approach for atomistic spin dynamics simulations are covered in this section, both regarding the derivation of the equation of motion of atomistic spins and the limitations and advantages this approach has, in terms of time scales and the sizes of simulation cells. In addition, it is discussed how all parameters of the atomistic Landau–Lifshitz equation can be calculated from DFT, how finite temperatures enter the theory, and how an efficient and practical implementation of these equations into software form should be done. Mechanisms of dissipation of energy and angular momentum between atomic spin, lattice, and electron reservoirs are also discussed in this part of the book.

4

The Atomistic Spin Dynamics Equation of Motion

From the information obtained in density functional theory (DFT), in particular the magnetic moments and the Heisenberg exchange parameters, one has the possibility to make a connection to atomistic spin dynamics. In this chapter, the essential features of this connection are described. It is also discussed under what length and time scales this approach is a relevant approximation. The master equation of atomistic spin dynamics is derived and discussed in detail. In addition, we give examples of how this equation describes the magnetization dynamics of a few model systems.

4.1 A few introductory comments

The stochastic Landau–Lifshitz–Gilbert equation

$$\frac{d\mathbf{m}_i}{dt} = -\gamma_L \mathbf{m}_i \times \left(\mathbf{B}_i + \mathbf{B}_i^{\text{fl}}\right) - \gamma_L \frac{\alpha}{m_i} \mathbf{m}_i \times \left[\mathbf{m}_i \times \left(\mathbf{B}_i + \mathbf{B}_i^{\text{fl}}\right)\right] \tag{4.1}$$

is the equation of motion for atomic moments, and the derivation of this equation is the main focus of this chapter and Chapter 5. In Eqn (4.1), $\mathbf{m}_i = \mathbf{m}_i(t)$ is the magnetic moment of an atom at site i in the lattice, as defined in Section 2.6. According to Eqn (4.1), the atomic moment at site i experiences an effective magnetic field $\mathbf{B}_i = \mathbf{B}_i(t)$, and a stochastic magnetic field $\mathbf{B}_i^{\text{fl}} = \mathbf{B}_i^{\text{fl}}(t)$. The first term of the equation is the precessional motion, while the second term describes the damping motion. Furthermore, γ is the gyromagnetic ratio, while the renormalized gyromagnetic ratio is

$$\gamma_L = \frac{\gamma}{(1 + \alpha^2)}, \tag{4.2}$$

and α is a scalar (isotropic) Gilbert damping constant. The renormalization of the gyromagnetic ratio is a consequence of expressing the Landau–Lifshitz–Gilbert equation in

Atomistic Spin Dynamics. Olle Eriksson, Anders Bergman, Lars Bergqvist, Johan Hellsvik. First Edition.
© Olle Eriksson, Anders Bergman, Lars Bergqvist, Johan Hellsvik 2017. First published in 2017 by Oxford University Press.

the same form as the Landau–Lifshitz equation. The relationship between the Landau–Lifshitz equation and the Landau–Lifshitz–Gilbert equation is analysed in detail in Section 4.6. The damping parameter and the more general forms of it will be discussed in detail in Chapter 6, especially with focus on how one can evaluate it from first-principles electronic structure theory as described in Chapters 1 and 2. The fluctuating, stochastic magnetic field $\mathbf{B}_i^{\mathrm{fl}}$ describes how temperature effects enter the theory of atomistic spin dynamics, in a Langevin dynamics approach. This is the topic of Chapter 5 and will not be discussed further in this chapter. We note also that, disregarding the Langevin forces $\mathbf{B}_i^{\mathrm{fl}}$, the form of Eqn (4.1) is identical to the micromagnetic Landau–Lifshitz equation (MMLL; L. Landau and Lifshitz, 1935), with the important difference being that, in the MMLL equation, the primary variable is a vector field describing the magnetization density, while Eqn (4.1) describes the dynamics of the natural building block of any material, atoms and their magnetic moments.

This and the following two chapters hence contain the derivation of the stochastic Landau–Lifshitz (SLL) equation of motion from first principles, methods for calculating of the damping parameter, and how to incorporate temperature. In the present chapter, the equations of motion for atomistic spin dynamics (ASD) will be worked out, while Chapter 5 covers finite temperature effects, and Chapter 6 presents techniques for the calculation of the Gilbert damping parameter from state-of-the-art electronic structure methods.

4.2   Spin dynamics from first principles

Spin dynamics from first principles is a relatively young development. In two pioneering papers from the mid-1990s, Antropov et al. (1995, 1996) developed in detail how the time evolution of the magnetization at finite temperatures can be calculated within DFT. The scheme was very extensive and included treatment of systems with substantial orbital contribution to the magnetization and opened up for simultaneous spin and molecular dynamics. Example calculations were done for γ-Fe, a crystalline phase of iron known for its complicated magnetic structure. Alternative derivations of the equation of motion for local spin magnetic moments in the adiabatic limit have also been communicated by Niu et al., (1999) Z. Qian and Vignale (2002), and Bhattacharjee, Nordström, et al. (2012). Constraining magnetic fields were used in the self-consistent calculations of adiabatic spin dynamics by Stocks et al. (1998), and Ujfalussy et al. (1999). Very recently, time-dependent DFT (TD-DFT) methods have been developed in order to investigate spin dynamics beyond the adiabatic approximation; this will be commented on in Chapter 12.

To date, the majority of theoretical studies of ASD take as their starting point a magnetic Hamiltonian (see e.g. G. Brown et al., 2001; Nowak et al., 2005; Skubic et al., 2008; Evans et al., 2014). The Heisenberg exchange and other parameters in the Hamiltonian (see Section 4.8) can either be calculated from first principles by means of DFT methods (Skubic et al., 2008), or be fitted to experimental data from, for example, neutron scattering and Raman spectroscopy measurements or to reproduce values of more aggregated properties such as the critical temperature and the spin wave stiffness. The former approach is the route taken in the UppASD software (http://www.physics.uu.se/uppasd).

4.3   Equations of motion for the spin and charge densities

Omitting the spin–orbit coupling terms, the Kohn–Sham Hamiltonian Eqn (1.31) reduces to

$$H_{\alpha\beta}^{KS} = -\frac{1}{2}\nabla^2 \delta_{\alpha\beta} + v_{\alpha\beta}^{eff}(\mathbf{r}, t) + \left\{\frac{1}{2c}\hat{\sigma} \cdot \mathbf{B}^{eff}(\mathbf{r}, t)\right\}_{\alpha\beta}, \tag{4.3}$$

where the index KS stands for Kohn–Sham. From this Hamiltonian, the equations of motion for the charge and spin densities, defined from the eigenstates of Eqn (4.3) as $n(\mathbf{r}, t) = \psi^\dagger(\mathbf{r}, t)\psi(\mathbf{r}, t)$ and $s(\mathbf{r}, t) = \psi^\dagger(\mathbf{r}, t)\hat{s}\psi(\mathbf{r}, t)$, respectively, can be derived in several ways. The most compact derivation proceeds over the Heisenberg equation of motion for operators. The alternative derivation by means of the Pauli–Schrödinger equation and its complex conjugate is more lengthy, but has the advantage that the complex conjugations of operators and wave functions are more transparent. In the following presentation, a mixture of the two approaches is used.

Suppressing the orbital index and the spin index, the Pauli–Schrödinger Kohn–Sham equation and its complex conjugate are written compactly as

$$i\frac{\partial \psi(\mathbf{r}, t)}{\partial t} = H^{KS}\psi(\mathbf{r}, t); \qquad -i\frac{\partial \psi^\dagger(\mathbf{r}, t)}{\partial t} = \left[H^{KS}\psi(\mathbf{r}, t)\right]^\dagger. \tag{4.4}$$

The time evolution of the charge density $n(\mathbf{r}, t) = \psi^\dagger(\mathbf{r}, t)\psi(\mathbf{r}, t)$ is straightforwardly calculated as

$$\begin{aligned}
\frac{\partial n(\mathbf{r}, t)}{\partial t} &= \frac{\partial}{\partial t}\left[\psi^\dagger(\mathbf{r}, t)\psi(\mathbf{r}, t)\right] = \psi^\dagger(\mathbf{r}, t)\frac{\partial \psi(\mathbf{r}, t)}{\partial t} + \frac{\partial \psi^\dagger(\mathbf{r}, t)}{\partial t}\psi(\mathbf{r}, t) \\
&= \frac{1}{i}\left\{\psi^\dagger(\mathbf{r}, t)H^{KS}\psi(\mathbf{r}, t) - [H^{KS}\psi(\mathbf{r}, t)]^\dagger \psi(\mathbf{r}, t)\right\} \\
&= \frac{1}{2i}\nabla \cdot \left\{[\nabla\psi^\dagger(\mathbf{r}, t)]\psi(\mathbf{r}, t) - \psi^\dagger(\mathbf{r}, t)\nabla\psi(\mathbf{r}, t)\right\} \\
&= -\nabla \cdot \mathbf{J}_p,
\end{aligned} \tag{4.5}$$

where the paramagnetic charge current density has been defined as

$$\mathbf{J}_p = \frac{1}{2i}\{\psi^\dagger(\mathbf{r}, t)[\nabla\psi(\mathbf{r}, t)(\mathbf{r})] - [\nabla\psi^\dagger(\mathbf{r}, t)]\psi(\mathbf{r}, t)(\mathbf{r})\}. \tag{4.6}$$

Similarly, the time evolution of the spin density $s(\mathbf{r}, t) = \psi^\dagger(\mathbf{r}, t)\hat{s}\psi(\mathbf{r}, t)$ proceeds as

$$
\begin{aligned}
\frac{\partial s(\mathbf{r}, t)}{\partial t} &= \frac{\partial}{\partial t}\left[\psi^\dagger(\mathbf{r}, t)\hat{s}\psi(\mathbf{r}, t)\right] = \psi^\dagger(\mathbf{r}, t)\hat{s}\frac{\partial \psi(\mathbf{r}, t)}{\partial t} + \frac{\partial \psi^\dagger(\mathbf{r}, t)}{\partial t}\hat{s}\psi(\mathbf{r}, t) \\
&= \frac{1}{2i}\left\{\psi^\dagger(\mathbf{r}, t)\hat{s}H^{\mathrm{KS}}\psi(\mathbf{r}, t) - \left[H^{\mathrm{KS}}\psi(\mathbf{r}, t)\right]^\dagger \hat{\sigma}\psi(\mathbf{r}, t)\right\} \\
&= \frac{1}{4i}\nabla \cdot \left\{[\hat{\sigma}\nabla\psi^\dagger(\mathbf{r}, t)]\psi(\mathbf{r}, t) - \psi^\dagger(\mathbf{r}, t)\hat{\sigma}\nabla\psi(\mathbf{r}, t)\right\} - \gamma s \times \mathbf{B}^{\mathrm{eff}} \\
&= -\nabla \cdot \mathbf{J}_s - \gamma s \times \mathbf{B}^{\mathrm{eff}},
\end{aligned}
\tag{4.7}
$$

where $\mathbf{B}^{\mathrm{eff}}$ is the magnetic part of the effective Kohn–Sham potential, see Eqn (4.3). The spin current density, a tensorial quantity formed from the direct product of orbital and spin space, is defined as

$$
\mathbf{J}_s = \frac{1}{4i}\{\psi^\dagger(\mathbf{r}, t)[\hat{\sigma}\nabla\psi(\mathbf{r}, t)] - [\hat{\sigma}\nabla\psi^\dagger(\mathbf{r}, t)]\psi(\mathbf{r}, t)\},
\tag{4.8}
$$

to analog the paramagnetic charge current density. We note that the definition of the spin current has been a matter of some debate. A profound difference between the charge and the spin current is that whereas electronic charge is a preserved quantity obeying a continuity equation, Eqn (4.5), spin is not necessarily conserved. Some authors (G. Sun and Xie, 2005) refer to the quantity defined by Eqn (4.8) as the linear spin current, and the precession of the spin as the angular spin current.

The divergence operator in the last part of Eqn (4.7) acts on the orbital space. The action of the spin currents on the magnetization is the mechanism behind spin-transfer torque (Ralph and Stiles, 2008), which has received much attention since the seminal works of Slonczewski (1996) and Berger (1996). The mechanism is also important as it contributes to the relaxation of the magnetization. To conclude this part, we note that, in compact form, the TD-DFT equations for the charge and spin densities, in the absence of stochastic damping mechanisms, are written as

$$
\frac{\partial n(\mathbf{r}, t)}{\partial t} = -\nabla \cdot \mathbf{J}_p,
\tag{4.9}
$$

$$
\frac{\partial s(\mathbf{r}, t)}{\partial t} = -\nabla \cdot \mathbf{J}_s - \gamma s \times \mathbf{B}^{\mathrm{eff}}.
\tag{4.10}
$$

If we omit the current term and integrate the expression for spin density over, for example, an atomic sphere at site i, the so-called atomic moment approximation (see Fig. 2.7), so that we can do the replacement $s(\mathbf{r}, t) \rightarrow \mathbf{m}_i(t)$, we arrive at the precessional term of Eqn (4.1). However, this requires a separation of the time-scale of electron motion from that of the atomic spins, an issue addressed in Chapter 2, Section 2.6. Before proceeding with this analysis, we first make a small comment on alternative ways of deriving Eqns (4.9) and (4.10). Wieser (2015) recently explored how equations for dissipative quantum dynamics can be derived if one takes as the starting point a non-Hermitian Hamiltonian

$$H = H_0 - i\lambda\Gamma \tag{4.11}$$

and calculates the time evolution of the expectation value of the density matrix through the Liouville or Heisenberg equation; H_0 is a Hermitian Hamiltonian of the quantum system, γ is a non-Hermitian Hamiltonian describing dissipation processes, and $\lambda \in \mathbb{R}_0^+$. The norm of the wave function was preserved by replacing γ with $\gamma - \langle\gamma\rangle$ (Wieser, 2015). Working with the Schrödinger equation

$$i\frac{d}{dt}|\psi(t)\rangle = (H - i\lambda\,[\Gamma - \langle\Gamma\rangle])\,|\psi(t)\rangle, \tag{4.12}$$

it is possible to arrive at the equation of motion for the expectation value of the spin operator,

$$\frac{d\langle\hat{\mathbf{s}}\rangle}{dt} = \langle\hat{\mathbf{s}}\rangle \times \mathbf{B}_{\mathrm{eff}} - \lambda\langle\hat{\mathbf{s}}\rangle \times \left(\langle\hat{\mathbf{s}}\rangle \times \mathbf{B}_{\mathrm{eff}}\right), \tag{4.13}$$

which yields both the precessional term, as found also in Eqns (4.9) and (4.10), as well as a damping term that appears as the second term on the right-hand side of Eqn (4.13).

4.4   Local coordinate systems and the adiabatic approximation

Atomistic spin dynamics simulations, using the SLL equation, are commonly pursued for the order of 10^4 to 10^9 atomic magnetic moments. A direct calculation of the spin polarized TD-DFT equations of motion for such large number of atoms is not yet feasible due to the excessive computational effort required, and Eqn (4.1) has to be used instead. For the case of a few atoms (<10), pioneering TD-DFT studies have very recently been pursued, which will be discussed as an outlook on ab initio spin dynamics, in Chapter 12. The step taken to avoid the complications with solving a time-dependent Kohn–Sham equation is a separation of slower and faster time scales, where the slower developing variables are regarded as frozen on time scales on which the faster degrees of freedom evolve. This was discussed loosely in Chapter 2 and is further elaborated upon here. The starting point for the adiabatic approximation is to assume that the time-independent Schrödinger equation holds at any instant of time for an effective potential that depends parametrically on time. For slow enough time evolution of the potential, the eigenvalues will depend on time in such a way that no levels cross and transitions to other levels do not occur. The most well-known example of the adiabatic approximation is the Born–Oppenheimer approximation, in which the time scales for electronic and ionic motion in a crystal or molecule are separated. The Born–Oppenheimer approximation relies on the difference in mass of heavy ions and light electrons and can be worked out in a systematic manner (Ashcroft and Mermin, 1976). For the dynamics of a spin polarized density matrix, the slow variable is the evolution in time of the local

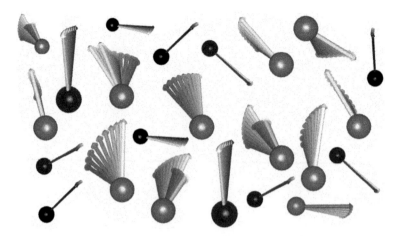

Figure 4.1 *A pictorial illustration of the dynamics in a two-component magnetic solid. Femtosecond-time-scale fast dynamics occur for the charge density and the magnitude of the magnetization, which both, in the local coordinate frame defined by Eqn (4.14), are diagonal in the density matrix. Transversal fluctuations of the magnetization correspond to the slower dynamics of the off-diagonal parts of the density matrix, a dynamics which on atomic length scales nevertheless is in the multiples of teraherz regime. Figure kindly provided by Dr. Diana Iuşan.*

direction of the magnetization, which typically is relevant for time scales of picoseconds or longer, while the dynamics of the electrons is many orders of magnitude faster (of the order of 0.1 fs as was discussed in Chapter 2, Section 2.6), as is schematically shown in Fig. 4.1.

Before giving the full details for establishing a connection between Eqns (4.9), (4.13), and (4.1), we outline how a straightforward time evolution of the TD-DFT equations may be obtained. We start by considering the density matrix in a general form, where due to, for example, a non-collinear magnetic configuration, the density matrix is non-diagonal and hence has components $\rho_{\alpha\beta}$ that are non-zero, Eqn (1.27). It is however possible to find a local coordinate system at any point in space and at any time, where a diagonal representation of the density matrix can be formed, by a unitary transformation (Sandratskii, 1998; Kübler, 2000):

$$\begin{pmatrix} \rho^{\uparrow}(\mathbf{r},t) & 0 \\ 0 & \rho^{\downarrow}(\mathbf{r},t) \end{pmatrix} = \begin{pmatrix} U_{\alpha\alpha}(\mathbf{r},t) & U_{\alpha\beta}(\mathbf{r},t) \\ U_{\beta\alpha}(\mathbf{r},t) & U_{\beta\beta}(\mathbf{r},t) \end{pmatrix} \begin{pmatrix} \rho_{\alpha\alpha}(\mathbf{r},t) & \rho_{\alpha\beta}(\mathbf{r},t) \\ \rho_{\beta\alpha}(\mathbf{r},t) & \rho_{\beta\beta}(\mathbf{r},t) \end{pmatrix} \begin{pmatrix} U_{\alpha\alpha}(\mathbf{r},t) & U_{\alpha\beta}(\mathbf{r},t) \\ U_{\beta\alpha}(\mathbf{r},t) & U_{\beta\beta}(\mathbf{r},t) \end{pmatrix}^{\dagger},$$

$$(4.14)$$

where $\rho^{\uparrow}$ and $\rho^{\downarrow}$ are the components of majority 'spin-up' and minority 'spin-down' densities, respectively, defined locally in space and time. In this equation, $U_{\alpha\beta}[\theta(\mathbf{r},t), \phi(\mathbf{r},t)]$ is a spin half-rotation matrix:

$$U_{\alpha\beta}(\theta(\mathbf{r},t),\phi(\mathbf{r},t)) = \begin{pmatrix} \exp(i\frac{\phi}{2})\cos(\frac{\theta}{2}) & \exp(-i\frac{\phi}{2})\sin(\frac{\theta}{2}) \\ -\exp(i\frac{\phi}{2})\sin(\frac{\theta}{2}) & \exp(-i\frac{\phi}{2})\cos(\frac{\theta}{2}) \end{pmatrix}. \tag{4.15}$$

The rotation matrix $U(\theta,\phi)$ is specified to diagonalize $\rho_{\alpha\beta}$, and hence the required angles θ, ϕ that define $U(\theta,\phi)$ and describe the direction of the magnetization can be evaluated from $\rho_{\alpha\beta}$ as

$$\tan\phi = -\frac{\mathrm{Im}(\rho_{\alpha\beta})}{\mathrm{Re}(\rho_{\alpha\beta})},$$

$$\tan\theta = \frac{2\sqrt{(\mathrm{Re}(\rho_{\alpha\beta}))^2 + (\mathrm{Im}(\rho_{\alpha\beta}))^2}}{(\rho_{\alpha\alpha} - \rho_{\beta\beta})}. \tag{4.16}$$

In the adiabatic limit, one can freeze the magnetization direction at each point in space, and via, for example, the local density approximation, obtain an effective potential in which the directions that specify 'spin up' and 'spin down' vary in space. A self-consistent calculation of the density matrix using Eqn (1.27) can be obtained from solutions to the time-independent Kohn–Sham equation, Eqn (1.26), generalized to the non-collinear case (Sandratskii, 1998; Kübler, 2000):

$$\left[-\frac{\nabla^2}{2}\mathbf{1} + \begin{pmatrix} V_{\alpha\alpha}^{\mathrm{eff}}(\mathbf{r}) & V_{\alpha\beta}^{\mathrm{eff}}(\mathbf{r}) \\ V_{\beta\alpha}^{\mathrm{eff}}(\mathbf{r}) & V_{\beta\beta}^{\mathrm{eff}}(\mathbf{r}) \end{pmatrix} \right] \begin{pmatrix} \psi_{i\alpha}(\mathbf{r}) \\ \psi_{i\beta}(\mathbf{r}) \end{pmatrix} = \epsilon_i \begin{pmatrix} \psi_{i\alpha}(\mathbf{r}) \\ \psi_{i\beta}(\mathbf{r}) \end{pmatrix}. \tag{4.17}$$

Note that, in this equation, the wave function is written in a general spinor form, the effective potential in Eqns (1.32) and (1.33) is written in matrix form, and $\mathbf{1}$ is the 2×2 identity matrix. For a collinear magnetization density, the off-diagonal elements of the effective potential, $V_{\alpha\beta}^{\mathrm{eff}}$ in Eqn (4.17), are equal to 0, and we can solve the Kohn–Sham equation separately for $\psi_{i\alpha}$ and $\psi_{i\beta}$, that is, we recover Eqn (1.26).

Schematically, the evolution in time of the fast and slow degrees of freedoms in a magnetic solid is shown in Fig. 4.1. In the adiabatic approximation of TD-DFT the complete set of equations for a new direction of the magnetization density is obtained from the density matrix evaluated from Eqn (4.17) and (1.27), together with Eqn (4.16). In presence of a driving field, that is, a time-dependent component of the field $\mathbf{B}^{\mathrm{eff}}$ in Eqn (4.9), new magnetization directions should be updated at each time step. The calculation of the evolution of the magnetization and charge densities from a time t to $t+\Delta t$ then becomes a two-step procedure. The first step is to solve Eqn (4.17) until self-consistency is obtained. This step involves a constraining field so that the atomic moment directions are fixed along some direction. The second step is to evolve the equation of motions for the slow degrees of freedom (i.e. the magnetization direction according to Eqn (4.9)) in a time interval Δt. This step requires the calculation of a local field, $\mathbf{B}^{\mathrm{eff}}$, which can be evaluated from Kohn–Sham wave functions. With the new directions for the atomic magnetic moments, Eqn (4.17) is solved anew, with an updated constraining field, and the process is iterated.

4.5 The atomic moment approximation and constraining field

In general, the magnetization density can show non-collinearity also on intra-atomic length scales, as exemplified by deviations from purely collinear magnetism in ferromagnetic fcc Pu (Nordström and Singh, 1996) and non-collinear deviations from an antiferromagnetic Néel structure in unsupported Cr monolayers (Sharma et al., 2007). For many materials and situations, it is reasonable to approximate the full non-collinearity of the density matrix by assuming that the magnetization density is locally collinear in the vicinity of an atom (Sandratskii, 1998; Kübler, 2000). In this atomic moment approximation, the volume one integrates over becomes an issue but, as will be discussed in Chapter 9, this is in practice seldom a problem (Andersen and Jepsen, 1984; Turek et al., 1997). Regardless of how the partitioning of space is done, an atomic magnetic moment can be defined for each lattice site i as the integral of the magnetization density over the region Ω_i,

$$\mathbf{m}_i = \int_{\Omega_i} \mathbf{m}(\mathbf{r}) \, dr. \qquad (4.18)$$

A general non-collinear magnetic configuration does not constitute a ground state within DFT. The iteration of Eqn (4.17) towards self-consistency for the electronic structure would by itself drift towards the ground state, for example, a collinear ferromagnetic state. To prevent this, constraining fields can be included. The technique used in practice is to use additional constraints in the form of Lagrangian multipliers (Dederichs et al., 1984; Schwarz and Mohn, 1984), in order to constrain some quantity, such as the total particle number of the unit cell, the l-projected charge inside a muffin-tin or atomic sphere, or the size of the magnetic moment (Mohn, 2006). For self-consistent field ASD, the directions of the atomic magnetic moments should be fixed during the calculation of the electronic structure. This can be achieved by the addition of a constraint:

$$\int_{\Omega_i} \mathbf{m}(\mathbf{r}, t) \times \mathbf{e}_i^{\mathrm{con}} \, dr = 0, \ \forall i. \qquad (4.19)$$

The cross-product form of the constraint ensures that the atomic magnetic moment has no components normal to the direction of the prescribed magnetization direction $\mathbf{e}_i^{\mathrm{con}}$. An important observation was done by (Stocks et al., 1998), who realized that the magnetic field that is used to constrain the direction of the magnetic moment can be used to construct the effective field that gives the correct precession torque on the local magnetic moment, a torque that is needed to perform self-consistent field calculation spin dynamics simulations. The technique, called the constrained local moment approach (CLM), has, for example, been used to investigate the canted magnetism of a finite Co chain along a Pt(111) surface step edge (Újfalussy et al., 2004). We will now discuss this method further with respect to how it works in practice, closely following the original discussion by Stocks et al. (1998).

Using the atomic moment approximation, introduced in Chapter 2 and in Eqn (4.18), we define a unit vector $\mathbf{e}_i$ of the local magnetization $\mathbf{e}_i = \mathbf{m}_i/|\mathbf{m}_i|$. Expressed in terms of the unit vectors, the equation of motion reads (Antropov et al., 1996)

$$\frac{d\mathbf{e}_i}{dt} = -2\mathbf{e}_i \times \mathbf{B}_i(\{\mathbf{e}_i\}), \tag{4.20}$$

where $\mathbf{B}_i(\{\mathbf{e}_i\})$ is the effective magnetic field on each site given by the expression

$$\mathbf{B}_i(\{\mathbf{e}_i\}) = -\frac{\partial E(\{\mathbf{e}_i\})}{\partial \mathbf{m}_i}, \tag{4.21}$$

where $E(\{\mathbf{e}_i\})$ is not an energy calculated from a parametrized magnetic Hamiltonian but instead a DFT energy functional. In the CLM approach, the Kohn–Sham equation, Eqn (4.17) is solved in the presence of constraining field on each atomic moment in order to ensure that the local magnetic moment is kept fixed using Eqn (4.19).

The constraint is achieved by adding a penalty energy functional to the normal energy functional, Eqn (1.20), which is thus generalized to noncollinear magnetic states (Section 1.5) such that the total energy functional becomes

$$E^{\mathrm{con}}\left[\rho(\mathbf{r}); \mathbf{m}(\mathbf{r}); \mathbf{B}^{\mathrm{con}}(\mathbf{r})\right] = E\left[\rho(\mathbf{r}); \mathbf{m}(\mathbf{r})\right] + \int \mathbf{m}(\mathbf{r}) \cdot \mathbf{B}^{\mathrm{con}}(\mathbf{r}). \tag{4.22}$$

The effective field appearing in the Kohn–Sham Hamiltonian for non-collinear magnetic states, Eqn (4.3), including the constraining field, then reads (Stocks et al., 1998)

$$\mathbf{B}_i^{\mathrm{eff}} = \mathbf{B}_i^{\mathrm{xc}}[\mathbf{m}(\mathbf{r})] + \mathbf{B}_i^{\mathrm{con}}(\mathbf{r}) = \mathbf{e}_i B_i^{\mathrm{xc}} + \mathbf{B}_i^{\mathrm{con}}(\mathbf{r}). \tag{4.23}$$

Self-consistency is enforced by the following three conditions:

$$n(\mathbf{r}) = \mathrm{Tr}[\hat{\rho}(\mathbf{r})], \tag{4.24}$$
$$\mathbf{m}(\mathbf{r}) = \mathbf{e}_i \left\{\mathbf{e}_i \cdot \mathrm{Tr}[\hat{\sigma}\rho(\mathbf{r})]\right\}, \tag{4.25}$$

$$\frac{\partial E^{\mathrm{con}}\left[\rho(\mathbf{r}); \mathbf{m}(\mathbf{r}); \mathbf{B}^{\mathrm{con}}(\mathbf{r})\right]}{\partial \mathbf{e}_i} = 0, \forall i. \tag{4.26}$$

Given that the atomic moment approximation was used from the very start, it is possible to use the following functional form for the constraining field:

$$\mathbf{B}_i^{\mathrm{con}}(\mathbf{r}) = c_i B^{\mathrm{xc}}(\mathbf{r}), \tag{4.27}$$

where c_i denotes a constraining factor. With the above constraints and functional form for the constraining field, the effective field to be used in conjunction with the equation of motion, Eqn (4.20), comes out as

$$\mathbf{B}_i = -\frac{c_i}{m_i} \int_{\Omega_i} m(\mathbf{r}) B^{\text{xc}}(\mathbf{r}) \, \mathrm{d}\mathbf{r}. \tag{4.28}$$

In order to find the constraining field according to the self-consistency conditions, an iterative procedure was proposed by Stocks et al. (1998): as the zeroth step of the iteration, a trial initial constraining field c_i^{in} is chosen; then, in subsequent iteration steps, the constraining field is obtained according to

$$\mathbf{c}_i^{\text{new}} = \mathbf{c}_i^{\text{in}} - \left(\mathbf{c}_i^{\text{in}} \cdot \mathbf{e}_i\right)\mathbf{e}_i - \left[\mathbf{e}_i^{\text{out}} - \left(\mathbf{e}_i^{\text{out}} \cdot \mathbf{e}_i\right)\mathbf{e}_i\right], \tag{4.29}$$

a procedure similar to schemes for charge density mixing to achieve self-consistency in normal density functional calculations. Eqn (4.29) guarantees that $\mathbf{e}_i^{\text{out}}$ converges to the constrained orientation and that $\mathbf{c}_i^{\text{in}}$ converges to $\mathbf{c}_i$.

4.6   Damping motion and relaxation

With the atomic moment approximation introduced in Section 4.5, and in Section 2.6, we can now focus on the master equation of ASD, Eqn (4.1). The first term on the right-hand side of this equation describes the conservative dynamics of an ideal spin system that does not exchange energy with its environment. If this were the only driving term of the magnetization dynamics, the system would, for example, not respond to changes in the temperature of its environment. Mechanisms describing dissipation of energy and angular momentum are found in the second term on the right-hand side of Eqn (4.1), and are referred to as damping. We will devote Chapter 6 to a detailed description on how to calculate the damping from first principles, but we make a shorter analysis here on its effects and the mechanisms that are responsible for it.

Landau and Lifshitz included a relaxation term in the equation of motion, already in their original paper (L. Landau and Lifshitz, 1935) on magnetization dynamics. They proposed a double-cross-product damping term in addition to the precession term, yielding the equation known as the Landau–Lifshitz equation:

$$\frac{\partial \mathbf{m}}{\partial t} = -\gamma \mathbf{m} \times \mathbf{B} - \frac{\lambda}{m}\mathbf{m} \times (\mathbf{m} \times \mathbf{B}). \tag{4.30}$$

The damping torque is seen from this equation to be perpendicular to the precession torque, and its relative contribution is determined by the parameter λ. To illustrate the effects of the precessional term and the damping contribution, we show in Fig. 4.2 the motion of a single spin exposed to an external magnetic field, $\mathbf{B}$. Note that, in this figure, we have replaced λ with $\gamma\alpha$ for reasons that will be clear below. The dynamics outlined in Fig. 4.2 is seen to be a spiralling motion where the spin eventually aligns with the field.

In classical mechanics, the friction force acting on a particle moving through a viscous media is to first-order approximation proportional to, and negative in direction to, its velocity (Goldstein et al., 2002). Gilbert (2004) introduced an analogous damping force for magnetization dynamics, thus creating the Landau–Lifshitz–Gilbert equation:

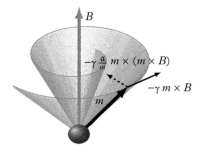

Figure 4.2 *Motion described by the Landau–Lifshitz equation. The atomic magnetic moment* **m** *(thick black arrow) revolves in damped precessional motion around the effective magnetic field* **B** *(thick grey arrow). The thin black arrow shows the tangent vector of the precessional motion* $-\gamma m_i \times B$, *and the dashed thin black arrow shows the tangent vector of the damping motion* $-\gamma \frac{\alpha}{m} m \times (m \times B)$. *For a scalar Gilbert damping* α *parameter, the damping motion is perpendicular to the precessional motion;* γ *is the gyromagnetic ratio. Figure kindly provided by Dr. Danny Thonig.*

$$\frac{\partial \mathbf{m}}{\partial t} = -\gamma \mathbf{m} \times \mathbf{B} + \frac{\alpha}{m} \mathbf{m} \times \frac{\partial \mathbf{m}}{\partial t}. \tag{4.31}$$

As for the Landau–Lifshitz damping torque, the Gilbert damping torque is perpendicular to the precession torque. In the limit of small damping, the solutions of the Landau–Lifshitz and the Landau–Lifshitz–Gilbert equations are close to each other. For larger damping, the discrepancy is significant. As will be shown in Section 4.7, the Landau–Lifshitz and the Landau–Lifshitz–Gilbert equations are identical in the case of isotropic damping, provided a renormalized gyromagnetic ratio is introduced. In the micromagnetic community, the Gilbert form of damping has been the more popular choice, and it was indeed shortcomings of the Landau–Lifshitz equation for damping that motivated Gilbert's research.

In Fig. 4.3, we illustrate, with a simple but realistic example, how the precession torque and the damping term of the SLL equation contribute to the time evolution of magnetism. The system under consideration is a thin film of a ferromagnetic material, a monolayer of Fe on a W(110) surface, where all magnetic interactions are described by Heisenberg exchange calculated using multiple scattering theory as described in Section 3.4 and as reported by Bergman et al. (2010). Initially, this ferromagnetic system has one of its atomic spins reversed to align antiparallel to its neighbours, as shown in the upper left part of Fig. 4.3, where the increased coloured contrast in the figure indicates the degree of deviation from ferromagnetic alignment. At $t = 0$ ps, it is only the spin in the middle of the simulation cell that has a reversed orientation. However, as time passes, other atomic spins that are in the vicinity of the original flipped atomic spin start to have their orientations deviate from collinear ferromagnetic order. This can be seen in, for example, the spin configuration at $t = 0.5$ ps, where a 'spin cloud' has developed. The simulations shown in Fig. 4.3 have been done without damping, and therefore energy and angular momentum do not leave the simulation cell. The time evolution of the spin cloud is therefore not the result of a global energy minimization but may be seen as a local energy minimization in which the exchange energy is spread out between more and more atomic spins. With time, the spin cloud continues to grow, but as it does so it becomes more and more diffuse, that is, with a less pronounced deviation from the ferromagnetic ground state. This is clear from the snapshots of the simulation at $t = 1$,

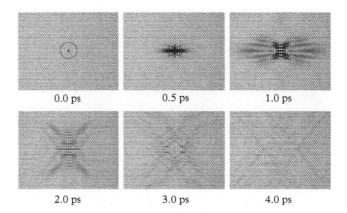

Figure 4.3 *Simulated snapshots of the relaxation of a single flipped spin in a monolayer of Fe on a W(110) surface. The first panel shows a single flipped spin in the centre of the ring, which is added for clarity. Here, no damping is included in the simulation, and the system does thus not relax to the equilibrium ferromagnetic state, but the excitation gets more and more smeared out across the system over time.*

2, and 3 ps, as these have less and less colour contrast as the spin cloud becomes larger and larger time over. An interesting feature that can be seen in Fig. 4.3 is also that the 'cloud' of distorted spins does not propagate isotropically across the sample; first, it extends in the horizontal (001) direction and, after some time, it extends more in the vertical (1$\bar{1}$0) direction. This is caused by the fact that, owing to the symmetry of the surface, the exchange interactions differ along the different surface directions. An effect that can be inferred from this anisotropic distribution of the spin cloud is that the group velocity of magnons on this surface would also behave anisotropically and depend on their propagation direction. The influence of damping is shown in Fig. 4.4. Here, all parameters are kept the same as for the simulation in Fig. 4.3, with the only difference being the presence of a Gilbert damping parameter of $\alpha = 0.01$. It may be noted that initially the results in Fig. 4.4 are similar to those in Fig. 4.3; for example, a spin cloud develops and grows with time. However, in Fig. 4.4, energy and angular momentum can dissipate from the system and, as a consequence, the system relaxes to its global energy minimum, that is, the ferromagnetic state. For this reason, the spin-cloud becomes more and more diffuse in time, as is clear in Fig. 4.4, and, eventually, it disappears completely.

We end this section with a short comment on Eqn (4.13), which naturally introduces a damping term in the equation of motion of spin operators. Hence, damping is seen to naturally emerge from a microscopic theory of quantum mechanical operators. If one adopts the atomic moment approximation, damping emerges in an equation of motion of atomic spins and is not an ad hoc interaction that simply is added to the Landau–Lifshitz equation.

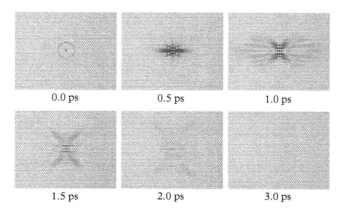

Figure 4.4 *Simulated snapshots of the relaxation of a single flipped spin in a monolayer of Fe on a W(110) surface. The first panel shows a single flipped spin in the centre of the ring, which is added for clarity. Here, a finite damping is included in the simulation, and the system relaxes to the equilibrium ferromagnetic state within a few picoseconds.*

4.7 The relation between the Landau–Lifshitz and the Landau–Lifshitz–Gilbert equations

The Landau–Lifshitz equation (Eqn (4.30)), and the Landau–Lifshitz–Gilbert equation (Eqn (4.31)), are closely related. In the case of isotropic damping, they are identical if a renormalized gyromagnetic ratio is introduced, as will be shown in this section. That the Landau–Lifshitz and the Landau–Lifshifz–Gilbert equations are not identical in the case of anisotropic damping is discussed by Steiauf and Fähnle (2005). The precession torque and the damping torque, whether in the Landau–Lifshitz form or the Gilbert form, preserve the magnitude of the magnetization **m**. This is easily proved by first observing that this is equivalent to the statement that

$$\frac{\partial \mathbf{m}^2}{\partial t} \equiv 0. \tag{4.32}$$

Since

$$\frac{\partial \mathbf{m}^2}{\partial t} = 2\mathbf{m} \cdot \frac{\partial \mathbf{m}}{\partial t}, \tag{4.33}$$

we should proceed with a scalar multiplication with **m** and its time derivative, here for the Landau–Lifshitz–Gilbert equation,

$$\mathbf{m} \cdot \frac{\partial \mathbf{m}}{\partial t} = -\gamma \mathbf{m} \cdot (\mathbf{m} \times \mathbf{B}) + \frac{\alpha}{m} \mathbf{m} \cdot \left(\mathbf{m} \times \frac{\partial \mathbf{m}}{\partial t} \right)$$

$$= -\gamma \mathbf{B} \cdot (\mathbf{m} \times \mathbf{m}) + \frac{\alpha}{m} \frac{\partial \mathbf{m}}{\partial t} \cdot (\mathbf{m} \times \mathbf{m}) = 0, \qquad (4.34)$$

which proves Eqn (4.32). The derivation using the Landau–Lifshitz equation is very similar.

We now proceed to establish a connection between the Landau–Lifshitz and the Landau–Lifshitz–Gilbert equations. Starting from the Landau–Lifshitz–Gilbert equation, the Landau–Lifshitz equation can be obtained after multiplication with $\mathbf{m}$ from the left:

$$\mathbf{m} \times \frac{\partial \mathbf{m}}{\partial t} = -\gamma \mathbf{m} \times (\mathbf{m} \times \mathbf{B}) + \frac{\alpha}{m} \mathbf{m} \times (\mathbf{m} \times \frac{\partial \mathbf{m}}{\partial t}) \qquad (4.35)$$

$$= -\gamma \mathbf{m} \times (\mathbf{m} \times \mathbf{B}) + \frac{\alpha}{m} \left[\mathbf{m}(\mathbf{m} \cdot \frac{\partial \mathbf{m}}{\partial t}) - \frac{\partial \mathbf{m}}{\partial t}(\mathbf{m} \cdot \mathbf{m}) \right].$$

Using now the relation $\mathbf{m} \cdot \frac{\partial \mathbf{m}}{\partial t} = 0$, derived in Eqn (4.34), we get

$$\mathbf{m} \times \frac{\partial \mathbf{m}}{\partial t} = -\gamma \mathbf{m} \times (\mathbf{m} \times \mathbf{B}) - \alpha m \frac{\partial \mathbf{m}}{\partial t}. \qquad (4.36)$$

This is substituted into the right-hand side of the Landau–Lifshitz–Gilbert equation:

$$\frac{\partial \mathbf{m}}{\partial t} = -\gamma \mathbf{m} \times \mathbf{B} - \frac{\alpha}{m} \left[\gamma \mathbf{m} \times (\mathbf{m} \times \mathbf{B}) + \alpha m \frac{\partial \mathbf{m}}{\partial t} \right]$$

$$= -\gamma \mathbf{m} \times \mathbf{B} - \frac{\alpha}{m} \gamma \mathbf{m} \times (\mathbf{m} \times \mathbf{B}) - \alpha^2 m \frac{\partial \mathbf{m}}{\partial t}. \qquad (4.37)$$

The terms containing $\frac{\partial \mathbf{m}}{\partial t}$ are collected on the left to give

$$(1 + \alpha^2) \frac{\partial \mathbf{m}}{\partial t} = -\gamma \mathbf{m} \times \mathbf{B} - \frac{\alpha}{m} \gamma \mathbf{m} \times (\mathbf{m} \times \mathbf{B}), \qquad (4.38)$$

$$\frac{\partial \mathbf{m}}{\partial t} = -\frac{\gamma}{(1 + \alpha^2)} \mathbf{m} \times \mathbf{B} - \frac{\alpha}{(1 + \alpha^2)m} \gamma \mathbf{m} \times (\mathbf{m} \times \mathbf{B})$$

in the Landau–Lifshitz form, with the relaxation expressed as a double cross product. Using the renormalized gyromagnetic ratio defined in Eqn (4.2) and expressing the Landau–Lifshitz relaxation parameter λ in units of the Gilbert damping parameter α,

$$\lambda = \frac{\gamma \alpha}{(1 + \alpha^2)} = \gamma_L \alpha, \qquad (4.39)$$

the Landau–Lifshitz equation can be written as

$$\frac{\partial \mathbf{m}}{\partial t} = -\gamma_{\mathrm{L}} \mathbf{m} \times \mathbf{B} - \gamma_{\mathrm{L}} \frac{\alpha}{m} \mathbf{m} \times (\mathbf{m} \times \mathbf{B}). \tag{4.40}$$

For isotropic damping, the Landau–Lifshitz and the Landau–Lifshitz–Gilbert equation differ only in the regard that $\gamma_{\mathrm{L}} \neq \gamma$. A common practice is to use the Landau–Lifshitz equation but specify the damping as a Gilbert damping parameter α. In the limit of vanishing damping, $\alpha \to 0$, the gyromagnetic ratios become equal, $\gamma_{\mathrm{L}} \to \gamma$. In studies of dynamics in the presence of a time-dependent external field, it is important that the correct gyromagnetic ratio γ_{L} is used during the simulation. In other cases, it is also possible to rescale the time after the simulation, corresponding to using a different gyromagnetic ratio.

4.8 The magnetic Hamiltonian

A critical aspect of ASD is establishing a practical expression for $\mathbf{B}_i$ in Eqn (4.1). The field $\mathbf{B}_i$ should naturally contain all relevant interactions that each atomic moment may experience, where the most important one naturally is the interatomic exchange interaction. Other terms like the Dzyaloshinskii–Moriya (DM) interaction, the anisotropy field, classical dipolar interactions, and an external Zeeman term may also be considered. The most efficient way to consider these interactions is to include them in an effective spin Hamiltonian, for which one may evaluate the field $\mathbf{B}_i$ from the expression

$$\mathbf{B}_i = -\frac{\partial \mathcal{H}}{\partial \mathbf{m}_i}. \tag{4.41}$$

As was described in Chapter 2, Section 2.6, it is possible to evaluate the interatomic exchange from first-principles theory, and provided spin–orbit interaction is considered, it is possible to also obtain accurate results for the DM interaction and the magnetic anisotropy energy (see Chapter 3, Section 3.3). Suitable parametrized forms of these interactions can then be used for the total spin Hamiltonian, which then becomes one of the critical aspects of the multiscale approach used when connecting DFT results to ASD simulations. This is much in the same way as quantum mechanical forces are used in ab initio molecular dynamics simulations to evolve in time the nuclear coordinates of atoms. Most magnetic Hamiltonians can be decomposed into the following terms:

$$\mathcal{H} = \mathcal{H}_{\mathrm{Heis}} + \mathcal{H}_{\mathrm{MAE}} + \mathcal{H}_{\mathrm{DM}} + \mathcal{H}_{\mathrm{dd}} + \mathcal{H}_{\mathrm{ext}}. \tag{4.42}$$

Here, the first term represents the interatomic Heisenberg exchange interaction, the second term represents the DM interaction, the third term represents the magnetocrystalline anisotropy energy (MAE), the magnetostatic dipole–dipole interaction is included in the fourth term, and the fifth term is the Zeeman energy.

From Eqn (4.17), the Kohn–Sham equation of any non-collinear configuration can be evaluated, and hence also its total energy. This can be done, for example, for a ferromagnetic host in which the magnetization direction of any of its atoms is rotated by a small amount. This could for instance be done from the rotation matrix in Eqn (4.14), possibly with a constraining field as described above, and any continuous rotation could be considered in this way. Hence, a rotation of any atoms' or any pair of atom's magnetization direction could be considered, and hence also the changes in total energy for these rotations. Adopting the atomic moment approximation, this then corresponds to energies of continuous rotations of atomic spins, something which is described parametrically in Eqn (2.19). Hence, it is possible to connect results from DFT to exchange interactions of the Heisenberg form, in particular for classical spins. This is indeed what was done by Liechtenstein, Katsnelson, and Gubanov et al., (1984, 1987), who used a multiple scattering formalism for the electronic structure (see also Chapter 3, Section 3.4). In addition, in these works, pair exchange was shown to be of the bilinear Heisenberg form. Hence, both the Heisenberg form of the exchange interaction, and a practical method of how to evaluate the strength of the exchange between any pair of atoms, were established. Relativistic or non-collinear generalization to this formalism yield slightly modified results, but the basic idea of all these generalizations is the same. Consequently, first-principles theory can be used to evaluate bilinear and biquadratic exchange parameters and the DM interaction, as well as exchange interactions at finite temperatures (Liechtenstein, Katsnelson, and Gubanov et al., 1984, 1987; Udvardi et al., 2003; Szilva et al., 2013).

An important development in recent years has been advances in methods to calculate effective exchange interactions for strongly correlated systems. Kvashnin et al. (2015) reported exchange interaction parameters for body-centred cubic Fe, hexagonal close-packed Gd, and antiferromagnetic NiO, calculating these values by using a numerical scheme that combines first-principles full-potential electronic structure theory with dynamical mean field theory (Grånäs et al., 2012). Formalisms to address the time evolution of exchange interactions in non-equilibrium correlated systems have been communicated (Secchi et al., 2013; Mentink and Eckstein, 2014; Mentink et al., 2015), with the latter two papers reporting ultrafast quenching and reversible control of the exchange interaction in Mott insulators. To account also for the interaction of orbital magnetic moments was one motivation for the study by Secchi et al. (2015).

In the remaining part of this section, we will formulate the mathematical expressions for all the various interactions in Eqn (4.42). The typically most important interaction is Heisenberg exchange, formulated in Eqn (2.19). Compared to other contributions to the internal energy, it is dominant. One of the consequences of this is that it is mainly the Heisenberg exchange that governs the temperature dependence of the magnetic system. One kind of interaction that has previously not been considered important but has recently been shown to be relevant in systems with reduced symmetry is the DM interaction (Dzyaloshinskii, 1957; Moriya, 1960), which can be written as

$$\mathcal{H}_{DM} = -\frac{1}{2}\sum_{i,j} \mathbf{D}_{ij} \cdot (\mathbf{m}_i \times \mathbf{m}_j), \tag{4.43}$$

where the coupling constant is referred to as the DM vector, $\mathbf{D}_{ij}$.[1] An important class of systems in which the DM interaction plays a role comprises thin metallic films, in which the surface breaks the inversion symmetry, as will be treated in detail in Chapter 9. For some magnetic materials, higher-order exchange couplings could also be of relevance. An example is the biquadratic exchange coupling

$$\mathcal{H}_{\text{bq}} = -\frac{1}{2} \sum_{i,j} \mathcal{J}_{ij}^{\text{bq}} (\mathbf{m}_i \cdot \mathbf{m}_j)^2, \tag{4.44}$$

which is important, for example, for some type II multiferroics such $RMnO_3$ compounds (Hayden et al., 2010; Mochizuki et al., 2010b) and CuO (Yablonskii, 1990; Pasrija and Kumar, 2013; Hellsvik et al., 2014), as well as for finite temperature exchange (Szilva et al., 2013). There also have been suggestions that a biquadratic form of the DM interaction, as well as multi-site interactions such as four-spin interactions (Heinze et al., 2011; Fedorova et al., 2015), sometimes needs to be added to the Hamiltonian, in order to obtain a good description of the energetics and dynamics of the system at hand. In addition to these pair interactions, a proper description of a spin system may often also have to contain single-site contributions to the energy and effective magnetic field. The prime example of this is MAE. For cubic bulk systems, the MAE is typically quenched, as an example, the MAE for body-centred cubic Fe is of the order of 1.4 μeV per atom. On the other hand, systems with lower symmetry and/or the presence of heavier elements, where the spin–orbit coupling is stronger, can have much larger MAE. The technologically important FePt alloy in the $L1_0$ structure has an MAE on the order of 1 meV per formula unit. Many magnetically hard materials, that is, those with large MAE, exhibit a uniaxial magnetic anisotropy.

The expressions for uniaxial anisotropy (Eqn (3.1)) and cubic anisotropy, (Eqn (3.2)), which in Chapter 3, Section 3.3, were stated in angles and directional cosines, can equivalently be formulated in terms of local magnetic moment unit vectors $\hat{\mathbf{e}}_i = \mathbf{m}_i/|\mathbf{m}_i|$. For uniaxial anisotropy the leading orders read

$$\mathcal{H}_{\text{uni}} = -\sum_i \left\{ K_i^1 (\hat{\mathbf{e}}_i \cdot \mathbf{e}_i^k)^2 + K_i^2 (\hat{\mathbf{e}}_i \cdot \mathbf{e}_i^k)^4 + \cdots \right\}, \tag{4.45}$$

where K_i^n is the uniaxial anisotropy constant of order n, and $\mathbf{e}_i^k$ is the axis of the uniaxial anisotropy. The first-order anisotropy term is typically dominant and is thus often the only anisotropy term that needs to be included in the Hamiltonian. As was the case with the magnetic exchange terms, there exist different conventions regarding the sign of the anisotropy terms. A minus sign in front of the summation, as is used in Eqn (4.45), means that a positive value of K^1 leads to a favoured magnetization axis, an easy axis, while a negative K^1 would infer a cost in energy to align the magnetic moments along the anisotropy axis, in which case one gets a hard axis, or easy plane, anisotropy. The

---

[1] The needed symmetry breaking is the lack of inversion symmetry for the magnetic ions in the material.

MAE is typically very weak in cubic systems but can be modelled if needed and then expressed as

$$\mathcal{H}_{\text{cub}} = -\sum_{i} \left\{ K_i^1 \left(e_{i,x}^2 e_{i,y}^2 + e_{i,x}^2 e_{i,z}^2 + e_{i,y}^2 e_{i,z}^2 \right) + K_i^2 \left(e_{i,x}^2 e_{i,y}^2 e_{i,z}^2 \right) + \cdots \right\}, \tag{4.46}$$

where, once again, care needs to be taken regarding the sign of the contributing terms.

The MAE, which is a relativistic, quantum mechanical property coming from spin–orbit coupling, is hence seen to provide coupling between the magnetic moments and the underlying lattice. A similar coupling comes from the classical magnetostatic interactions between the atomic (magnetic) dipole moments. As is the case with classical dipoles, magnetic moments interact in such a way as to minimize the magnetic stray field from the system. The result is a set of weak but long-range dipole–dipole interactions which give rise to a demagnetizing field in the system at hand. In micromagnetics, that is, studies of magnetization dynamics on length scales larger than what is typically accessible by atomistic magnetization dynamics, the demagnetizing field is of critical importance. Two important effects that stem from the demagnetizing field are worth mentioning here. First, since it strives to decrease the stray field from a sample, the field favours magnetization patterns that forms closed loops of the magnetic field inside the system. This is the driving force behind the formation of magnetic domains in bulk systems. Another effect is that, for systems without spherical symmetry, the demagnetizing field causes an additional magnetic anisotropy that depends on the shape of the system. This is called shape anisotropy and, as a rule of thumb, shape anisotropy prefers that magnetic moments lie along directions of the material that are long. As a well-known example, shape anisotropy prefers in-plane magnetization patterns in thin samples.

For the typical length scale considered with ASD, dipolar interactions are much weaker than other interactions and can thus often be neglected or at most be considered in the form of an additional shape-anisotropy contribution that can be added to the onsite magnetocrystalline anisotropy discussed above. Nevertheless, there might be instances where dipolar effects play a more important role in the atomistic magnetization dynamics and then there are several possible ways of treating them. For finite sizes, the demagnetizing field can be calculated by 'brute force', that is, by simply considering the full dipole–dipole interaction matrix

$$\mathcal{H}_{\text{dd}} = -\frac{\mu_0}{4\pi} \sum_{i,j,j \neq i} \frac{1}{r_{ij}^3} \left[3 \left(\mathbf{m}_i \cdot \hat{\mathbf{r}}_{ij} \right) \left(\mathbf{m}_j \cdot \hat{\mathbf{r}}_{ij} \right) - \mathbf{m}_i \cdot \mathbf{m}_j \right], \tag{4.47}$$

where μ_0 is the magnetic constant.

5

Spin Dynamics at Finite Temperature

In Chapter 4, we presented a microscopic derivation of the Landau–Lifshitz equation and its connection to ab initio results, such as that provided by density functional theory. All the analysis of Chapter 4 was done by considering the temperature $T = 0$ K. Since most magnetic phenomena of interest are observed at finite temperature, it is important to generalize the analysis presented above to incorporate the effects of finite temperature. In the discussion of Eqn (4.1), it was mentioned briefly that finite temperature effects are incorporated in the stochastic field $\mathbf{B}_i^{\mathrm{fl}}$. Details of the coupling between temperature and the stochastic field comprise the main topic of this chapter.

5.1 Langevin dynamics

The study of equilibrium properties of quantum mechanical and classical spin models is a well-established field of condensed matter and statistical physics. The physics that take place in the vicinity of phase transitions has for long time been of particular interest (Fisher, 1967). A large body of theoretical work has been presented, with important concepts such as finite size scaling, universality classes, and the renormalization group (Fisher, 1974; Cardy, 1996). For less technical introductions to these topics, the papers by Stanley (1999) and Delamotte (2004) are suggested reading. For atomic spin dynamics, the theory of dynamic critical phenomena (Hohenberg and Halperin, 1977) is relevant, as it encompasses phenomena such as critical slowing down (K.Chen and Landau, 1994) and the relaxation processes of frustrated systems (A. Young, 1998). The thermal equilibrium properties of a spin system modelled by a classical Hamiltonian follow Boltzmann statistics (Plischke and Bergersen, 1994). The Boltzmann statistics is an important starting point also for simulations of atomistic spin dynamics at finite temperatures. Following up on their study of static behaviour, K. Chen and D. P. Landau (1994) performed spin dynamics simulations on the Heisenberg model.

At finite temperature, atomic magnetic moments experience not only the deterministic precession torque and the damping torque connected to the effective field $\mathbf{B}_i$ but also stochastic torques due to thermal fluctuations. Stochastic torques can be included as fluctuating fields $\mathbf{B}_i^{\mathrm{fl}}$ in the full stochastic Landau–Lifshitz equation. As will be described in this chapter, the power of fluctuating torques needs to be connected to the

Atomistic Spin Dynamics. Olle Eriksson, Anders Bergman, Lars Bergqvist, Johan Hellsvik. First Edition.
© Olle Eriksson, Anders Bergman, Lars Bergqvist, Johan Hellsvik 2017. First published in 2017 by Oxford University Press.

mechanisms for damping. That fluctuations and relaxations are intimately connected is a very important result of equilibrium statistical physics and is expressed in the fluctuation–dissipation theorem (Callen, 1951; Kubo, 1966; Plischke and Bergersen, 1994).

The first investigations on how to include fluctuations in magnetization dynamics were undertaken by William Fullet Brown, J. (1963), who let a stochastic torque contribute to both the precession and the damping motion. To find out a relation between the strength of the fluctuating field and the rate of damping, he formulated the Fokker–Planck equation (Risken, 1989) for the stochastic version of the Landau–Lifshitz equation. The result is an equation that couples the amplitude of the stochastic field $\mathbf{B}_i^{\mathrm{fl}}$ to the damping parameter α and the temperature T. The form of this relation is similar to that between the fluctuating forces and the diffusion constant in Einstein's theory of Brownian motion (Einstein, 1906). Choosing to include the fluctuating fields only in the precession term, Kubo and Hashitsu (1970) worked out a formalism similar to, but not identical to, Brown's expressions. Brown and Kubo were modelling single-domain magnetic particles and single spins, respectively. It was proposed by Antropov and Harmon (1996) that the same formalism would be applicable to atomic magnetic moments.

The stochastic Landau–Lifshitz equation, which includes stochastic fields in both the precession term and the damping term, in Chapter 4, as was given in Eqn (4.1). This equation can be classified as a Langevin equation, with multiplicative noise introduced as Langevin forces (van Kampen, 2007). The term 'multiplicative' implies that the Langevin forces enter the equation with coefficients depending on the system variables. This indeed is the case for the stochastic Landau–Lifshitz equation, as $\mathbf{B}_i^{\mathrm{fl}}(t)$ is a noise term that occurs in a cross product or double cross product with the system variable $\mathbf{m}_i$. Stochastic differential equations and related aspects of stochastic processes are covered in a textbook by van Kampen (2007). As will be discussed in Section 5.4.2, for every Langevin equation, there is an equivalent Fokker–Planck equation. The Fokker–Planck equation and techniques for how to solve it are discussed in detail in the monograph by Risken (1989).

The fluctuating fields are caused by a large number of weakly coupled microscopic events and follow, due to the central limit theorem, a Gaussian distribution. In principle, the noise can be correlated both in orbital space and in spin space. In the following, it is assumed that the components $B_{i,\mu}^{\mathrm{fl}}$, with $\mu = \{x, y, z\}$, are uncorrelated and that the fluctuating fields for different atomic magnetic moments $\mathbf{m}_i$ are uncorrelated. Assuming furthermore that the average of the fluctuating field is 0, the first two moments of the fluctuating field are given by the expressions

$$\langle B_{i,\mu}^{\mathrm{fl}}(t)\rangle = 0, \quad \langle B_{i,\mu}^{\mathrm{fl}}(t)B_{j,\nu}^{\mathrm{fl}}(s)\rangle = 2D\delta_{\mu\nu}\delta_{ij}f(t-s), \tag{5.1}$$

respectively. Note that the correlations between spins are described by the autocorrelation function $f(t,s)$, which, for the present discussion, is expressed in the form $f(t-s)$. The first moment, $\langle B_{i,\mu}^{\mathrm{fl}}(t)\rangle = 0$, is the average value of the stochastic process and the second moment, $\langle B^{\text{fl}}_{i,\mu}(t)B^{\text{fl}}_{j,\nu}(s)\rangle$, is its variance, whose magnitude is described by D. Eqn (5.1) hence describes correlations that are local in space and time. A short, but finite, correlation in time can be modelled, for example, by exponentially correlated coloured noise (Milshtein and Tret'yakov, 1994):

$$\langle B^{\text{fl}}_{i,\mu}(t)\rangle = 0, \quad \langle B^{\text{fl}}_{i,\mu}(t)B^{\text{fl}}_{j,\nu}(s)\rangle = \frac{b^2}{2a}e^{(-a|t-s|)}. \tag{5.2}$$

Fluctuations that have finite correlation time are characterized by a noise power that depends on frequency and are for this reason referred to as 'coloured'. This can be inferred through the Wiener–Khintchine theorem (Risken, 1989), which states that the power spectrum of a stationary fluctuating process can be calculated as the Fourier transform of the autocorrelation function

$$S(\omega) = \int_{-\infty}^{\infty} f(t')e^{i\omega t'}\, dt'. \tag{5.3}$$

If the stochastic process is very fast in comparison to the deterministic part of the motion, the autocorrelation function can be approximated with the product $f(t') = q\delta(t')$, where q is amplitude, and $\delta(t')$ is the Dirac δ-function. That noise with zero correlation time has a flat power spectrum, completely lacking dependence on frequency, is easily shown, since

$$S(\omega) = \int_{-\infty}^{\infty} q\delta(t')e^{i\omega t'}\, dt' = 2q. \tag{5.4}$$

This kind of noise is called white noise, as it has the same power for all frequencies. For the reason that coloured noise is more difficult to handle mathematically than white noise, the assumption of noise with zero correlation time is often used in description of fluctuations in physical systems. The Langevin forces that play a role in atomistic spin dynamics are caused by microscopic events that, according to the discussion in Chapter 2, Section 2.6, can have time scales of the order of 0.1 fs.

In Fig. 5.1, we show schematically how the spin system interacts both with the reservoir of electrons as well as the lattice, and we give estimates of time scales for how information between the different reservoirs is exchanged. In a non-equilibrium situation, the temperature of the three different reservoirs may not be the same. How the temperature of each reservoir develops as the system equilibrates is normally described by the so-called three-temperature model (see e.g Beaurepaire et al., 1996), which is discussed in more detail in Chapter 11, where we describe ultrafast magnetization phenomena. We note here that the division of a material into three reservoirs is, in principle, complex. This is especially so concerning the division into a reservoir of atomic spins and an electron reservoir, since it is the electrons themselves that form the atomic spins. This has to be remembered when dealing with ultrafast phenomena in magnetization dynamics (see Chapter 11).

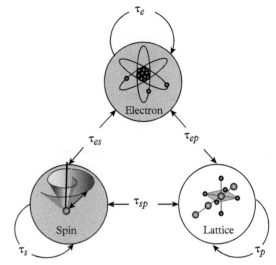

Figure 5.1 *Schematic flow of energy between spin reservoirs, lattice reservoirs, and electron reservoirs. Time scales are $\tau_{ep} \approx 1$ ps, $\tau_{sp} \approx 100$ ps, and $\tau_{es} \approx 0.1$ fs. Figure kindly provided by Dr. Danny Thonig.*

5.2 Stochastic differential equations

Using the Einstein convention, with summation over repeated indices implied without writing the summation sign, the general form of a multidimensional Langevin equation may be written as

$$\frac{\partial X_i}{\partial t} = h_i(\{X_i\}, t) + g_{ij}(\{X_i\}, t)\Gamma_j(t). \tag{5.5}$$

This is a multidimensional stochastic differential equation of the Ito form with indices $i, j = 1, 2, \ldots, n$, where n is the dimension of the equation. The large symbol X_i denotes the multidimensional stochastic variable, and t is the time variable. The first term on the right-hand side, $h_i(\{X_i\}, t)$, accounts for the deterministic drift; the second term represents stochastic diffusion.

To solve a stochastic differential equation, the techniques for solving ordinary differential equations must be extended with techniques for stochastic processes. The reason is that the integral forms of stochastic differential equations are not directly integrable in the standard Riemann or Lebesgue sense. Stating the Ito stochastic differential equation in integral equation form,

$$X_i(t + \tau)$$
$$= X_i(t) + \int_t^{t+\tau} h_i(\{X(t')\}, t')\, dt' + \int_t^{t+\tau} g_{ij}(\{X(t')\}, t')\Gamma_j(t')\, dt', \tag{5.6}$$

we observe that the last term on the right-hand side contains the Langevin force $\Gamma_j(t)$, which is not a continuous function but a stochastic variable. To proceed, the integral which includes the stochastic variable $\Gamma_j(t')$ is replaced with an integral containing the increment $dW_j(t') = \int_0^t \Gamma_j(t')\,dt'$. This transforms the second integral into a Riemann–Stieltjes integral, so that

$$X_i(t+\tau)$$
$$= X_i(t) + \int_t^{t+\tau} h_i(\{X(t')\},t')\,dt' + \int_t^{t+\tau} g_{ij}(\{X(t')\},t')\,dW_j(t'). \qquad (5.7)$$

The distribution of the increment $dW_j(t')$ has the same form as the distribution of the Langevin force does. In the case of Gaussian white noise in Eqn (5.1), the moments of $dW_j(t')$ are given by

$$\langle dW_{i,\mu}\rangle = 0, \quad \langle dW_{i,\mu}\,dW_{j,\nu}\rangle = 2D\tau\delta_{\mu\nu}\delta_{ij}. \qquad (5.8)$$

The integral form of the stochastic differential equation is still not a uniquely defined expression, as the stochastic calculus requires a prescription on how the last term on the right-hand side of Eqn (5.7) should be evaluated. In informal wording, it must be specified for which times t', and with which weights, the term $g_{ij}(\{X(t')\},t')$ should be multiplied with the increment $dW_j(t')$. Stochastic integrals can be calculated as the mean square limit of a sum over n terms with $n \to \infty$. In the notation of Kloeden and Platen (1992), with w denoting specific values of the stochastic variable W, the expression for the integral is

$$I(\lambda) = (\lambda) \int_0^\tau f(t,w)\,dW_t(w) \qquad (5.9)$$

$$= \lim_{n\to\infty} \sum_{j=1}^n \left\{ (1-\lambda)f\left(t_j^{(n)},w\right) + \lambda f\left(t_{j+1}^{(n)},w\right)\right\}\left\{W_{t_{j+1}^{(n)}}(w) - W_{t_j^{(n)}}(w)\right\},$$

with evaluation points $t_j^{(n)}$ for partitions $0 = t_1^{(n)} < t_2^{(n)} < \cdots < t_{n+1}^{(n)} < \tau$, for which

$$\delta^{(n)} = \max_{1\le j\le n}\left(t_{j+1}^{(n)} - t_j^{(n)}\right) \to 0 \qquad \text{when} \qquad n \to \infty. \qquad (5.10)$$

The parameter λ determines the weights of $f\left(t_j^{(n)},w\right)$ and $f\left(t_{j+1}^{(n)},w\right)$. The two common choices for the stochastic integral are the Ito and Stratonovich integrals. The Ito integral is defined by the choice $\lambda = 0$, and the Stratonovich integral by the choice $\lambda = 1/2$. For Langevin equations of the form of Eqn (5.5), it is often, but not always, the Stratonovich calculus that gives the solution that is related to a physical process (Kloeden and Platen, 1992; van Kampen, 2007). Both definitions are reminiscent of the standard Riemann integral in so far that they are the $n \to \infty$ limit of a sum of n terms. In the Ito integral, only the value $f\left(t_j^{(n)},w\right)$ of the integrand is included. In the Stratonovich integral, the value $f\left(t_{j+1}^{(n)},w\right)$ is also used, with the values at $t_j^{(n)}$ and $t_{j+1}^{(n)}$ contributing with equal weight.

In terms of a physical process, the Ito definition describes the situation where the stochastic contribution during the time interval $\left[t_{j+1}^{(n)}, t_j^{(n)}\right]$ does not affect the deterministic contribution during the same time interval.

For many physical processes it is more appropriate to use the Stratonovich calculus. Here, the stochastic contribution for the time interval $\left[t_j^{(n)}, t_{j+1}^{(n)}\right]$ does affect the deterministic contribution for the same time interval. At an intermediate time $\tilde{t}$, with $t_j^{(n)} < \tilde{t} < t_{j+1}^{(n)}$, the system variable $x_{\tilde{t}}$ can have a different value in the presence of the stochastic force than it would have had in the absence of the stochastic force. The deterministic evolution of the system variable from time $\tilde{t}$ and onwards depends on the value $x_{\tilde{t}}$ and it is clear that the presence of a stochastic force can affect the deterministic evolution. This contribution to the otherwise deterministic drift term is called noise-induced drift, or alternatively, spurious noise. The calculation of the noise-induced drift is not trivial but is well described in textbooks on stochastic calculus (Risken, 1989; Kloeden and Platen, 1992).The expression for the drift coefficient with noise-induced drift is needed for some explicit numerical schemes for integration of the stochastic Landau–Lifshitz equation and is also needed together with the diffusion constant in order to formulate the Fokker–Planck equation for the stochastic Landau–Lifshitz equation. The expressions for the drift coefficient with noise-induced drift included, $D_i(\{x_i\}, t)$, and the diffusion coefficient $D_{ij}(\{x_i\}, t)$ of the multidimensional Langevin equation are written as

$$D_i(\{x_i\}, t) = h_i(\{x_i\}, t) + D g_{kj}(\{x_i\}, t) \frac{\partial g_{ij}(\{x_i\}, t)}{\partial x_k}, \tag{5.11}$$

$$D_{ij}(\{x_i\}, t) = D g_{ik}(\{x_i\}, t) g_{jk}(\{x_i\}, t), \tag{5.12}$$

respectively, where D is the noise amplitude of the stochastic process specified by Eqn (5.8). It turns out that the Stratonovich integral, by construction, takes account for the noise-induced drift. This implies that the integrands can be used as they are with no explicit addition of a noise-induced part to the drift coefficient. It can be shown (García-Palacios and Lázaro, 1998) that, for the stochastic Landau–Lifshitz equation, with Gaussian white noise, the Ito calculus gives a solution that fails to reproduce the Boltzmann distribution for the energies of the magnetic moments in the effective field. Instead, the Stratonovich definition of the stochastic integral should be used.

5.3   Finite difference approximations to stochastic differential equations and the choice of stochastic calculus

The schemes that are used to solve ordinary differential equations can be extended to be applicable to the case of stochastic differential equations. It is important to ensure that the stochastic counterpart to an ordinary differential equation integrator gives a solution that converges to the kind of stochastic calculus that has been chosen to apply to the stochastic Landau–Lifshitz that is investigated. We will briefly discuss this issue here, and save a detailed discussion on numerical techniques to integrate the Landau–Lifshitz or stochastic Landau–Lifshitz equations for Chapter 7, Section 7.4. As an example, the Euler integrator will be discussed. This is the simplest form of an explicit scheme where time is discretized as $t_0 < t_1 < t_2 < \cdots < t_n$, and the solutions are obtained as the sequences $X_0 < X_1 < X_2 < \cdots < X_n$. The stochastic extension of the Euler scheme for an ordinary differential equation reads

$$X_i(t + \tau) = X_i(t) + h_i(\{X_i\}, t)\tau + g_{ik}(\{X_i\}, t)dW_k, \qquad (5.13)$$

where τ is the time step, and the moments of the stochastic increments are given by Eqn (5.8). To obtain the value $X_i(t + \tau)$, the Euler scheme uses only the value of $X_i(t)$ at a time t_i. By comparing this with the definition of the Stratonovich and the Ito integrals, it seems that the Euler scheme corresponds to the solution given by the Ito integral, as the definition of the Ito integral uses the value of the integrand only at $\left(t_j^{(n)}\right)$. It can be proved (McShane, 1974; Rümelin, 1982) that the Euler solution converges to the Ito integral in the quadratic mean. The Euler scheme can still be used to obtain the numerical solution in the Stratonovich sense if the integrands $h_i(\{X_i\}, t)$ and $g_{ik}(\{X_i\}, t)$ are substituted with the Stratonovich calculus drift (Eqn (5.11)) and diffusion (Eqn (5.12)) coefficients. The Euler scheme that converges in the quadratic mean to the Stratonovich solution is thus written as

$$X_i(t + \tau) = X_i(t) + D_i(\{X_i\}, t)\tau + D_{ik}(\{X_i\}, t)dW_k. \qquad (5.14)$$

5.4 Fluctuation–dissipation relations for the stochastic Landau–Lifshitz equation

A fluctuation–dissipation relation (FDR) that specifies how noise power and the dissipation coefficient are related can be derived in different ways, using classical or quantum statistics. In this section, we present how an FDR for the stochastic Landau–Lifshitz equation can be derived, either by means of the Fokker–Planck equation or from a general formulation of the fluctuation–dissipation theorem. We also outline how quantum effects can be incorporated.

We will, in Sections 5.4.1–5.4.3, make use of index notation for the vector algebra in which the subscripts i, j, k, l, m, n are indices for the Cartesian spin space coordinates x, y, z. Note that these includes the indices i, j, which elsewhere in this book are used as site indices. In index notation, the cross product is expressed as $(\mathbf{A} \times \mathbf{B})_i = \epsilon_{ijk}A_jB_k$, and the BAC-CAB formula for the double cross product, $\mathbf{A} \times (\mathbf{B} \times \mathbf{C}) = \mathbf{B}(\mathbf{A} \cdot \mathbf{C}) - \mathbf{C}(\mathbf{A} \cdot \mathbf{B})$, is written $\epsilon_{ijk}\epsilon_{klm} = \delta_{km}\delta_{ln} - \delta_{kn}\delta_{lm}$, or explicitly for vectors $\mathbf{A}, \mathbf{B}, \mathbf{C}$ as $\epsilon_{ijk}\epsilon_{klm}A_jB_kC_m = B_iA_lC_l - C_iA_lB_l$.

5.4.1 The stochastic Landau–Lifshitz equation in the form of the Langevin equation

To write the stochastic Landau–Lifshitz equation in the form of a general Langevin equation, the terms containing $\mathbf{B}_i$ and $\mathbf{B}_i^{\text{fl}}$ are separated:

$$\frac{d\mathbf{m}}{dt} = -\gamma \mathbf{m} \times \mathbf{B} - \gamma \frac{\alpha}{m} \mathbf{m} \times (\mathbf{m} \times \mathbf{B})$$
$$-\gamma \mathbf{m} \times \mathbf{B}^{\text{fl}} - \gamma \frac{\alpha}{m} \mathbf{m} \times (\mathbf{m} \times \mathbf{B}^{\text{fl}}). \qquad (5.15)$$

After some straightforward algebra, the terms $h_i(\{\mathbf{m}_i\}, t)$ and $g_{ik}(\{\mathbf{m}_i\}, t)$, defined in Eqn (5.5), can be identified as follows:

$$h_i = -\gamma \epsilon_{ijk} m_j B_k - \gamma \frac{\alpha}{m_i} (m_i m_k B_k - m^2 B_i), \qquad (5.16)$$

$$g_{ik} = -\gamma \epsilon_{ijk} m_j - \gamma \frac{\alpha}{m_i} (m_i m_k - \delta_{ik} m_i m_k). \qquad (5.17)$$

5.4.2 The Fokker–Planck equation

For every Langevin equation there is a corresponding Fokker–Planck equation. The Fokker–Planck equation governs the time evolution of the probability distribution of the states (Risken, 1989). The Fokker–Planck equation for the probability density $P(\{x_i\}, t)$ is expressed in terms of the drift coefficient and the diffusion coefficient:

$$\frac{\partial P(\{x_i\}, t)}{\partial t} = \left[-\frac{\partial}{\partial x_i} D_i(\{x_i\}, t) + \frac{\partial}{\partial x_i x_j} D_{ij}(\{x_i\}, t) \right] P(\{x_i\}, t). \qquad (5.18)$$

The drift coefficient and the diffusion coefficient are here the ones given by Eqns (5.11) and (5.12). Following García-Palacios and Lázaro (1998), the terms of the Fokker–Planck equation are regrouped to yield a form of the equation that will be suitable for the stochastic Landau–Lifshitz equation. First, the drift and diffusion coefficients are substituted in Eqn (5.18) to give

$$\frac{\partial P}{\partial t} = -\frac{\partial}{\partial x_i} \left[\left(h_i + D g_{jk} \frac{\partial g_{ik}}{\partial x_j} \right) P \right] + \frac{\partial^2}{\partial x_i \partial x_j} \left[(D g_{ik} g_{jk}) P \right]. \qquad (5.19)$$

By taking the derivatives in the second term on the right-hand side of the equation, and by permuting dummy indices j and k, we obtain the expression

$$\frac{\partial P}{\partial t} = -\frac{\partial}{\partial x_i} \left\{ \left[h_i - D g_{ik} \left(\frac{\partial g_{jk}}{\partial x_j} \right) - D g_{ik} g_{jk} \frac{\partial}{\partial x_j} \right] P \right\}, \qquad (5.20)$$

where the last equality sign defines the probability current $\mathcal{J}$, so that the Fokker–Planck equation can be written in the compact form $\partial P/\partial t = -\partial \mathcal{J}_i/\partial x_i$ which constitutes a continuity equation for the probability density.

The expressions for h_i and g_{ik} can be substituted into three terms in the square brackets of the right-hand side of the Fokker–Planck equation in the form of Eqn (5.20). The first term in square brackets in Eqn (5.20) simply gives the contribution $h_i P$. To evaluate the second term, we first calculate

$$\frac{\partial g_{ik}}{\partial m_j} = \frac{\partial}{\partial m_j}\left[-\gamma \epsilon_{ijk} m_j - \gamma \frac{\alpha}{m}(m_i m_k - m^2 \delta_{ik} m^2)\right]$$
$$= -\gamma \epsilon_{ijk} - \gamma \frac{\alpha}{m}\delta_{ij} m_k + \gamma \frac{\alpha}{m}\delta_{jk} m_i, \tag{5.21}$$

and, specifically by setting i to j and noting that $\delta_{ii} = 3$, we obtain

$$\frac{\partial g_{jk}}{\partial m_j} = -\gamma \frac{\alpha}{m} 2 m_k. \tag{5.22}$$

Inserting this in the second term of Eqn (5.20), we find after simplification, that the term vanishes, that is,

$$g_{ik}\left(\frac{\partial g_{jk}}{\partial x_j}\right) P = 0. \tag{5.23}$$

For the third term of Eqn (5.20), we obtain

$$g_{ik} g_{jk} \frac{\partial P}{\partial m_j}$$
$$= \left[-\gamma \epsilon_{ijk} m_j - \gamma \frac{\alpha}{m}(m_i m_k - m^2 \delta_{ik} m^2)\right]\left[-\gamma \epsilon_{ijk} m_j - \gamma \frac{\alpha}{m}(m_i m_k - m^2 \delta_{ik} m^2)\right]\frac{\partial P}{\partial m_j}$$
$$= -\gamma^2 (1 + \alpha^2)\left[\mathbf{m} \times \left(\mathbf{m} \times \frac{\partial P}{\partial \mathbf{m}}\right)\right]_i. \tag{5.24}$$

Eventually, the Fokker–Planck equation for the stochastic Landau–Lifshitz equation, in the form of a continuity equation for the probability distribution, can be written as

$$\frac{\partial P}{\partial t} = -\frac{\partial}{\partial \mathbf{m}} \cdot \left\{\left[-\gamma \mathbf{m} \times \mathbf{B} - \gamma \frac{\alpha}{m}\mathbf{m} \times (\mathbf{m} \times \mathbf{B}) + D\gamma^2(1 + \alpha^2)\frac{\mathbf{m}}{m} \times \left(\mathbf{m} \times \frac{\partial}{\partial \mathbf{m}}\right)\right] P\right\}. \tag{5.25}$$

In equilibrium, the time derivative of the probability distribution P_0 should vanish, that is, $\frac{\partial P_0}{\partial t} = 0$, and the probability distribution should be a Boltzmann distribution that is, $P_0(\{\mathbf{m}_i\}) = e^{-\beta \mathcal{H}(\{\mathbf{m}_i\})}$, where $\beta = (k_B T)^{-1}$, and $\mathcal{H}(\{\mathbf{m}_i\})$, is a magnetic Hamiltonian of the type expressed in Eqn (4.42). The spatial derivative in the last term of the

Fokker–Planck equation acts on the probability distribution. By recalling Eqn (4.41), the derivative can be swiftly calculated:

$$\frac{\partial P_0}{\partial \mathbf{m}} = \frac{\partial \left\{ e^{-\beta \mathcal{H}(\mathbf{m})} \right\}}{\partial \mathbf{m}} = -\beta \frac{\partial \mathcal{H}(\mathbf{m})}{\partial \mathbf{m}} e^{-\beta \mathcal{H}(\mathbf{m})} = \beta \mathbf{B} P_0. \tag{5.26}$$

The last equation is inserted into the Fokker–Planck equation, resulting in an equation from which it is possible to extract the relations between D, α, and T, as is required for thermodynamic consistency:

$$\frac{\partial P_0}{\partial t} \tag{5.27}$$

$$= -\frac{\partial}{\partial \mathbf{m}} \left\{ -\frac{\gamma}{\beta} \mathbf{m} \times \frac{\partial P_0}{\partial \mathbf{m}} - \frac{\gamma}{\beta} \frac{\alpha}{m} \mathbf{m} \times \left(\mathbf{m} \times \frac{\partial P_0}{\partial \mathbf{m}} \right) + D\gamma^2 (1 + \alpha^2) \mathbf{m} \times \left(\mathbf{m} \times \frac{\partial P_0}{\partial \mathbf{m}} \right) \right\}.$$

The first term is divergence-free, as can be established by the straightforward calculation

$$\frac{\partial}{\partial \mathbf{m}} \cdot \left[\frac{\gamma}{\beta} \mathbf{m} \times \frac{\partial P_0}{\partial \mathbf{m}} \right] = \frac{\gamma}{\beta} \partial m_i (\epsilon_{ijk} m_j \partial m_k P_0) = 0, \tag{5.28}$$

and we are left with

$$\frac{\partial P_0}{\partial t}$$

$$= -\frac{\partial}{\partial \mathbf{m}} \left\{ \left[D\gamma^2 (1 + \alpha^2) - \frac{\gamma}{\beta} \frac{\alpha}{m} \right] \mathbf{m} \times \left(\mathbf{m} \times \frac{\partial P_0}{\partial \mathbf{m}} \right) \right\}, \tag{5.29}$$

so that a stationary solution is found when

$$D\gamma^2 (1 + \alpha^2) - \frac{\gamma}{\beta} \frac{\alpha}{m} = 0. \tag{5.30}$$

Solving for D and substituting for β gives the relation

$$D = \frac{\alpha}{1 + \alpha^2} \frac{k_B T}{\gamma m}. \tag{5.31}$$

This equation is the Einstein-type FDR for the stochastic Landau–Lifshitz equation in the Gilbert form. The noise power, D, turns out to be frequency independent, precisely as the Wiener–Khintchine theorem had already revealed. As m is in the denominator, it follows that large magnetic moments are less susceptible to fluctuations than small magnetic moments are. This derivation has been the traditional one, involving the Fokker–Planck equation. Had the corresponding derivation instead been pursued for the stochastic Landau–Lifshitz equation, the expression

$$D = \alpha \frac{k_{\mathrm{B}} T}{\gamma m} \qquad (5.32)$$

would have been obtained. We note that the difference in the noise amplitudes for the two stochastic differential equation s is due to the renormalized gyromagnetic ratio used in the stochastic Landau–Lifshitz on the Gilbert form.

5.4.3 Fluctuation–dissipation relations with quantum corrections

The FDRs is Eqns (5.31) and (5.32) were derived using classical statistics and are asymptotically valid for higher temperatures. For lower temperatures, the quantized energy levels of quantum statistics are necessary. Following Woo et al. (2015), the fluctuation–dissipation ratio $\eta = \eta(T)$ is defined as

$$\eta = \frac{D \gamma m}{\alpha}. \qquad (5.33)$$

For the case of classical statistics, this ratio is nothing but $\eta(T) = k_{\mathrm{B}} T$. Using quantum statistics, Woo et al. (2015) derived the following quantum fluctuation–dissipation theorem (QFDR):

$$
\begin{aligned}
\eta(T) &= \frac{1}{N} \sum_k \frac{\hbar \omega}{\exp(\hbar \omega_k / k_{\mathrm{B}} T - 1)} \\
&= \int_0^\infty \frac{\hbar \omega}{\exp(\hbar \omega_k / k_{\mathrm{B}} T - 1)} g_m(\omega, T) \, d\omega,
\end{aligned} \qquad (5.34)
$$

where $g_m(\omega, T)$ is the magnon density of states at temperature T. In the limit of high temperatures $\hbar \omega \ll k_{\mathrm{B}} T$, the classical limit of Eqn (5.31) is recovered. For low temperatures, the Bloch $T^{3/2}$ law is obtained.

5.5 Conservation properties of the Landau–Lifshitz equation

The Landau–Lifshitz equations for interacting spin systems are coupled non-linear differential equations. It is only for a few particular cases that exact closed analytical solutions exist, such as for a two-spin system, and for coherent rotation of coupled spins in an external magnetic field. In general, it is necessary to resort to analytical approximation techniques or to schemes for numerical simulation of the equations of motion. Analytical solutions are rare, not only for the trajectories in time of the individual magnetic moments, but also for the trajectories in time of the average magnetization and higher moments (García-Palacios and Lázaro, 1998). An interesting development is the analytical (but not closed) solution for moments, magnetization recently derived for the stochastic Landau–Lifshitz equation (Ma and Dudarev, 2011). With a focus on slightly longer length scales than is typically the case for ASD simulations, an advanced treatise on non-linear magnetization dynamics has been provided by Bertotti et al. (2009).

For small amplitude motion where the spins oscillate around a ground-state spin configuration, it is possible to linearize the Landau–Lifshitz equation to yield a system of linear equations for which analytical or numerical solutions for the eigenfrequencies and eigenmodes can be straightforwardly obtained. This approach has successfully been developed in classical theories for magnetic resonance (Kittel, 1948; Keffer and Kittel, 1952; Lax and Button, 1962; Gurevihvc and Melkov, 1996), as well as quantum mechanical theories for spin waves (Holstein and Primakoff, 1940; van Kranendonk and Van Vleck, 1958; Achiezer et al., 1968; Kittel and Fong, 1987) and will be discussed further in Chapter 8. Temperature can be introduced into this framework in the form of temperature-dependent sublattice magnetization and coupling constants.

Even though analytical solutions of the Landau–Lifshitz equation exist only for rare cases, there are a number of conservation relations that one can derive analytically, that apply also for general situations that involve numerical simulations. To start with, we note that solutions to the stochastic Landau–Lifshitz equation can be grouped into three different classes with regard to the damping parameter and the temperature.

1. $\alpha = 0, T = 0$. The equation of motion for the spin systems is then the Landau–Lifshitz equation, with conservative dynamics. The internal energy U of the magnetic system is preserved, also in the presence of an external field. No direct channel for transfer between Zeeman energy and exchange energy is allowed. A direct channel between anisotropy energy and exchange energy is possible.

2. $\alpha > 0$, $T = 0$. The equation of motion for the spin systems is then the Landau–Lifshitz equation, with Landau–Lifshitz dissipative dynamics. Energy is transferred out from the magnetic system. The dynamics have a Lyapunov structure, which means that, in the case that the external magnetic field does not depend on time, the free energy is monotonously decreasing (D'Aquino et al., 2005).

3. $\alpha > 0$, $T \geq 0$. The equation of motion for the spin systems is then the stochastic Landau–Lifshitz equation, with dissipative dynamics in the presence of a heat bath. This is the standard equation that is solved in atomistic spin dynamics simulations. Averaging over heat baths is essential, as the finite size of the simulation cells makes the evolution of the spin system sensitive to random-walk-like drift of the total angular momentum.

Atomistic spin dynamics simulations are commonly performed for the third class in the list above, involving damped dynamics at finite temperature, since this is the situation that most closely resembles an experimental situation. Many of the exact relations apply only to the conservative dynamics (Class 1). An integrator for this class should ideally preserve all or nearly all of these conservation laws identically. For small damping, where $\alpha \ll 1$ (Class 2 or 3), the dynamics can be regarded as a perturbation to $\alpha = 0$ dynamics (Class 1). Therefore, a solver that performs well for the first class is likely to be a good choice also for the second or third class.

As expressed in Eqn (4.32), the lengths of individual spins are constants of motion of the Landau–Lifshitz equation. This property carries over to the stochastic Landau–Lifshitz equation (Class 3), for which preservation of spin length is the *only* conserved quantity. In $\alpha = 0$ (Class 1) dynamics, conservation laws can exist also for the total angular momentum. If the magnetic Hamiltonian only includes isotropic exchange interactions, the total angular momentum is a conserved property, as no external torque acts on the system. In precession in an external magnetic field, the total angular momentum associated with the magnetic moments precesses. This is the essence of gyroscopic precession: angular momentum precesses with the tangent vector for the motion of the tip of the angular momentum vector parallel to the applied torque. The total angular momentum is here not a preserved quantity, however, the projection of the angular momentum in the direction of the applied field is a constant of motion.

At zero temperature and $\alpha > 0$ (Class 2), an important relation for the energy dissipation holds for the case of a time-independent external magnetic field, namely, that the internal energy is a decreasing function of time (D'Aquino et al., 2005). This property can be derived by first observing that the internal magnetic energy U in a volume V can be expressed as

$$
\begin{aligned}
\frac{dU}{dt} &= \int_V \left(\frac{\partial U}{\partial \mathbf{m}} \cdot \frac{\partial \mathbf{m}}{\partial t} + \frac{\partial U}{\partial \mathbf{B}} \cdot \frac{\partial \mathbf{B}}{\partial t} \right) d\mathbf{r} \\
&= \int_V \left(-\mathbf{B} \cdot \frac{\partial \mathbf{m}}{\partial t} + \mathbf{m} \cdot \frac{\partial \mathbf{B}}{\partial t} \right) d\mathbf{r}.
\end{aligned}
\tag{5.35}
$$

From the Landau–Lifshitz equation, we have

$$
\frac{\partial \mathbf{m}}{\partial t} = -\gamma \mathbf{m} \times \left(\mathbf{B} - \frac{\alpha}{m} \frac{\partial \mathbf{m}}{\partial t} \right),
\tag{5.36}
$$

which can be used to evaluate the terms in Eqn (5.35). From scalar multiplication of the Landau–Lifshitz equation with $\mathbf{B} - \frac{\alpha}{m} \frac{\partial \mathbf{m}}{\partial t}$, and properties of the scalar triple product, it can be established that $\frac{\partial \mathbf{m}}{\partial t} \cdot (\mathbf{B} - \frac{\alpha}{m} \frac{\partial \mathbf{m}}{\partial t}) = 0$. This identity can be substituted into the expression for internal energy to give

$$
\frac{dU}{dt} = \int_V \left(-\alpha \left| \frac{\partial \mathbf{m}}{\partial t} \right|^2 + \mathbf{m} \cdot \frac{\partial \mathbf{B}}{\partial t} \right) d\mathbf{r}.
\tag{5.37}
$$

For the case that the applied magnetic field is constant in time, the expression simplifies to

$$
\frac{dU}{dt} = -\int_V \alpha \left| \frac{\partial \mathbf{m}}{\partial t} \right|^2 < 0,
\tag{5.38}
$$

that is, the internal energy is a decreasing function.

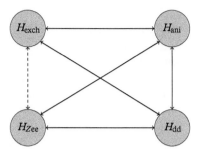

Figure 5.2 *The flow of energy between different parts of magnetic interactions for* α = 0 *dynamics. The five full lines indicate energy channels which are allowed, and the dashed line indicates the forbidden channel between Heisenberg exchange (exch) and Zeeman energy (zee); ani, stands for anisotropy; dd, stands for dipole–dipole.*

Even for the case of conservative dynamics, energy can however be transferred within the magnetic system, between the different contributions to the spin Hamiltonian. The time derivative of different parts of the magnetic energy can be calculated from the commutator of each contribution and the total magnetic Hamiltonian (Safonov and Neal Bertram, 2001). The commutator between Heisenberg exchange energy and the magnetocrystalline anisotropy is, in general, not 0. This implies that, for systems with only magnetic anisotropy and Heisenberg exchange, energy can be transferred from (to) the magnetocrystalline anisotropy energy to (from) the exchange energy in the case of conservative dynamics as well. This transfer of energy is accompanied by a transfer of angular momentum. The commutator of the Heisenberg part of the interatomic exchange energy and the Zeeman energy is always 0. Consequently, for systems where these two terms dominate the spin Hamiltonian, energy cannot be exchanged between these terms.

The flow of energy within a conservative magnetic system can be visualized as a graph with four vertices (shown in Fig. 5.2), all of which are mutually connected by, in total, six edges. Out of the six energy channels, some channels are always closed (e.g. exchange energy ↔ Zeeman energy), whereas some of the others can be opened or closed depending on geometry, interactions, and the magnetic configuration at an instant of time. If all the four energy types are connected, directly or over the other sorts of energy, the amount of energy of different types will evolve over time, reflecting the nature of how the magnetic moments precess.

5.6   Finite temperature exchange

So far, we have discussed finite temperature effects via a noise term (or the stochastic magnetic field $\mathbf{B}_i^{fl}$) in Eqn (4.1). In many applications, it is sufficient to only consider finite temperature effects on this level of approximation, while, for example, the size of the magnetic moment or the parameters of the spin Hamiltonian used to evaluate the local field, $\mathbf{B}_i$, in Eqn (4.1) are obtained from DFT calculations at $T = 0$ K. On this level of approximation, two parameters of Eqn (4.1) depend on temperature. One is the stochastic field whose strength depends on temperature according to Eqns (5.1) and (5.31), while the other is $\mathbf{B}_i$, which depends on temperature via the particular spin configuration at each time step of the simulations. There are, however, examples where this is a less accurate approximation and it is desirable to include finite temperature effects also in the evaluation of all parameters of the spin Hamiltonian that are used to calculate $\mathbf{B}_i$ in Eqn (4.1). This involves primarily the size of the magnetic moment and the exchange parameters. We outline shortly one method to do this, as proposed by Szilva et al. (2013). The idea is to couple iteratively the information obtained from finite temperature spin dynamics with electronic structure theory in the following way. First, from Eqn (2.23), one obtains exchange parameters that are used to perform a finite temperature spin-dynamics simulation using Eqn (4.1). At this temperature, the atomistic spins form a non-collinear configuration, and once equilibration has been realized in the simulations, information about the degree of non-collinearity is brought back to the electronic structure part of the problem. Typically, this is done by a few parameters that characterize the non-collinear magnetic state, for example, the angles θ and ϕ in Eqn (4.16). With

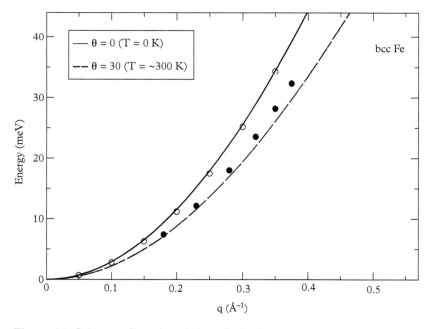

Figure 5.3 *Spin wave dispersion relation calculated at room temperature and zero temperature in body-centred cubic (bcc) Fe, as a function of wavevector q and plotted in a direction from Γ to H. The open circles represent magnetization measurements taken at 4.2 K (data from Pauthenet, R., (1982), Experimental verification of spin-wave theory in high fields,* J. Appl. Phys., 53, 8187–92, *and Pauthenet, R., (1982), Spin-waves in nickel, iron, and yttrium-iron garnet,* J. Appl. Phys., 53, 2029–31), *and the filled circles refer to room temperature neutron scattering measurement data (from Lynn, J. W., (1975), Temperature dependence of the magnetic excitations in iron.* Phys. Rev. B, *11, 2624–37). Reprinted figure with permission from Szilva, A., Costa, M., Bergman, A., Szunyogh, L., Nordström, L., and Eriksson, O.,* Phys. Rev. Lett., *111, 127204, 2013. Copyright 2013 by the American Physical Society.*

these parameters (angles), the electronic structure problem is solved anew (where one needs to consider Eqn (1.31) for a non-collinear description of the Kohn–Sham equation). From the solution of this new electronic structure calculation, both the magnetic moment and the Heisenberg exchange parameters are recalculated and passed on to the stochastic spin dynamics equation, Eqn (4.1). The whole procedure can then be iterated until self-consistency.

We note that Szilva et al. (2013) argued that a bilinear form of the spin Hamiltonian is not sufficient in the case of non-collinear magnetism and that a smaller biquadratic term needs to be added, such that the effective spin Hamiltonian becomes

$$\mathcal{H}^Q = -\frac{1}{2}\sum_{ij}^{i\neq j}\mathcal{J}'_{ij}\mathbf{n}_i\mathbf{n}_j - \frac{1}{2}\sum_{ij}^{i\neq j} B_{ij}\left(\mathbf{n}_i\mathbf{n}_j\right)^2 . \tag{5.39}$$

Note that a revised bilinear parameter $\mathcal{J}'_{ij}$ is introduced here, which dominates the problem, and it is in general different from the exchange parameter of Eqn (2.23). The parameter B_{ij} gives the strength of the biquadratic term and may also be calculated from first-principles theory (Szilva et al., 2013). We illustrate in Fig. 5.3 the effects of finite temperature on the exchange parameters for body-centred cubic Fe, using magnon excitation energies taken from the work by Szilva et al. (2013). Note that the figure contains experimental data at low temperature (4.2 K) and room temperature, as well as theoretical data from $T = 0$ K and room temperature. For simplicity, the magnon spectra shown in Fig. 5.3 are calculated with a so-called adiabatic approach (Kübler, 2009; described in detail in Chapter 9), using temperature-dependent exchange parameters. Note that, with increasing temperature, the magnon dispersion softens, an effect which is both observed experimentally as well as found in the theory. One may also note that, on a quantitative level, the agreement between theory and experiment is rather good, which naturally gives credence to the method proposed by Szilva et al. (2013).

5.7   Some final comments

In Sections 5.1, 5.2, and 5.4 we outlined details on how finite temperature effects can be incorporated as a fluctuating field in the stochastic Landau–Lifshitz equation (Eqn (4.1)), and how the size of the fluctuating field is connected to the temperature of the atomistic spins, as well as the damping parameter of the materials (Eqns (5.1) and (5.31)). We have also outlined that interatomic exchange parameters in general should depend on temperature, and we have suggested one way to calculate such temperature effects. Sections 5.1, 5.2, and 5.4 also contained discussions of how finite temperature effects necessarily lead to a stochastic differential equation. Practical details of the implementation of the finite temperature effects of Eqn (4.1), together with general practicalities of numerical solutions to this equation, will be given in Chapter 7. Before entering into that discussion we will outline in Chapter 6 how one can evaluate from ab initio theory the strength of the damping parameter of Eqn (4.1).

6

The Damping Term, from First Principles

In Chapters 1–5, we described the basic principles of density functional theory, gave examples of how accurate it is to describe static magnetic properties in general, and derived from this basis the master equation for atomistic spin dynamics; the stochastic Landau–Lifshitz (or stochastic Landau–Lifshitz–Gilbert) equation. However, one term was not described in these chapters, namely the damping parameter. This parameter is a crucial one in the stochastic Landau–Lifshitz (or stochastic Landau–Lifshitz–Gilbert) equation, since it allows for energy and angular momentum to dissipate from the simulation cell. The damping parameter can be evaluated from density functional theory and the Kohn–Sham equation, and it is possible to determine its value experimentally. This chapter covers in detail the theoretical aspects of how to calculate the damping parameter. Chapter 8 is focused, among other things, on the experimental detection of damping via ferromagnetic resonance.

6.1 Background

Any effect that dissipates energy and angular momentum from the magnetic system to the environment is denoted 'damping' in the context of magnetization dynamics. The most common damping mechanism is scattering due to phonons and magnons, at elevated temperature, with associated spin-flip scattering, but scattering against lattice imperfections and the generation of eddy currents may also be important. Without any damping processes present, the total energy is preserved in the magnetic system. It is clear that damping has an important impact on the magnetization dynamics properties, and reliable determination from both experiments and first-principles calculations is important. The main driving force for damping, that is, the transfer of angular momentum from the spin system, comes from spin–orbit coupling. Normally, first-principles calculations of damping only deal with this dominant mechanism, but it is important to remember that there are other mechanisms. Conduction electrons from the environment can transfer angular momentum to the spin system and vice versa, and this is the basic mechanism behind spin-transfer torque, which can lead to damping or

Atomistic Spin Dynamics. Olle Eriksson, Anders Bergman, Lars Bergqvist, Johan Hellsvik. First Edition.
© Olle Eriksson, Anders Bergman, Lars Bergqvist, Johan Hellsvik 2017. First published in 2017 by Oxford University Press.

anti-damping of magnetic moments. Moreover, a dynamic magnetic system radiates electromagnetic waves, which are governed by Maxwell's equations and which also may involve angular momentum transfer. A consistent phenomenological formulation of the damping in the Landau–Lifshitz (LL) equation originates from the pioneering thesis work of Thomas Gilbert in 1956 (Gilbert, 2004). The Landau–Lifshitz–Gilbert (LLG) equation, Eqn (4.31), introduced in Chapter 4, is the starting point for most of the magnetization the dynamics studies. It is written here in the slightly more general form

$$\frac{d\mathbf{m}}{dt} = -\gamma \mathbf{m} \times \mathbf{B}_{\text{eff}} + \frac{1}{m}\mathbf{m} \times \left[\tilde{\alpha}(\mathbf{m}) \cdot \frac{d\mathbf{m}}{dt} \right], \tag{6.1}$$

where the second term on the right-hand side of Eqn (6.1) is the damping term written as a 3×3 damping tensor $\tilde{\alpha}$. Very often, the damping tensor is assumed to be isotropic and scalar, replacing the full tensor with a scalar parameter quantity $\alpha = \alpha_{11}$, where α_{11} is a diagonal element of the tensor. A simple symmetry analysis of the damping tensor and its dependence on the magnetization direction is discussed in Section 6.6. This chapter is devoted to the calculation of the damping, approaching it theoretically from first principles. The experimental background and methods, that is, ferromagnetic resonance, for measuring damping are given in Chapter 8.

Two distinctly different approaches for calculating damping from first principles have emerged: one using the breathing Fermi surface model and the torque correlation model due to Kambersky (Kambersky, 1970, 1976; Kunes and Kambersky, 2002; Kamberský, 2007; Fähnle et al., 2005, 2006; Gilmore et al., 2007), and one using scattering theory from linear response (Brataas et al., 2008; Starikov et al., 2010; Ebert et al., 2011; Bhattacharjee, Nordström, et al., 2012; Mankovsky et al., 2013).

6.2   The breathing Fermi surface

The Fermi surface shape determines many of the properties of a metal: electric, magnetic, thermal, and optical. The shape and occupation number variation at the Fermi edge determines directly the electric current as an example. Moreover, parallel sheets in the Fermi surface can give rise to nesting, causing magnetic instabilities driving the system to a spin-spiral state or a spin density wave (as found in Cr). Damping can also be related to change of the Fermi surface shape in a model introduced by Kambersky (1976), the so-called breathing Fermi surface, which was later modified to account for non-collinear magnetism by Steiauf and Fähnle (2005). The model describes the transfer of energy and angular momentum from the electrons to the lattice through electron scattering processes due to spin–orbit coupling. As discussed in Section 2.3, the energies of the Bloch states $\epsilon_{\mathbf{k}i}$ (where i is the band index, and $\mathbf{k}$ is the wavevector) depend on the magnetic moment's direction, denoted by $\mathbf{e} = \mathbf{m}/m$. If the direction of the magnetic moment changes, so that $\mathbf{e}' = \mathbf{e} + \delta\mathbf{e}$, not only does the Fermi surface shape change and 'breathes', but the occupation numbers near the Fermi level also change, as illustrated in Fig. 6.1.

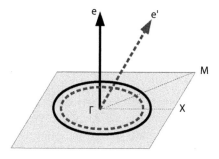

Figure 6.1 *Schematic picture of breathing Fermi surface model for damping. In grey, the Brillouin zone for a square, two-dimensional lattice is shown, along with the symmetry points* Γ, X, *and* M. *Due to the spin–orbit coupling, when the moment with direction* e *changes to a new direction* e′, *the occupation numbers change and thus the shape and size of the Fermi surface, illustrated as a ellipse with full (dashed) line before (after) the orientation change. Figure redrawn from Thonig, D., (2013), 'Magnetization Dynamics and Magnetic Ground State Properties from First Principles', PhD thesis, Martin Luther University, Halle.*

The occupation number of each Bloch state n_{ki} in equilibrium is equal to the Fermi–Dirac equilibrium occupation number f_{ki}. The adiabatic approximation (see Chapter 4), describes slow magnetization dynamics where the electronic system in any instant is in its ground state with respect to the magnetic configuration, and it follows that the occupation numbers of the electrons instantly adapt to the new configuration. The motivation behind this approximation is that the electron scattering processes appear on a time scale that is shorter than the characteristic time scale for magnetization dynamics and, during this time, the moments are precessing without damping:

$$\dot{e} = T(\{e'\}), \tag{6.2}$$

where **T** is the torque acting on a magnetic moment and is defined as

$$T(\{e'\}) = -\frac{2\mu_B}{\hbar}\frac{dE(\{e'\})}{de} \times e, \tag{6.3}$$

where E is the total energy that depends on the magnetic configuration $\{e'\}$. In reality, the scattering processes require a finite time, and the adiabatic approximation is a bit too drastic. The breathing Fermi surface model is designed to describe a situation that is slightly non-adiabatic but close enough to the adiabatic situation so that the adiabatic wave function and eigenvalues are used. The deviation from the strict adiabatic condition is accounted for by introducing a non-adiabatic occupation number n_{ki} that has a time lag to f_{ki}, using the relaxation time model

$$\frac{dn_{ki}(t)}{dt} = -\frac{1}{\tau_{ki}}[n_{ki}(t) - f_{ki}(t)], \tag{6.4}$$

where τ_{ki} is the relaxation time of each electronic state. No statement of the physical origin of the scattering process is made here, and different processes will have different relaxation times. Moreover, only scattering among states close to the Fermi surface is considered. An approximate solution to Eqn (6.4), assuming that the relaxation times are much smaller than the characteristic time scale for moment rotations, is given by

$$n_{ki}(t) = f_{ki}(t) - \tau_{ki}\frac{df_{ki}(t)}{dt}. \tag{6.5}$$

From Eqn (6.5), it is clear that changes in the occupation numbers by scattering causes a transfer of energy and angular momentum from the electrons sub-system. This is the main source for the damping microscopically.

From density functional theory (see Chapter 1), combined with the so-called force theorem, one may obtain a simple expression for the difference of the total energy functional for two different moment directions, **e** and **e′**, that reads

$$\Delta E = \sum_{ki} n_{ki}\epsilon_{ki} - \sum_{ki} n'_{ki}\epsilon'_{ki}. \tag{6.6}$$

The first and second sums of the equation correspond to eigenvalues evaluated with the magnetization along **e** and **e′**, respectively. This approximation of energy differences is surprisingly accurate and eases the calculation of the damping parameter, as will now be described. First, we note that damping in the LLG equation can be seen as an additional field $\tilde{B}_{eff}$ characterized by variation of total energy with respect to variation in moment direction (Fähnle et al., 2005):

$$\tilde{B}_{eff} = -\frac{1}{M}\frac{\partial E}{\partial \mathbf{e}} = -\frac{1}{M}\sum_{ki} n_{ki}\left[\{\mathbf{e'}(t)\}\right]\frac{\partial\epsilon_{ki}\left[\{\mathbf{e'}(t)\}\right]}{\partial \mathbf{e}}, \tag{6.7}$$

where we have used Eqn (6.6) for the infinitesimally small differences between **e** and **e′**. Combining Eqns (6.5) and (6.7) and comparing the result with the LLG equation of Eqn (6.1), the elements of the damping tensor $\tilde{\alpha}(\mathbf{m})$ can be expressed as

$$\tilde{\alpha}_{lm} = -\frac{\gamma}{m}\sum_{ki}\tau_{ki}\frac{\partial f_{ki}}{\partial\epsilon_{ki}}\frac{\partial\epsilon_{ki}}{\partial e_l}\frac{\partial\epsilon_{ki}}{\partial e_m}. \tag{6.8}$$

As pointed out by Fähnle et al. (2005), the damping tensor expression has many similarities with the Drude expression for the electric conductivity tensor $\hat{\sigma}$ in the semi-classical approximation

$$\hat{\sigma}_{lm} = -q^2\sum_{ki}\tau_{ki}\frac{\partial f_{ki}}{\partial\epsilon_{ki}}\frac{\partial\epsilon_{ki}}{\partial k_l}\frac{\partial\epsilon_{ki}}{\partial k_m}, \tag{6.9}$$

where q is the elementary charge. A common approximation in practical calculations is to assume a state-independent relaxation time, that is, $\tau_{ki} \equiv \tau$, resulting in the simplified damping tensor expression

$$\frac{\tilde{\alpha}_{lm}}{\tau} = -\frac{\gamma}{m}\sum_{ki}\frac{\partial f_{ki}}{\partial\epsilon_{ki}}\frac{\partial\epsilon_{ki}}{\partial e_l}\frac{\partial\epsilon_{ki}}{\partial e_l m}. \tag{6.10}$$

Assuming that the main temperature dependence of damping and conductivity is coming from the relaxation time, it can be concluded, due to the similarities between the expressions, that damping and conductivity have similar temperature dependence in this level of approximation. More specifically, from ferromagnetic resonance experiments, it has been shown that the temperature dependence of the damping has two contributions: one proportional to the conductivity, and the other one proportional to the resistivity. Within the level of approximations that has been introduced so far, in particular using the relaxation time approximation, which is not able to describe interband transitions, only the conductivity-like term persists within the breathing Fermi surface model. This has the consequence that numerically calculated values of α are orders of magnitude off from experimental values, if typical relaxation times from conductivity measurements are used. Clearly, the description for calculating Gilbert damping needs to be improved and some of the approximations done in the breathing Fermi surface model should be lifted. For simplicity, the expressions in this system were derived for a collinear magnetized system, but it would be relatively straightforward to generalize them to a non-collinear case (Fähnle et al., 2006).

6.3   The torque correlation model

The main drawback of the breathing Fermi surface model is the explicit assumption of relaxation time, together with a failure to describe interband scattering events. However, one model that is both a generalization of and extension to the breathing Fermi surface model and which remedies most of the shortcomings is Kamberský et al.'s torque correlation model (Gilmore et al., 2007; Kamberský, 2007) which has been shown to be equivalent to the Kubo formalism (Brataas et al., 2008; Bhattacharjee, Nordström, et al., 2012; Ebert et al., 2011; Mankovsky et al., 2013). The starting point for the torque correlation model, following the derivation by Steiauf and Fähnle (2005), is the realization that the energy derivatives with respect to moment direction in the general expression of the damping tensor elements, Eqn (6.8), can be obtained from the Hellmann–Feynman theorem:

$$\frac{\partial \epsilon_{ki}}{\partial \mathbf{e}} = \langle \psi_{ki} | \frac{\delta \hat{H}}{\delta \mathbf{e}} | \psi_{ki} \rangle. \tag{6.11}$$

For tidying up the notation, a transverse torque operator $\hat{T}$ is introduced and defined as

$$\langle \psi_{ki} | \frac{\delta \hat{H}}{\delta \mathbf{e}} | \psi_{ki} \rangle = \langle \psi_{ki} | \hat{T} | \psi_{ki} \rangle = T_{ki}, \tag{6.12}$$

with

$$\hat{T} = \left[\sigma, \hat{H}_{so} \right] = \mathbf{e} \times \sum_j \xi_j(r) \hat{l}_j \times \hat{s}_j(\mathbf{e}), \tag{6.13}$$

where σ and $\hat{H}_{so}$ are the Pauli spin matrices and spin–orbit Hamiltonian, respectively, with the angular momentum operator $\hat{\mathbf{l}}$ and the spin operator $\hat{\mathbf{s}}$. The torque operator quantifies transitions between states induced by spin–orbit torque and has been shown to have both intraband and interband transitions, where the first term often dominates. Combining Eqns (6.8) and (6.12), we can finally write an expression for the damping tensor elements in a form suitable for first-principles calculations (Gilmore et al., 2007); this form is normally referred to as the torque correlation model:

$$\hat{\alpha}_{lm} = \frac{\gamma}{m} \sum_{kij} \mathbf{T}_{kij}^{l} \mathbf{T}_{kij}^{m} W_{kij}, \qquad (6.14)$$

where W_{kij} is a 'weighting factor' of the scattering events defined from the derivative of the Fermi–Dirac function with respect to energy and eigenstates (or bands) to the Kohn–Sham equation, as represented by the slightly more general form of a spectral function, A. The derived equation reads as follows:

$$W_{kij} = -\int \frac{\partial f_{k,i}}{\partial \epsilon} A_{ki}(\epsilon_{ki}, \Gamma) A_{kj}(\epsilon_{kj}, \Gamma) \, d\epsilon, \qquad (6.15)$$

where the spectral functions have a Lorentzian shape and are centred at ϵ_{ki} with a width $\Gamma = h/\tau$ (h is the Plank constant) that is proportional to the scattering rate against defects, for example, the electron–phonon coupling, as illustrated in Fig. 6.2.

The torque correlation model captures two different scattering events: scattering within a single band $i = j$, and between two different bands $i \neq j$. The overlap of the spectral functions is proportional (inverse) to the scattering time for intraband (interband) scattering (Kambersky, 1976; Gilmore et al., 2007). Qualitatively, it then follows that intraband contributions match conductivity-like terms, while interband contributions match resistivity-like terms.

The torque correlation model still requires knowledge about the scattering rate to determine the damping, however, it has been found in calculations, for example, as

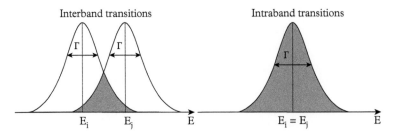

Figure 6.2 *Interband and intraband transitions in the torque correlation model. The larger the overlap is between two states, with their corresponding Lorentzians, the larger is the transition rate. The width Γ is proportional to the electron–phonon coupling. Figure kindly provided by Danny Thonig.*

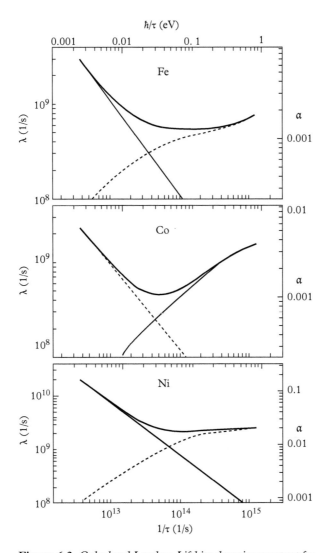

Figure 6.3 *Calculated Landau–Lifshitz damping constant for Fe, Co, and Ni from the torque correlation model (right-hand scale). Thick solid curves give the total damping parameter while dotted curves give the intraband contributions, and dashed lines give the interband contributions. Values for* λ *equal to the damping rate,* λ = γmα, *are given in SI units. The right axis is the equivalent Gilbert damping parameter* α, *and the top axis is the full width half maximum of the electron spectral functions. Reprinted figure with permission from Gilmore, K., Idzerda, Y. U., and Stiles, M. D., Phys. Rev. Lett., 99, 027204, 2007. Copyright 2007 by the American Physical Society.*

shown in Fig. 6.3, that the damping has a unique minimum with respect to the scattering rate. The minimum is often interpreted as the final calculated damping from the torque correlation model and, even if the exact value does depend on the scattering rate, quantitatively, it is often a great improvement, compared to the breathing Fermi surface model. The downwards sloping lines in Fig. 6.3 correspond to intraband contributions, and the upwards dashed lines correspond to interband contributions. In Fig. 6.3, λ denotes the scattering rate, which is proportional to the damping as $\lambda = \gamma m \alpha$.

6.4   The linear response formulation

As has been pointed out several times, the main weak point of both the breathing Fermi surface model and the torque correlation model is their dependence on relaxation time or the scattering rate. Brataas et al. (2008) have shown that the relaxation time approximation can be lifted by describing Gilbert damping by means of scattering theory and that this approach is equivalent to a linear response formulation or Kubo formalism. The development gives a formal basis for describing Gilbert damping and electric transport within a unified, parameter-free theory of disordered alloys (Starikov et al., 2010). Within the coherent potential approximation (CPA) for disordered alloys, it is straightforward to include potential scattering due to chemical, magnetic, and thermal disorders. Once again starting with the LLG equation, we can obtain the time derivative of the magnetic energy as follows (Brataas et al., 2008; Starikov et al., 2010; Ebert et al., 2011; Mankovsky et al., 2013):

$$\dot{E}_{mag} = \mathbf{B}_{eff} \cdot \frac{d\mathbf{m}}{dt} = \frac{m}{\gamma} \dot{\mathbf{m}} [\tilde{\alpha}(\mathbf{m}) \dot{\mathbf{m}}]. \tag{6.16}$$

If the system is described by an underlying time-dependent Hamiltonian $\hat{H}(t)$, then the energy dissipation is given by $\dot{E}_{dis} = \langle d\hat{H}/dt \rangle$. Expanding the Hamiltonian linearly around the equilibrium magnetization, with $\mathbf{m}_0 + \mathbf{u}(t)$, one can write

$$\hat{H} = \hat{H}_0(\mathbf{m}_0) + \sum_i u_i(t) \frac{\partial}{\partial u_i} \hat{H}(\mathbf{m}_0), \tag{6.17}$$

where $i = x, y, z$ is a Cartesian index. Using the more compact notation of implicit summation over repeated indices, we can rewrite Eqn (6.17) as

$$\hat{H} = \hat{H}_0(\mathbf{m}_0) + u_i \partial_{u_i} \hat{H}(\mathbf{m}_0). \tag{6.18}$$

The rate of energy change to lowest (first) order then equals $\dot{E} = \dot{u}(t) \times \langle \partial_i \hat{H} \rangle$, with

$$\langle \partial_i \hat{H} \rangle = \langle \partial_i \hat{H} \rangle_0 + \int_{-\infty}^{\infty} d\tau' \chi_{ij}(t - t') u_j(t'), \tag{6.19}$$

where $\langle \ldots \rangle_0$ denotes the equilibrium expectation value and χ is the generalized exchange interaction tensor

$$\chi_{ij}(t - t') = -\frac{i}{\hbar}\theta(t - t')\langle \partial_i\hat{H}(t), \partial_j\hat{H}(t')\rangle_0. \tag{6.20}$$

The generalized exchange interaction tensor is a retarded correlation function in the interaction picture for the time evolution, and the leading and higher-order dampings, that is, moments of inertia (Bhattacharjee, Nordström, et al., 2012), can be extracted from it. In order to proceed, a couple of approximations need to be introduced. If we are interested in the adiabatic limit of the magnetization dynamics, that is, the Gilbert damping, then we need to impose sufficiently slow magnetization dynamics such that $u_j(t') \approx u_j(t) + (t - t')\dot{u}_j(t)$. Moreover, neglecting longitudinal fluctuations such that $\mathbf{m}^2 = 1$, and $\dot{\mathbf{m}} \cdot \mathbf{m} = 0$, the rate of change of the magnetic energy can be expressed as

$$\dot{E} = \lim_{\omega \to 0} i\partial_\omega \chi_{ij}(\omega)\dot{u}_i\dot{u}_j. \tag{6.21}$$

Here, $\chi_{ij}(\omega) = \int_{-\infty}^{\infty} d\tau \chi_{ij}(\tau)\exp(i\omega\tau)$, that is, the frequency-dependent generalized exchange interaction tensor. Now scattering theory is used as a basis for expanding the correlation function, Eqn (6.20). Expanding the Hamiltonian in a free-electron part and a scattering potential and making use of linear response theory, the rate of change of the magnetic energy in the low temperature adiabatic limit can finally be expressed as (Brataas et al., 2008)

$$\dot{E} = -\pi\hbar \sum_{nm} \sum_{ij} \dot{u}_i\dot{u}_j \langle \psi_n|\partial_i\hat{H}|\psi_m\rangle\langle \psi_m|\partial_j\hat{H}|\psi_n\rangle \times \delta(E_F - E_n)\delta(E_F - E_m), \tag{6.22}$$

where n and m denote eigenstates of the system. Identifying $\dot{E}$ with Eqn (6.16), an explicit expression of the Gilbert damping tensor $\hat{\alpha}$ is obtained, or equivalently, the relaxation rate tensor $\hat{G} = \hat{\alpha}\gamma m$. In practice, there are several advantages to expressing the sum over eigenstates $|\psi_n\rangle$ in terms of the retarded one-electron Green's function

$$\mathrm{Im}\,G^+(E_F) = -\pi \sum_n |\psi_n\rangle\langle \psi_n|\delta(E_F - E_n), \tag{6.23}$$

which leads to a Kubo–Greenwood-like equation for the damping tensor elements by combining Eqns (6.16), (6.22), and (6.23):

$$\alpha_{ij} = -\frac{\hbar\gamma}{\pi m}\mathrm{Tr}\langle \partial_i\hat{H}\mathrm{Im}\,G^+(E_F)\partial_j\hat{H}\mathrm{Im}\,G^+(E_F)\rangle, \tag{6.24}$$

where $\langle \ldots \rangle$ denotes the configurational average in the case of disorder. Identifying $T_i = \partial_i\hat{H}$ with the component of the magnetic transverse torque operator, Eqn (6.12), the Kubo–Greenwood-like equation for α is expressed as a torque–torque correlation function. However, in contrast to the standard torque correlation model of damping, Eqn (6.14), which describes the electronic structure in terms of Bloch states, here the electronic structure is instead represented by electronic Green's functions. Within scattering theory, it is convenient to introduce a scattering path operator $P_{xx'}^{nm}(E)$ that describes transfer of an electronic wave coming in at a site m into a wave going out from site n, with all possible intermediate scattering events being accounted for. Using this formalism, we have finally reached a compact expression for the damping parameter suitable for implementation in electronic structure programs (Ebert et al., 2011; Mankovsky et al., 2013):

$$\alpha_{ii} = \frac{g}{\pi M_{\text{tot}}} \sum_n \text{Tr} \langle \hat{\mathbf{T}}_{0i} \hat{\mathbf{P}}_{0n} \hat{\mathbf{T}}_{ni} \hat{\tau}_{n0} \rangle, \tag{6.25}$$

where g is the Landé factor.

6.5   Inclusion of disorder and temperature effects

To treat disorder of various kinds, for instance atomic disorder, thermal disorder, and magnetic disorder, it is convenient to describe the electronic structure within the CPA of alloy theory. An important technical detail while carrying out the configurational averages is to account for vertex corrections (Butler, 1985; Carva et al., 2006), which correspond to terms such as $\langle \partial_i \hat{H} \text{Im} G^+ \partial_j \hat{H} \text{Im} G^+ \rangle - \langle \partial_i \hat{H} \text{Im} G^+ \rangle \langle \partial_j \hat{H} \text{Im} G^+ \rangle$. For conductivity calculations, it has been shown that vertex corrections are crucial for obtaining quantitative agreement with experiments. Similar conclusions are expected for damping calculations due to the similarities of the expressions.

   In order to include thermal vibrations, which are an important mechanism for electron scattering at finite temperature, the most general way is to add an electron–phonon self-energy when calculating the Green's function G^+ in Eqn (6.24). However, this complicates the calculations and implementation and has rarely been used in practice; instead, simplified approaches have been developed. Liu et al. (2011) proposed the so-called 'frozen thermal lattice disorder,' and Ebert et al. (2011) developed a similar technique, which makes a restriction to the elastic scattering processes in a quasi-static picture with atom displacements from the equilibrium position due to thermal fluctuations. This structure can be modelled using an analogy to alloy theory (Ebert et al., 2015); in this approach, a discrete set of displacements are chosen in such a way that the obtained configuration reproduces the root mean square displacements $\sqrt{\langle u^2 \rangle_T}$ of the Debye model, at a given temperature T, $\langle u^2 \rangle_T = \frac{1}{4} \frac{3\hbar^2}{\pi^2 m k_B \Theta_D} \left[\frac{\Phi(\Theta_D/T)}{\Theta_D/T} + \frac{1}{4} \right]$. Similarly, spin fluctuations at finite temperature due to magnons can be accounted for by tilting the magnetic moments away from the quantization axis (z-axis) such that the root mean square average tilting $\langle \theta \rangle_T$ correponds to average magnetization along the z-axis ($M_z = \langle M(T) \rangle$). Typically, $M(T)$ is calculated from Monte Carlo simulations or assuming a simplified functional dependence on temperature. In Fig. 6.4, the inclusion of thermal displacements and spin fluctuations are schematically illustrated.

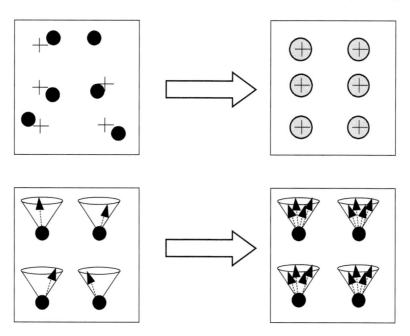

Figure 6.4 *Schematic illustration of the alloy analogy model for atomic displacements due to thermal fluctuations of phonons (upper panel) and from spin fluctuations from magnons (lower panel). Atoms are marked as filled circles, with the magnetic moment as an arrow. The left side illustrates the real situation, with '+' denoting the equilibrium positions, and the right side shows the condition after averaging over several configurations via coherent potential approximation. Reprinted figure with permission from Ebert, H., Mankovsky, S., Chadova, K., Polesya, S., Minár, J., and Ködderitzsch, D., Phys. Rev. B, 91, 165132, 2015. Copyright 2015 by the American Physical Society.*

In Fig. 6.5, damping calculations for Fe, Co, and Ni are displayed using linear response and including temperature effects from atomic displacements (Mankovsky et al., 2013). Qualitatively, the calculations capture the overall temperature dependence found from ferromagnetic resonance experiments; however, quantitatively, the calculated values seem to underestimate the experimental values somewhat. One reason for the disagreement could be due to neglected temperature effects of spin fluctuations. Other sources of errors are additional forms of scattering in the experiments, for instance, from surface effects of the thin films used in the experiment, or from unwanted impurities or other lattice defects. At low temperature, $T < 100$ K, an increase in damping is found both in experiments and calculations. The behaviour can be explained via the torque correlation model, where a transition from low temperature conductivity-like to high temperature resistivity-like behaviour is expected, changing from dominating intraband transitions to dominating interband transitions.

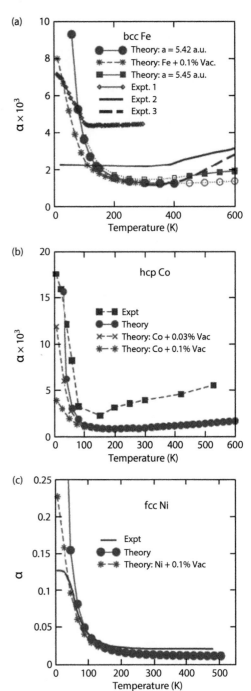

Figure 6.5 *Calculated Gilbert damping of Fe, Co, and Ni as a function of temperature, including atomic displacements due to thermal fluctuations. (a) Body-centred (bcc) Fe for two different lattice constants: a = 5.42 Å, and a = 5.45 Å. (b) Hexagonal close-packed (hcp) Co without and with two different concentrations of vacancies (Vac). (c) Face-centred cubic (fcc) Ni. Reprinted figure with permission from Mankovsky, S., Ködderitzsch, D., Woltersdorf, G., and Ebert, H., Phys. Rev. B, 87, 014430, 2013. Copyright 2013 by the American Physical Society. Experimental data mainly from Bhagat, S. M., and Lubitz, P., (1974), Temperature variation of ferromagnetic relaxation in the 3d transition metals, Phys. Rev. B, 10, 179–85, and Heinrich, B., and Frait, Z., (1966), Temperature dependence of the FMR linewidth of iron single-crystal platelets, Phys. Status Solidi (b), 16, K11-K14, via ferromagnetic resonance.*

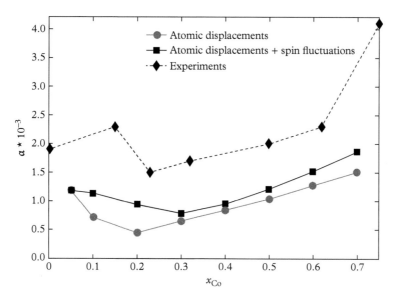

Figure 6.6 *Room temperature Gilbert damping* α *of the body-centred cubic random alloy* $Fe_{1-x}Co_x$, *where x is the Co concentration. Calculations are performed using linear response without and with spin fluctuations at room temperature. Experimental values are adapted from Oogane, M., Wakitani, T., Yakata, S., Yilgin, R., Ando, Y., Sakuma, A., and Miyazaki, T., (2006), Magnetic damping in ferromagnetic thin films,* Jpn J. Appl. Phys., *45, 3889–91.*

   In random alloys, additional scattering for the electrons is possible due to the random distribution of atoms of different size and mass. Many of the technologically important materials for spintronics and magnonics are transition metal alloys, like Fe-Ni alloys, in particular permalloy ($Fe_{0.19}Ni_{0.81}$) and Fe Co alloys, due to the low damping these materials possess. In Fig. 6.6, calculated and measured Gilbert damping of Fe Co random alloys is shown. It is immediately noticed that the trends are similar but the experimental values are higher than calculations. There could be several reasons for this discrepancy. First of all, the samples in the experiments are thick films that may have imperfections as well as extrinsic contributions to the damping, in addition to the intrinsic Gilbert damping, while the calculations are performed in ideal bulk, crystalline conditions. Second, the calculations may underestimate the disorder effect since they are based on single-site CPA. However, a more sophisticated treatment using non-local CPA does not change the results significantly. Moreover, at room temperature, it is clear that additional scattering provided from spin fluctuations on top of the atomic displacements increases the damping, which would improve the agreement between experiment and theory.

6.6   Symmetry analysis of the damping tensor

As pointed out in Section 6.1, the damping term is generally a tensor that is written as

$$\tilde{\alpha}(\mathbf{m}) = \begin{pmatrix} \alpha_{xx} & \alpha_{xy} & \alpha_{xz} \\ \alpha_{yx} & \alpha_{yy} & \alpha_{yz} \\ \alpha_{zx} & \alpha_{zy} & \alpha_{zz} \end{pmatrix}. \tag{6.26}$$

Following the same arguments as presented by Fähnle, Drautz, Singer, Steiauf and Berkov (2005), we see that using a general tensor formulation of the damping has the consequence that in the general case, the damping torque in Eqn (6.1), $H_{\mathrm{damp}} = (1/m)\mathbf{m} \times (\tilde{\alpha} \cdot d\mathbf{m}/dt)$ is not parallel to $d\mathbf{m}/dt$. A Gilbert equation with a scalar damping α is obtained only for the special case when $d\mathbf{m}/dt$ is an eigenvector for the tensor $\tilde{\alpha}(\mathbf{m})$, with the corresponding eigenvalue α. If the general LLG equation, Eqn (6.1), is multiplied by $\mathbf{m}$, one obtains $\mathbf{m} \cdot d\mathbf{m}/dt = 0$, that is, only perpendicular (transversal) components of $\mathbf{m}$ change in time. Choosing a local frame of reference such that the moment is aligned along the z-axis, that is, $\mathbf{m} = m\mathbf{e}_z$, we obtain $d\mathbf{m}/dt = (dm_x/dt, dm_y/dt, 0)$. Moreover, if the damping tensor $\tilde{\alpha}$ takes the diagonal form, then

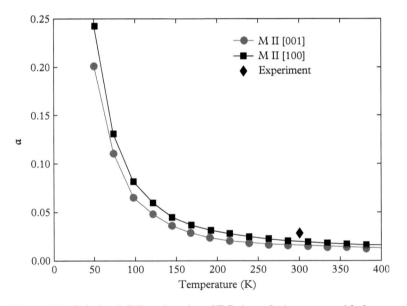

Figure 6.7 *Calculated Gilbert damping of FePt in an L10 structure with the magnetization axis parallel either to the [001] (z) direction (M II [001]) or the [100] (x) direction (M II [100]). Room temperature experimental value from ferromagnetic resonance from Chen, Z., Yi, M., Chen, M., Li, S., Zhou, S., and Lai, T., (2012), Spin waves and small intrinsic damping in an in-plane magnetized FePt film, Appl. Phys. Lett., 101, 224402.*

$$\tilde{\alpha}(\mathbf{m}) = \begin{pmatrix} \alpha_\perp & 0 & 0 \\ 0 & \alpha_\perp & 0 \\ 0 & 0 & \alpha_\| \end{pmatrix}, \tag{6.27}$$

with $\alpha_\perp$ and $\alpha_\|$ denoting the transversal and longitudinal damping, respectively. If the longitudinal components are set to 0, that is, moment size is not allowed to change, then the damping term can be reduced to a scalar $\alpha = \alpha_\perp$, and the standard LLG equation is recovered.

It can be shown that the general damping expressions from either the torque correlation model (Eqn (6.14)), or linear response (Eqn (6.24)), gives the form of $\tilde{\alpha}$ as in Eqn (6.27) if the magnetization is aligned parallel to a threefold or fourfold symmetry axis of the system. In the case of scalar damping, the LLG and the LL equations are easily transformed into each other (see Chapter 4), but that is no longer the case if tensorial damping is employed.

In the general case, the damping not only has a tensorial form but also is dependent on the magnetization direction $\mathbf{m}$. Since the driving mechanism for damping is the spin–orbit interaction that couples the spin to the crystallographic axis, the electronic structure and therefore the total energy depends on the magnetization direction (see Chapters 2 and 3). Even if the magnetization direction is chosen in such a way that the damping is diagonal (Eqn (6.27)), the value of α will depend on $\mathbf{m}$ and give rise to anisotropic damping. Take FePt in the L1$_0$ structure as an example, as it is a material with a large influence of spin–orbit effects, as revealed by the large uniaxial magnetocrystalline anisotropic energy. For this compound, the energy difference between the magnetization direction along the z-axis and in the xy-plane is $\sim$2.7 meV, with the z-axis being the easy axis. As a consequence, the damping is different, depending on the magnetization direction, as shown in Fig. 6.7. Even if the difference in α for different magnetization directions is distinguishable, this anisotropy is not expected to play a major role in practical simulations using the LLG equation.

7

Implementation

In Chapters 4–6, the underlying theory for atomistic spin dynamics (ASD) has been presented. In this chapter, we will present the technical aspects of ASD, in particular, how the method can be implemented in actual computer software. This involves the calculation of effective fields and the creation of neighbour lists for setting up the geometry of the system of interest as well as choosing a suitable integrator scheme for the stochastic Landau–Lifshitz (or stochastic Landau–Lifshitz–Gilbert (SLLG)) equation. We also give examples of extraction and processing of relevant observables that are common output from simulations. ASD simulations can be a computationally heavy tool but it is also very well adapted for modern computer architectures like massive parallel computing and/or graphics processing units (GPUs), and we provide examples of how to utilize these architectures in an efficient manner. We use the software UppASD as example, but the discussion could be applied to any other ASD software.

7.1   UppASD

The theory presented so far forms a solid basis for which any ASD implementation can stand on. In this chapter, we will present in detail how such an implementation can be performed. The presentation will at parts be quite technical since it is meant to be used as a guide for those interested in writing their own ASD software, or for those who aim to understand in detail the existing software and possibly make modifications to it. The algorithms described are currently implemented in the UppASD software (http://www.physics.uu.se/uppasd).

In short, the working parts in the ASD method can be divided into three main parts. Since the purpose is to solve and integrate the Landau–Lifshitz–Gilbert (LLG) or SLLG equations of motion, the first part is to calculate the effective field $\mathbf{B}_i$ acting on each individual magnetic moment $\mathbf{m}_i$. Once the effective field is calculated, the SLLG equation needs to be integrated to give a new set of moments. The integration of the equation of motion comprises the second part of the ASD implementation. In principle, combining the two first parts is actually all that is needed to perform a proper ASD simulation, but in order to analyse the spin dynamics, sampling of relevant physical observables is needed. The construction and extraction of the wanted observables from the simulated moment trajectories is thus the third part that needs to be present in an ASD implementation.

Atomistic Spin Dynamics. Olle Eriksson, Anders Bergman, Lars Bergqvist, Johan Hellsvik. First Edition.
© Olle Eriksson, Anders Bergman, Lars Bergqvist, Johan Hellsvik 2017. First published in 2017 by Oxford University Press.

7.2 The effective magnetic field

In order to be able to solve or integrate the equations of motions as formulated by the SLLG or LLG equations, the effective magnetic field $\mathbf{B}_i$ needs to be calculated. The field acting on each magnetic moment, $\mathbf{m}_i$, can be obtained as the partial derivative of the energy with respect to the moment, that is, $\mathbf{B}_i = -\partial \mathcal{H} / \partial \mathbf{m}_i$, as stated in Eqn (4.41).

For the absolute majority of cases, that is except when considering a single moment or macrospin, the effective field $\mathbf{B}_i$ depends both on the actual moment $\mathbf{m}_i$ and on the interaction between $\mathbf{m}_i$ and other moments $\mathbf{m}_j$ present in the system. The coupling between moments has to be described according to a Hamiltonian that is relevant and accurate to the problem at hand. Given that the atomistic LLG equation relies on the concept of localized magnetic moments, a variant of the Heisenberg Hamiltonian is usually a good choice for the Hamiltonian. The different terms that may be considered in the Hamiltonian were presented in Section 4.8. Since the Heisenberg Hamiltonian and additional terms are defined explicitly in closed terms of interacting moments, the partial derivatives needed to obtain the effective field can readily be calculated analytically.

7.2.1 Neighbour lists

Before going into detail about the Hamiltonian, we will briefly discuss the concept of neighbour lists. Regardless of the actual form of the Hamiltonian, the constituent terms typically consist of either single-site terms or pair interactions between neighbouring moments. In the case of pair interactions, they are, in general, short-range, compared to the size of the simulated system.[1] Due to their locality, the lists of interacting atoms are called neighbour lists. The ability to create and use such lists in a clever way is important, as it enables computationally efficient calculations of the effective fields. Since short-range interactions play an important role in many applications of computational physics and related disciplines, the creation of neighbour lists is a well-studied subject. As an example, in molecular dynamics (MD) simulations, which in many ways are closely connected to ASD simulations, the positions of the ions change over time, and the neighbour lists do thus have to be updated regularly. The consequence is that the creation of neighbour lists in MD can actually take a significant part of the simulation time if not managed efficiently. In ASD, the moments are considered to be fixed on a lattice, and the neighbour lists only have to be constructed once for each considered system.

Even so, the neighbour mapping in ASD can benefit from using efficient algorithms. The perhaps most obvious approach of neighbour mapping is to loop over all moments in the system and then, for each moment, loop over all other moments to see if they are neighbours. The number of operations of that approach scales as $\mathcal{O}(N^2)$, where N is the number of moments in the system; thus, for very large systems, the neighbour mapping

---

[1] One should, however, note that there is an important exception to this stated short-rangedness, and that is when considering classical magnetostatic dipole–dipole interactions. This particular case of interaction will also be mentioned and discussed later in this chapter but, since it is often not very important for systems where ASD simulations can be considered to be the optimal choice of method, we will not focus on dipole–dipole interactions in this treatise.

Algorithm 7.1 Brute-force neighbour list creation

---

for $i \leq N$ **do**
    for $j \leq N$ **do**
        if i, j neighbours **then**
            $k \leftarrow k + 1$
            $neighbourlist(i, k) \leftarrow j$
        end if
    end for
end for

---

can take a very long time even if it is only done once. This straight-forward scheme for neighbour list creation is explicitly expressed in pseudo code in Algorithm 7.1. Since large systems in ASD simulations are typically constructed as a repetition of identical but smaller cells, such as the primitive or conventional unit cell of a crystalline magnetic material, a more efficient algorithm can be obtained by use of the translational symmetry. In the present implementation of neighbour list creation in UppASD, we currently use a method based on partitioning into these cells. Consider a system for which the unit cell contains N_{cell} atoms. By repetition of this unit cell N_x times along the x-direction, N_y times along the y-direction, and N_z times along the z-direction, a simulation cell with $N = N_{\text{cell}}N_xN_yN_z$ moments is obtained. The global index i of each moment $\mathbf{m}_i$ can be expressed as a function $i = f(i_{\text{cell}}, i_x, i_y, i_z)$ where i_{cell} runs over the moments of the unit cell, and the indices i_x, i_y, i_z determine to which repetition of the unit cell the moment belongs. This one-to-one mapping makes it possible to refer to any neighbour pair of magnetic moments either through indices $\{i, j\}$ or through indices $\{i_{\text{cell}}, i_x, i_y, i_z; j_{\text{cell}}, j_x, j_y, j_z\}$. Using this cell- and index-based algorithm, as outlined in Algorithm 7.2 the neighbour-mapping problem reduces in computational complexity to $\mathcal{O}(NN_{\text{cell}})$. Since the size of the unit cell is typically on the order of a few atoms, the gain in efficiency with this latter algorithm is large, compared to the technically simpler

---

Algorithm 7.2 Cell-based neighbour list creation

---

for $i_{\text{cell}} \leq N_{\text{cell}}$ **do**
    for each neighbour vector r_{ij} **do**
        $j_{\text{cell}}, j_x, j_y, j_z \leftarrow r_{ij}$
    end for
    for each i_x, i_y, i_z **do**
        $k \leftarrow k + 1$
        $i \leftarrow f(i_{\text{cell}}, i_x, i_y, i_z)$
        $j \leftarrow f(j_{\text{cell}}, i_x + j_x, i_y + j_y, i_z + j_z)$
        $neighbourlist(i, k) \leftarrow j$
    end for
end for

---

approach described earlier. The algorithm also relies on it being possible to calculate the index list function $f(i_{\text{cell}}, i, i_x, i_y, i_z)$ without effort, which is the case here as it comes naturally with the construction of the system from the repetition of the unit cell. In addition to translational symmetry, point-group symmetries are used in UppASD to set up the system, so that the input files need only to contain one instance of couplings which are equivalent with respect to point-group symmetry.

To compare, we can consider a one-dimensional chain of moments, the Heisenberg chain, aligned along the z-direction. In the case of only nearest-neighbour interactions, each moment i has two finite couplings: $\mathbf{r}_+ = (0, 0, 1)$, and $\mathbf{r}_- = (0, 0, -1)$. If the brute-force algorithm is used to find the correct neighbour moments, one has to loop over the full length of the chain for every moment and at each step see if the distance between the considered moments corresponds to $\mathbf{r}_+$ or $\mathbf{r}_-$. If, on the other hand, one uses the cell-based algorithm, where, for this chain, $N_{\text{cell}} = N_x = N_y = 1$, then the neighbour list is readily obtained for each moment by adding the index of the nearest moment before $(\mathbf{r}_- \implies j_z = -1)$ and after $(\mathbf{r}_+ \implies j_z = +1)$ to the neighbour list of each considered moment.

7.2.2   Contributions to the effective field

From a Heisenberg Hamiltonian, an analytic expression for the contribution to the effective magnetic field from the Heisenberg exchange interactions is readily calculated as

$$\mathbf{B}_i^{\text{Heis}} = -\frac{\partial \mathcal{H}_{\text{Heis}}}{\partial \mathbf{m}_i} = \frac{1}{2} \sum_j \mathcal{J}_{ij} \mathbf{m}_j + \frac{1}{2} \sum_j \mathcal{J}_{ji} \mathbf{m}_j = \sum_j \mathcal{J}_{ij} \mathbf{m}_j. \tag{7.1}$$

Similarly, the contributions from the Dzyaloshinskii–Moriya and biquadratic exchange interactions to the effective field read

$$\mathbf{B}_i^{\text{DM}} = -\frac{\partial \mathcal{H}_{\text{DM}}}{\partial \mathbf{m}_i} = -\sum_j \mathbf{D}_{ij} \times \mathbf{m}_j, \tag{7.2}$$

$$\mathbf{B}_i^{\text{bq}} = -\frac{\partial \mathcal{H}_{\text{bq}}}{\partial \mathbf{m}_i} = 2 \sum_j \mathcal{J}_{ij}^{\text{bq}} (\mathbf{m}_i \cdot \mathbf{m}_j) \mathbf{m}_j. \tag{7.3}$$

The contributions to the effective magnetic field from magnetocrystalline anisotropy are obtained, for the case of single-site uniaxial anisotropy, Eqn (4.45), as

$$\mathbf{B}_i^{\text{uni}} = \sum_i 2K_i^1 (\hat{\mathbf{e}}_i \cdot \hat{\mathbf{e}}_i^k) \hat{\mathbf{e}}_i + \sum_i 4K_i^2 (\hat{\mathbf{e}}_i \cdot \hat{\mathbf{e}}_i^k)^3 \hat{\mathbf{e}}_i, \tag{7.4}$$

and, for cubic anisotropy, Eqn (4.46), as

$$\mathbf{B}_i^{\text{cub}} = \sum_i 2K_i^1 \left[e_{ix}(e_{iy}^2 + e_{iz}^2)\hat{\mathbf{e}}_x + e_{iy}(e_{ix}^2 + e_{iz}^2)\hat{\mathbf{e}}_y + e_{iz}(e_{ix}^2 + e_{iy}^2)\hat{\mathbf{e}}_z \right] +$$

$$\sum_i 2K_i^2 \left[e_{ix}e_{iy}^2 e_{iz}^2 \hat{\mathbf{e}}_x + e_{iy}e_{iz}^2 e_{ix}^2 \hat{\mathbf{e}}_y + e_{iz}e_{iy}^2 e_{ix}^2 \hat{\mathbf{e}}_z \right]. \tag{7.5}$$

The demagnetizing field that results from the dipole–dipole interactions in the magneto-static Hamiltonian reads

$$\mathbf{B}_i^{\mathrm{dd}} = \frac{\mu_0}{2\pi} \sum_{j,j\neq i} \frac{1}{r_{ij}^3} \left[3\hat{\mathbf{r}}_{ij}\left(\mathbf{m}_j \cdot \hat{\mathbf{r}}_{ij}\right) - \mathbf{m}_j \right], \qquad (7.6)$$

where μ_0 is the magnetic constant. The long-range behaviour of the dipolar interactions thus comes from a $\frac{1}{r^3}$ dependence and, since the decay of the interaction strength is not exponential, the summation over spins, defined in Eqn (7.6), needs to be evaluated for the whole system. This results in a computational cost for the demagnetizing field that scales as $\mathcal{O}(N^2)$ if the number of considered spins is N. That scaling can be contrasted with the $\mathcal{O}(N)$ scaling that is needed to evaluate short-range exchange interactions and onsite anisotropies. Since N is large for systems where dipolar effects are important, the evaluation of the demagnetizing field completely dominates the computational cost and effectively limits the simulation size and time that are possible to consider. Luckily, there exist several more efficient methods to treat dipolar effects and calculate the demagnetizing field. Many of these methods were originally conceived for particle simulations but they also lend themselves to being implemented for both micromagnetic (Berkov et al., 2006) and atomistic magnetization dynamics simulations. For regular and finite geometries fast Fourier transform (FFT) methods can be employed in order to reduce the computational cost for evaluating the demagnetizing field. For periodic systems, the problem can be treated by Ewald summation (Ewald, 1921), or by using a fast multipole method (Greengard and Rokhlin, 1987).

7.3   Spin-transfer torque

In addition to external and internal magnetic fields, external electrical currents can have a large impact on magnetization dynamics (Berger, 1996; Slonczewski, 1996). In fact, this phenomenon of current-induced magnetization dynamics have recently attracted considerable attention, since it is in many cases much more efficient to drive important magnetodynamical processes with currents instead of fields (Ralph and Stiles, 2008). The driving mechanism behind the current-induced magnetization dynamics comes from the fact that, if a spin polarized current, which occurs naturally when an electric current is driven through a magnetic material, passes through a region of a material where the magnetization has a spatial variation, then the current and the localized magnetic moments in that region strive to align to each other. In that process, there is a transfer of angular momentum between the spin polarized charge carriers in the current and the magnetic moments. The resulting effect is often referred to as spin-transfer torque (STT). In order to model STT in atomistic, as well as in micromagnetic, magnetization dynamics simulations, the SLLG equation of motion, Eqn (4.1), needs to be augmented with additional terms. In the case of a varying magnetization in the system, that is a finite and well-defined gradient of the magnetization, then the STT can be included into the SLLG by introducing two additional terms that depend on the magnetization gradient as follows (S. Zhang and Li, 2004; Schieback et al., 2007):

$$\frac{d\mathbf{m}_i}{dt} = -\gamma_L \mathbf{m}_i \times \left(\mathbf{B}_i + \mathbf{B}_i^{\text{fl}}\right) - \gamma_L \frac{\alpha}{m_i} \mathbf{m}_i \times \left[\mathbf{m}_i \times \left(\mathbf{B}_i + \mathbf{B}_i^{\text{fl}}\right)\right]$$

$$+ \frac{u_i}{m_i} \frac{1 + \alpha\beta}{1 + \alpha^2} \mathbf{m}_i \times (\mathbf{m}_i \times \nabla\mathbf{m}_i) - u_i \frac{\alpha - \beta}{1 + \alpha^2} \mathbf{m}_i \times \nabla\mathbf{m}_i, \tag{7.7}$$

where $u_i = j_i P g / m_i$ is a site-dependent prefactor that depends on the local current density j_i and the spin polarization of the current P, and β is the so-called adiabaticity parameter.

7.4 Numerical integration of the Landau–Lifshitz and stochastic Landau–Lifshitz equations

Once the framework for calculations of the effective field that is acting on each magnetic moment is in place, the equations of motions can be integrated. As described in Chapters 4–5, the stochastic Landau–Lifshitz equation, Eqn (4.1), is valid for magnetic moments $\mathbf{m}_i$ with a constant length. It describes the motion as a combination of one precessional, conservative contribution and one dissipative or damping contribution, as illustrated in Fig. 4.2. The relative influence of the two contributions is determined by the damping constant α. The equation of motion yields a seemingly simple set of first-order differential equations where all contributions to the coupling between different moments $\mathbf{m}_i$ are contained in the magnitude and direction of the site-resolved effective field $\mathbf{B}_i$. However, the SLLG equation for an interacting spin system of N spins constitute a set of $3N$-coupled non-linear differential equations. Fully implicit integrators require solving $3N$-dimensional non-linear equation systems in each time step, wherefore the computational effort increases dramatically with the number of spins.

Since we here focus on the numerical solution of the equation of motions for atomistic spin dynamics situations, it is valuable to identify the characteristic time scales on which this dynamics takes place. From Eqns (4.1), (4.41), and (7.1), it is apparent that, in exchange-dominated systems, the precessional frequency of a magnetic moment out of equilibrium depends on the sum of the exchange interactions and the magnitude of the magnetic moment. For the kinds of systems that are typically relevant for ASD studies, the resulting precessional frequency is of the order of tens of terahertz. This can be exemplified with ASD simulations for the prototype ferromagnet bcc Fe, with exchange parameters taken from the work by Tao et al. (2005); these simulations show a precessional frequency of 81 THz for individual atomic moments. As a good rule of thumb, with a stable and efficient solver, it is necessary to evaluate the SLLG equations about 5–10 time steps per precessional period (Mentink et al., 2010). That means that a suitable time step for ASD simulations is typically of the order of 0.1–1.0 femtoseconds. However, the length of the time step used is a parameter that always should be tested for convergence, since the maximum possible time step depends on the material, the magnetic configuration, and, as will be discussed in this chapter, the temperature. We emphasize that, even though the SLLG equation might be evaluated using sub-femtosecond intervals, it does not necessarily describe the magnetization dynamics accurately on those small time scales since, in those cases, fast electronic processes which are not explicitly present in ASD/SLLG models play important roles (as discussed in Section 2.6). On the other hand, the typical phenomena that may be considered for ASD studies range from a few picoseconds, which is enough for many ultrafast switching scenarios or for investigating high energy exchange magnons with terahertz frequencies, to several nanoseconds for slower switching processes or low energy magnons in the gigahertz regime.

The development of reliable and efficient general purpose integrators for the LLG and SLLG equations is an active field of research. In addition to conservation properties, the transparency of a method, its applicability to general geometries and symmetries, and how it can be efficiently parallelized in software for modern computer architectures are all important factors to be taken under consideration when choosing a method for numerical simulation of the LLG or SLLG equations. In the case of finite temperature, it is also important that the stochastic differential equation (SDE)-extended flavours of the integrators converge to the Stratonovich solution of the SLLG equation. With the introduction of the stochastic fields, the problem of solving deterministic ordinary differential equations (ODEs) is replaced by solving SDEs. From a mathematical point of view, treating SDEs is a more complex task than treating their deterministic ODE counterparts. This is, for instance, reflected in analytical and numerical studies (Rümelin, 1982; Kloeden and Platen, 1992), where it has been shown that moving towards higher-order methods is not always as beneficial for SDEs as for ODEs because, despite the increased complexity of the methods, the possible size of stable time steps does not necessarily increase as well.

7.4.1 Properties of integrators

The properties and performance of an integrator can be characterized and investigated in various ways. For Hamiltonian dynamics, the total energy is a constant of motion. This is the case of $\alpha = 0$ dynamics for the LLG equation, as discussed in Chapter 5, Section 5.5. Some of the solvers to be discussed here preserve the individual spin lengths and the total energy exactly and independently of the step size h.[2] At the same time, there will be global and local truncation errors in the time trajectories of the system's individual degrees of freedom. For the case of ODEs, all of the methods described in Sections 7.4.4–7.4.8 are of order 2, that is, their local truncation error depends on the step size as $\mathcal{O}(h^3)$. How large the errors become in a specific application can be investigated in numerical analysis for integrable systems, for which comparisons between the numerical and analytical solutions are possible, as well as for large and complex systems for which no analytical solutions are available. Given that we in the end are interested in statistical quantities such as means and moments (see Section 7.5), it should be emphasized

---

[2] There will, in an implementation of the solver in computer software, always be rounding errors present, due to the finite precision of real numbers in computer arithmetics.

that high local order is not the most critical quality criteria. Instead, it is more important to preserve qualitative features such as correct energy behaviour, symplecticity, and reversibility.

The corresponding analyses of the local and global properties of an SDE are more involved, and concepts such as weak convergence and mean square order convergence are sometimes introduced (see e.g. McShane, 1974; Rümelin, 1982; Kloeden and Platen, 1992; Milstein et al., 2002; Milstein and Tretyakov, 2004). For the case of SDEs, the methods described in Sections 7.4.4–7.4.8 are of weak order 1 and mean square order 1/2. The performance of the integrators at finite temperature can be tested in various ways. Relevant test quantities are properly averaged observables measured for a system in thermal equilibrium. The temperature of the spin system can be measured by means of the method developed by Ma et al. (2010) and be compared with the temperature set by the heat bath. For a one-dimensional spin chain, an analytical expression for the mean energy per spin is available (Shubin and Zolotukhin, 1936; Fisher, 1964), allowing for comparisons between the mean energies measured in numerical simulations using different integrators and the exact result (Mentink et al., 2010).

7.4.2   Overview of stochastic Landau–Lifshitz integrators

The phase space of a classical spin system is $(S^2)^N$, where N is the number of spins, and $S^2 = \{S_i \in R^3; |S_i|^2 = 1\}$. Due to the constraint on the length of the spin and, in the absence of dissipation and temperature, the spin system constitutes a non-canonical Hamiltonian system, which means that the system is Hamiltonian but not expressed in the canonical position momentum variables. In contrast to the case for canonical systems, for which a host of stable integrators are known, the pursuit to develop stable integrators preserving non-canonical Hamiltonian structures (or, equivalently, non-canonical symplectic structures) is a much harder and more recent field of mathematics and scientific computing.

Depending on the complexity and order of a prescribed solver, it is possible to find a suitable trade-off between computational cost per iteration, and the largest possible time step length. As an example, the simple explicit Euler solver, which was presented in Chapter 5 as Eqn (5.13), only needs one evaluation of the effective field $\mathbf{B}_i$ at each time step. That means that, per time step, the Euler method is fast but, unfortunately, the error of the solver decreases only linearly with a decrease of the time step. In order to achieve stable solutions for realistic ASD simulations, time steps of the order of 10^{-3} fs are needed if an Euler solver is employed. A better behaviour can be obtained by moving towards explicit Runge–Kutta methods, such as, for example, the Heun method.

The preferred choice of preference for time-stepping integration of Hamiltonian systems is use of the so-called geometrical integrators, which are constructed so that they preserve identically some or all of the structures and the constants of motion of the system. An important example is the implicit midpoint (IMP) method, which benefits from unconditional linear stability, exact conservation of spin lengths, preservation of reversibility, and a rather straightforward implementation for the SLLG equation, with the one main disadvantage being the computational effort in the limit of a large number of spins.

The symplectic integrators, which comprise a subclass of the geometric integrators, approximate the exact time evolution of the Hamiltonian system. A formal and detailed discussion on symplectic integrators goes beyond the present chapter, and the reader is referred to textbooks on geometric integration, such as, for example, the one by Hairer et al. (2006), and to literature specifically targeting spin systems (Steinigeweg and Schmidt, 2006; McLachlan et al., 2014) and numerical methods for stochastic systems preserving symplectic structure (Milstein et al., 2002). Symplectic integrators have outstanding properties also for weakly damped Hamiltonian systems; for an example, see the work by (Modin and Söderlind, 2011), which is of high relevance for ASD simulations in which the damping typically is weak. For canonical Hamiltonians, there exist numerous symplectic integrators, such as for instance, the Verlet method, which is by far the most popular choice for integration in MD simulation. Geometrical integrators based on splitting techniques (Frank et al., 1997; Omelyan et al., 2001; Tsai et al., 2005; Steinigeweg and Schmidt, 2006; Ma et al., 2008; Ma and Dudarev, 2011) requires the splitting to be designed by hand (Steinigeweg and Schmidt, 2006), wherefore they are not directly transferable to generic lattice classes and Hamiltonians and in that regard are less suitable for a general purpose implementation of ASD. A very recent development is the spherical midpoint (SMP) method (McLachlan et al., 2014) for the SLLG equation. While the IMP method applied to spin systems is not symplectic, the SMP method is and so has improved accuracy and superior qualitative behaviour, as shown by McLachlan et al. (2014).

Methods for semi-implicit integration of the stochastic Landau–Lifshitz equation (Baňas, 2005), for instance, the three-time semi-implicit method developed by Serpico et al. (2001) and the semi-implicit scheme B(SIB) method (Mentink et al., 2010), benefit from sharing conservation qualities known from implicit integrators, such as linear stability, reversibility, and preservation of spin lengths, with the lower computational effort known from explicit integrators.

In the following, a selection of integrators will be presented in detail. The discussion will be pursued for the SLLG equation in its dimensionless and normalized form, which is convenient both for analysing the mathematical properties of the solvers and for implementation.

7.4.3   The dimensionless and normalized SLLG equation

The terms on the left- and right-hand sides of the SLLG equation (Eqn (4.1)) are, when using the SI system, in units of joules per tesla. This is because the magnetic moment $\mathbf{m}_i$ is in joules per tesla (Bohr magnetons), the magnetic field is in teslas, and the gyromagnetic ratio is radians per second per tesla. A magnetic reference field strength B_0 is introduced to define the dimensionless magnetic fields $\mathbf{B}'_i$ and $\mathbf{b}'_i$, through $\mathbf{B} = B_0\mathbf{B}'_i$, and $\mathbf{B}^{\text{fl}} = B_0\mathbf{b}'_i$. Together with the gyromagnetic ratio, the reference field strength provides a dimensionless rescaled time $t' = \gamma B_0 t$. The size of the reference magnetic field B_0 is arbitrary. With the unitary value, $B_0 = 1$ T, the conversion factor $(\gamma B_0)^{-1} = 5.7$ ps, so that one unit of dimensionless time $\Delta t' = 1$ corresponds to $\Delta t = 5.7$ ps. For the remaining part of the chapter, the primes on t', $\mathbf{B}'_i$, and on $\mathbf{b}'_i$ will be dropped, so that

t denotes dimensionless time, and $\mathbf{B}_i$ and $\mathbf{b}_i$ denote dimensionless magnetic fields. By dividing Eqn 4.1 by $\gamma B_0 |\mathbf{m}_i|$, a dimensionless and normalized SLLG equation can now swiftly be obtained and written in the form

$$\frac{d\mathbf{X}^i}{dt} = -\mathbf{X}^i \times [\mathbf{B}^i(\mathbf{X}) + \mathbf{b}^i] - \alpha \mathbf{X}^i \times \left\{ \mathbf{X}^i \times [\mathbf{B}^i(\mathbf{X}) + \mathbf{b}^i] \right\}, \quad i = 1, \ldots, N, \quad (7.8)$$

where $\mathbf{X}^i = (X^{i,x}, X^{i,y}, X^{i,z})^{\mathrm{T}}$ are unitary 3-vectors, and $\mathbf{X} = (\mathbf{X}^{1^{\mathrm{T}}}, \ldots, \mathbf{X}^{n^{\mathrm{T}}})^{\mathrm{T}}$ is an N-tuple of spins written as a $3N$-long column vector; $\mathbf{B}^i = (B^{i,x}, B^{i,y}, B^{i,z})^{\mathrm{T}}$, and $\mathbf{b}^i = (b^{i,x}, b^{i,y}, b^{i,z})^{\mathrm{T}}$, are the effective and stochastic magnetic fields, respectively. A convenient starting point, both in order to discuss and derive mathematical properties and to formulate the integrators, is to rewrite the SLLG equation as its differential form. Using compact notation, we write

$$d\mathbf{X}^i = \mathbf{X}^i \times \mathbf{a}_i(\mathbf{X})\, dt + \mathbf{X}^i \times \mathbf{f}(\mathbf{X}) \circ d\mathbf{W}^i(t),$$

$$\mathbf{X}^i(0) = \mathbf{x}_0^i, \quad |\mathbf{x}_0^i| = 1, \ i = 1, \ldots, N, \quad (7.9)$$

where $\mathbf{a}_i(\mathbf{X}), \mathbf{X} \in \mathbb{R}^{3N}$, is a three-dimensional column vector defined by

$$\mathbf{a}_i(\mathbf{X}) = -\mathbf{B}^i(\mathbf{X}) - \alpha \mathbf{X}^i \times \mathbf{B}^i(\mathbf{X}), \quad (7.10)$$

and $\mathbf{f}(\mathbf{x})$, $\mathbf{x} \in \mathbb{R}^3$, is a 3×3 matrix such that

$$\mathbf{f}(\mathbf{x})\mathbf{y} = -\sqrt{2D'}\mathbf{y} - \alpha\sqrt{2D'}\mathbf{x} \times \mathbf{y} \quad (7.11)$$

for any $y \in \mathbb{R}^3$; $\mathbf{W}^i(t) = (W_x^i(t), W_y^i(t), W_z^i(t))^{\mathrm{T}}$, $i = 1, \ldots, N$, is a 3-vector of independent standard Wiener processes, and $\circ$ denotes that Stratonovich calculus is used. In $\mathbf{f}(\mathbf{x})$ the strength D' of the stochastic field is included. In dimensionless variables (cf. Eqn (5.32)), it is given by

$$D_i' = \frac{\alpha}{1 + \alpha^2} \frac{k_B T}{m_i B_0}, \quad (7.12)$$

with the index i indicating that D_i' depends on the size of the magnetic moment m_i. The integrators yield approximate solutions of the SDEs at discrete points in time for a time interval $[0, t_{\text{sim}}]$. In general, adaptive time steps can be used but, in the following, we use, for simplicity, a uniform discretization of the time interval in K steps; this discretization corresponds to a time step $h = t_{\text{sim}}/K$; $\mathbf{X}_k^i$, $i = 1, \ldots, N$, denotes the approximate solution $\mathbf{X}^i(t_k)$, $i = 1, \ldots, N$, to the stochastic Landau–Lifshitz equation at time t_k, $k = 1, \ldots, K$. The initial value at $t = 0$ is written $\mathbf{X}_0^i = \mathbf{x}_0^i$, $i = 1, \ldots, N$.

7.4.4 Heun with projection

Numerical simulations of the SLLG equations have often been pursued with explicit integrators such as the predictor–corrector Heun method and higher-order explicit Runge–Kutta schemes. These benefit from simplicity of implementation, and have been proven to converge to the Stratonovich interpretation of the stochastic Landau–Lifshitz (Rümelin, 1982). This means that one does not need to introduce the noise-induced drift term in the SDE equation; it can be integrated as it stands in the form of an Ito equation. For this reason, and also the circumstance that yet higher-order stochastic Runge–Kutta schemes do not necessarily converge to a higher order (Rümelin, 1982), as would be the case for ODEs, the stochastic Heun integrator has been one of the most popular methods among the explicit schemes. When the Heun method is applied to a Cartesian coordinate representation of the SLLG equation, it fails to preserve the size of the individual magnetic moments. Therefore, in implementations of the Heun scheme for ASD, an additional normalization step is needed, where the length of each spin is normalized by projection on the unit sphere. With this projection step included, the method can for clarity (Mentink et al., 2010) be referred to as Heun with projection (HeunP), in order to distinguish it from the standard Heun method. A drawback with HeunP is that, when using the projection step, the Heun method does not preserve the total angular momentum. Neither the Heun method nor HeunP preserves the internal energy of the system. For the dimensionless SLLG equation, one time step in HeunP consists of three operations:

$$\mathcal{X}_k^i = \mathbf{X}_k^i + h\mathbf{X}_k^i \times \mathbf{a}_i(\mathbf{X}_k) + h^{1/2}\mathbf{X}_k^i \times \mathbf{f}(\mathbf{X}_k^i)\xi_{k+1}^{i,j}, \tag{7.13}$$

$$\tilde{\mathbf{X}}_{k+1}^i = \mathbf{X}_k^i + h[\mathbf{X}_k^i \times \mathbf{a}_i(\mathbf{X}_k) + \mathcal{X}_k^i \times \mathbf{a}_i(\mathcal{X}_k)]/2$$

$$+ h^{1/2}[\mathbf{X}_k^i \times \mathbf{f}(\mathbf{X}_k^i)\xi_{k+1}^{i,j} + \mathbf{X}_k^i \times \mathbf{f}(\mathcal{X}_k^i)\xi_{k+1}^{i,j}]/2, \tag{7.14}$$

$$\mathbf{X}_{k+1}^i = \tilde{\mathbf{X}}_{k+1}^i / |\tilde{\mathbf{X}}_{k+1}^i|, \tag{7.15}$$

where $\xi_{k+1}^{i,j}$, where $j = 1, 2, 3$, and $i = 1, \ldots, N$, and $k = 1, \ldots, K$, is an independent, identically distributed random variable, commonly normal distributed as $\xi_k^{i,j} \sim \mathcal{N}(0, 1)$. At first, the predictor $\mathcal{X}_k^i$ is calculated for all spins, to yield $\mathcal{X}_k = (\mathcal{X}_k^{1\mathrm{T}}, \ldots, \mathcal{X}_k^{N\mathrm{T}})^{\mathrm{T}}$; $\mathcal{X}_k$ is then used when calculating the corrector $\tilde{\mathbf{X}}_{k+1}^i$. The magnitude of $\tilde{\mathbf{X}}_{k+1}^i$ will, in general, drift away from unity, wherefore it is necessary to perform a projection step in which the normalized corrector $\mathbf{X}_{k+1}^i$ is calculated. Note that the very same increment $d\mathbf{W}_k^i$ (i.e. the same pseudorandom number $\xi_{k+1}^{i,j}$, where $j = 1, 2, 3$, and $i = 1, \ldots, N$) should be used for the predictor and the corrector. All the three operations are explicit, that is, for each of them, the variables on the left-hand side of the equations do not occur on the right-hand side, which implies that they can be performed independently for each spin. The computational complexity is linear in the number N of spins, and the integrator can be straightforwardly parallelized, as will be discussed in Section 7.7.

7.4.5   The geometric Depondt–Mertens method

As is often the case when solving ODEs, solvers that connect naturally to the dynamics of the problem at hand, in this case, that every update should be a rotation of the spin vectors, can exhibit significantly improved properties. This certainly holds true also for ASD simulations, where the need for renormalization can be removed and the general properties of the Heun solver be enhanced, if the actual dynamics of the ASD problem is considered (Visscher and Feng, 2002; Depondt and Mertens, 2009). It has already been illustrated in Fig. 4.2 that the dynamics for the LLG and SLLG equations is a combination of two rotational motions: one precessional, and one damping. These two rotations can in turn be considered as a single rotation around some axis, as can easily be seen by factorizing Eqn (7.8) as

$$\frac{d\mathbf{X}^i}{dt} = -\mathbf{X}^i \times \left([\mathbf{B}^i(\mathbf{X}) + \mathbf{b}^i] - \alpha\{\mathbf{X}^i \times [\mathbf{B}^i(\mathbf{X}) + \mathbf{b}^i]\} \right) \tag{7.16}$$

$$= -\mathbf{X}^i \times \mathbf{B}^i_{\text{rot}}(\mathbf{X}, \mathbf{b}^i), \quad i = 1, \dots, N.$$

The rotation of a vector, here the magnetic moment, around another vector, here determined by the precessional and damping fields, can be expressed by means of a Rodrigues rotation (Rodrigues, 1840), which means that the integrated motion can be written as

$$\mathbf{X}^i_{k+1} = \mathbf{R}^i_k \mathbf{X}^i_k, \tag{7.17}$$

where $\mathbf{R}^i_k$ is the Rodrigues rotation matrix for moment i which, following Depondt and Mertens (2009), can be expressed as

$$\mathbf{R}^i_k = \begin{pmatrix} b_x^2 u + \cos\omega & b_x b_y u - b_z \sin\omega & b_x b_z u + b_y \sin\omega \\ b_x b_y u + b_z \sin\omega & b_y^2 u + \cos\omega & b_y b_z u - b_x \sin\omega \\ b_x b_z u - b_y \sin\omega & b_y b_z u + b_x \sin\omega & b_z^2 u + \cos\omega \end{pmatrix}. \tag{7.18}$$

where $(b_x, b_y, b_z) = \mathbf{B}^i_{\text{rot}}/|\mathbf{B}^i_{\text{rot}}|$ are the coordinates of the unit vector along the local "rotational" field $\mathbf{B}^i_{\text{rot}}$ that now is the sum of both the precessional and damping fields, $\omega = |\mathbf{B}^i_{\text{rot}}|h$ is the precession angle, and $u = 1 - \cos\omega$. The novelty of Depondt and Mertens's (2009) method was the use of this rotational form in combination with the Heun method to integrate the stochastic Landau–Lifshitz equation. Since the lengths of the magnetic moments are conserved by the geometry of the solver, a much increased stability, compared to that obtained by HeunP, is achieved, and this stability allows for longer time step sizes.

7.4.6   The IMP method

In common with the explicit integrators, and in contrast to the splitting methods, the IMP method allows for an implementation of an SLLG integrator which can be used for general geometries, symmetries, and interactions (D'Aquino et al., 2005, 2006). The IMP method can be written compactly in the form

$$\mathbf{X}^i_{k+1} = \mathbf{X}^i_k + h\frac{\mathbf{X}^i_k + \mathbf{X}^i_{k+1}}{2} \times \mathbf{a}_i \left(\frac{\mathbf{X}_k + \mathbf{X}_{k+1}}{2} \right)$$

$$+ h^{1/2} \frac{\mathbf{X}^i_k + \mathbf{X}^i_{k+1}}{2} \times \mathbf{f} \left(\frac{\mathbf{X}^i_k + \mathbf{X}^i_{k+1}}{2} \right) \xi^{i,j}_{k+1}, \tag{7.19}$$

where $\xi^{i,j}_{k+1}$, where $j = 1, 2, 3$, and $i = 1, \ldots, N$, and $k = 1, \ldots, K$, is an independent, identically distributed random variable. The fact that $\mathbf{a}_i$ depends not only on $\mathbf{X}_k$ but also on $\mathbf{X}_{k+1}$ reveals that the method is implicit (cf. HeunP, where $\mathbf{a}_i$ in the predictor (corrector) depends on $\mathbf{X}_k$ ($\mathbf{X}_k$ and χ_k)). The IMP method can diverge for the case that $\xi^{i,j}_k \sim \mathcal{N}(0, 1)$ is used as a random variable (Milstein et al., 2002). A remedy is to cut off the tail of the Gaussian distributed random variable. This can be achieved by letting $\xi^{i,j}_k$ be distributed as an auxiliary variable ξ_h defined by

$$\xi_h = \begin{cases} \zeta, |\zeta| \leq A_h \\ A_h, \zeta > A_h \\ -A_h, \zeta < -A_h \end{cases} , \tag{7.20}$$

where $\xi \sim \mathcal{N}(0, 1)$, and $A_h = \sqrt{2|\ln h|}$ (Milstein et al., 2002; Milstein and Tretyakov, 2004).

The IMP method possesses numerous advantageous mathematical properties (see e.g. the listing in McLachlan et al., 2014). Of particular importance for the LLG and SLLG equation is that it conserves identically the length of individual spins, as illustrated in Fig. 7.1. This property can be proved by a shift of $\mathbf{X}^i_k$ to the left-hand side of Eqn (7.19), and subsequent multiplication with $\mathbf{X}^i_{k+1} + \mathbf{X}^i_k$, to give

$$(\mathbf{X}^i_{k+1} + \mathbf{X}^i_k) \cdot (\mathbf{X}^i_{k+1} - \mathbf{X}^i_k)$$

$$= h(\mathbf{X}^i_{k+1} + \mathbf{X}^i_k) \cdot \frac{\mathbf{X}^i_k + \mathbf{X}^i_{k+1}}{2} \times \mathbf{a}_i \left(\frac{\mathbf{X}_k + \mathbf{X}_{k+1}}{2} \right). \tag{7.21}$$

By calculating the scalar product on the left-hand side and cycling the factors in the scalar triple product on the right-hand side, the equation can be simplified to

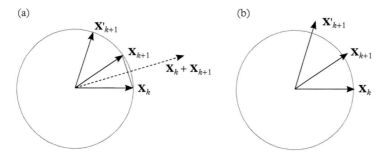

Figure 7.1 *Precession of a spin on the unit sphere xy-plane equator, for a magnetic field along z. (a) The implicit midpoint method preserves the magnitude of each spin $\mathbf{X}_k$ exactly, irrespectively of the size of the time step h when evolving the system over time $t_k \rightarrow t_{k+1}$, where $k = 1, \ldots, K$ steps of the time interval. (b) The Heun method does not preserve the magnitude of each spin $\mathbf{X}_k$. How much the spin changes its length depends on the step size h, and on the angle of the spin to the magnetic field.*

$$|\mathbf{X}_{k+1}^i|^2 - \mathbf{X}_{k+1}^i \cdot \mathbf{X}_k^i + \mathbf{X}_k^i \cdot \mathbf{X}_{k+1}^i - |\mathbf{X}_k^i|^2 \tag{7.22}$$

$$= \frac{h}{2}\mathbf{a}_i\left(\frac{\mathbf{X}_k + \mathbf{X}_{k+1}}{2}\right) \cdot \left[(\mathbf{X}_k^i + \mathbf{X}_{k+1}^i) \times (\mathbf{X}_{k+1}^i + \mathbf{X}_k^i)\right],$$

$$|\mathbf{X}_{k+1}^i|^2 - |\mathbf{X}_k^i|^2 = 0, \tag{7.23}$$

which shows that the length of each spin $\mathbf{X}_k^i$ is exactly preserved when evolving the system over time $t_k \rightarrow t_{k+1}$. For the case of isotropic spin Hamiltonians, and $\alpha = 0$, the IMP method conserves the total spin and the internal energy (Mentink et al., 2010). For $\alpha \geq 0$, the IMP method respects strictly the condition of non-increasing internal energy as expressed by Eqn (5.38) (D'Aquino et al., 2005). The IMP method is, however, not symplectic when applied to spin systems, since the symplectic structure is not canonical.

7.4.7   The McLachlan–Modin–Verdier SMP method

Very recently, McLachlan et al. (2014) developed the SMP method for the LLG equation. Extended to the SLLG equation, it can be written in the form

$$\mathbf{X}_{k+1}^i = \mathbf{X}_k^i + h \frac{\mathbf{X}_k^i + \mathbf{X}_{k+1}^i}{|\mathbf{X}_k^i + \mathbf{X}_{k+1}^i|} \times \mathbf{a}_i\left(\frac{\mathbf{X}_k^1 + \mathbf{X}_{k+1}^1}{|\mathbf{X}_k^1 + \mathbf{X}_{k+1}^1|}, \ldots, \frac{\mathbf{X}_k^N + \mathbf{X}_{k+1}^N}{|\mathbf{X}_k^N + \mathbf{X}_{k+1}^N|}\right)$$

$$+ h^{1/2} \frac{\mathbf{X}_k^i + \mathbf{X}_{k+1}^i}{|\mathbf{X}_k^i + \mathbf{X}_{k+1}^i|} \times \mathbf{f}\left(\frac{\mathbf{X}_k^i + \mathbf{X}_{k+1}^i}{|\mathbf{X}_k^i + \mathbf{X}_{k+1}^i|}\right) \xi_{k+1}^{ij}, \tag{7.24}$$

where $\xi_{k+1}^{i,j}$, where $j = 1, 2, 3$, and $i = 1, \ldots, N$, and $k = 1, \ldots, K$, is an independent, identically distributed random variable distributed as for the IMP method. The spherical midpoint method preserves many structural properties of the LLG equation. In particular, it is a symplectic integrator, which means that it yields the exact solution of a modified Hamiltonian system. Since the modified Hamiltonian function is exactly preserved, the original (exact) Hamiltonian is nearly preserved for arbitrary long time intervals. Applied to a non-linear perturbation of a spinning top (a single-spin system), the SMP method has a bounded energy error and produces periodic trajectories, whereas, with the IMP method the energy error drifts, and the trajectories are non-periodic (McLachlan et al., 2014, in press).

7.4.8 Mentink's SIB method

The solvers described above have in common that they are either explicit methods, or fully implicit methods. However, for the large systems typically considered for ASD simulations, the cost of solving the set of linear equations via a fully implicit method rapidly outgrows any possible gain from relaxing the time step requirements. A LLG solver that has been shown to be able to combine the stability from an implicit method yet limit the computational cost is the semi-implicit midpoint method, as introduced by Mentink et al. (2010). With this scheme, the implicit considerations only affect the magnetic moments locally, and the need to solve system wide sets of non-linear equations is thus negated. The SIB integrator reads

$$\mathcal{X}_k^i = \mathbf{X}_k^i + h\frac{\mathbf{X}_k^i + \mathcal{X}_k^i}{2} \times \mathbf{a}_i(\mathbf{X}_k) + h^{1/2}\frac{\mathbf{X}_k^i + \mathcal{X}_k^i}{2} \times \mathbf{f}(\mathbf{X}_k^i)\xi_{k+1}^{i,j} \tag{7.25}$$

$$\mathbf{X}_{k+1}^i = \mathbf{X}_k^i + h\frac{\mathbf{X}_k^i + \mathbf{X}_{k+1}^i}{2} \times \mathbf{a}_i\left(\frac{\mathbf{X}_k + \mathcal{X}_{k+1}}{2}\right)$$

$$+ h^{1/2}\frac{\mathbf{X}_k^i + \mathbf{X}_{k+1}^i}{2} \times \mathbf{f}\left(\frac{\mathbf{X}_k^i + \mathbf{X}_{k+1}^i}{2}\right)\xi_{k+1}^{i,j}, \tag{7.26}$$

where $\xi_{k+1}^{i,j}$, where $j = 1, 2, 3$, and $i = 1, \ldots, N$, and $k = 1, \ldots, K$, is an independent, identically distributed random variable distributed as for the IMP method. As the description 'semi-implicit' indicates, there is still a degree of implicitness in the SIB method but, as compared to that in the IMP method, it has been drastically reduced. Instead of the full $\mathbf{X}_{k_1}$ vector being implicit, the implicitness is now in the 'predictor' $\mathcal{X}_k^i$ (first operation) and in the 'corrector' $\mathbf{X}_{k+1}^i$ (second operation), both of which are calculated for each spin separately. This has the important consequence that the computational effort is reduced, from the need to solve in each time step a system of $3N$-dimensional non-linear equations, to solve instead two 3-dimensional linear systems of equations. In effect this renders a computational effort on par with explicit solvers. It is important, as for the Heun solver, that $\mathcal{X}_k^i$ has to be calculated first for all spins before $\mathbf{X}_{k+1}$ is computed. A central property of the SIB method is that both operations preserve the length of individual spins exactly, which can be proven similarly as for the IMP method. For a two-spin system with only Heisenberg interaction and $\alpha = 0$, the SIB method preserves total energy and total spin, just like the IMP method (Mentink et al., 2010). This conservation property extends also to some larger systems, for example, a bipartite lattice with only nearest-neighbour Heisenberg interaction.

7.4.9 Comparison of solvers

The decision of which kind of solver to use for integrating the SLLG equation of motion must be based on a balance between the computational cost of the solver and its numerical stability. As mentioned above, implicit schemes inherently have better numerical properties than explicit schemes, which allows for a relaxation of the time step length. However, for the large systems typically considered for atomistic spin-dynamics simulations, the cost of solving the set of linear equations that follows a fully implicit method rapidly outgrows any possible gain from relaxing the time step requirements. Since the number of simulated moments in ASD simulations typically ranges between a few thousand to millions of moments, implicit solvers are thus normally not even considered as an alternative.

Regarding explicit or semi-implicit solvers, empirical experience have shown that geometric properties, such as conserving the norms of the evolved moments, are indeed valuable with regards to solver efficiency. According to the discussion on solvers in this chapter, this singles out the Depondt–Mertens and the SIB schemes as suitable solvers, and they have both been extensively used for ASD simulations. As a comparison between different explicit and semi-implicit schemes, Fig. 7.2 shows the calculated norm of the deviation from a very accurate simulation (obtained for a time step of $h = 10^{-18}$ s using the Depondt–Mertens solver) when simulating the time evolution of a bcc model system, which is perturbed from equilibrium position in a zero temperature and zero-damping situation, with different solvers at a more reasonable time step of $h = 10^{-16}$ s. In Fig. 7.2 it is apparent that the Euler solver has a clearly lower accuracy than the HeunP, Depondt–Mertens, and SIB solvers do. The regular HeunP solver shows larger errors than the SIB and Depondt–Mertens schemes do, which is a clear example that it is indeed a drawback to not have the proper geometry-conserving properties that HeunP is lacking. The differences between the Depondt–Mertens and SIB schemes are small and, for this system, it could be said that they in principle behave with very similar accuracies, a trend noted in a majority of the performed ASD simulations. Since the numerical properties between the latter two schemes are on par with each other, the choice between the two solvers can instead be determined by other factors, such as simplicity of implementation, and the computational cost for each solver step. Also here, the two schemes are comparable, without large differences in the computational cost. Regarding implementation details, it can be argued that the Depondt–Mertens scheme may be easier than the SIB method to implement and generalize.

7.4.10 Random number generation, and statistics

A key component for simulations of the SLLG equation is the generation of the stream of pseudorandom numbers that form the stochastic field. The random numbers have

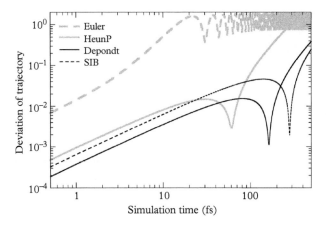

Figure 7.2 *Calculated error for trajectories predicted for a single moment for a bulk body-centred cubic model system. The error is calculated as the deviation from a Depondt–Mertens simulation with a time step of $h = 10^{-18}$s, and the time step used for the trajectories being compared is $h = 10^{-16}$s; Depondt, the Depondt–Mertens solver; Euler, the Euler solver; HeunP, the Heun method with projection; SIB, the semi-implicit midpoint solver.*

a few important criteria that need to be fulfilled, namely, they should be uncorrelated and follow a normal distribution. In addition, the generation of random numbers should preferably be fast, since it is a very common operation in the ASD procedure. For finite temperature simulations, three random numbers need to be drawn for the generation of the stochastic field for each magnetic moment, one or two times per time step. For this purpose, a good, efficient, and long-period pseudorandom number generator is needed. Most programming languages come with built-in random number generators (RNGs), but the quality of these may vary and, for that reason, it is often advisable to use external, third-party implementations of well-established RNGs. RNG algorithms differ vastly in complexity but, regardless, seemingly minor errors in an implementation of a RNG can totally destroy its desired properties. To make matters worse, a faulty RNG can be hard to detect and thus it is strongly advised not to use untested or self-written RNGs (Ferrenberg et al., 1992). The UppASD method uses an implementation of the Mersenne Twister (MT) RNG (Matsumoto and Nishimura, 1998), which is generally considered to be a good choice for most scientific computing applications, albeit there exist other methods that have statistical properties and efficiency that are on par with, or better than, the MT algorithm.

Another important issue to consider when generating random numbers is which distribution they should follow. Most RNGs generate numbers with a uniform distribution on the interval [0, 1]. However, the stochastic field entering into the stochastic Landau–Lifshitz equation is typically chosen to follow a Gaussian distribution with a zero mean and a variance according to Eqn (5.8).[3] For the SLLG equation in its differential and dimensionless form expressed in Eqn (7.9), the variance is unitary, given that the amplitude of the noise power is included in the coefficient defined by Eqn (7.11).

To achieve this distribution, a transformation from the uniform is needed. There are several reported transformations of this kind, including the Box–Muller transform (Box and Muller, 1958) and the ziggurat method (Marsaglia and Tsang, 2000). While the choice of transformation method may seem as an insignificant detail of an ASD implementation, it can actually be observed that using the latter of the two can significantly speed up the random number generation. Without going into detail, the main reason for the improved efficiency is that, in both these algorithms, some of the generated random numbers need to be discarded and that, for the ziggurat method, the share of the random numbers that need to be thrown away is significantly smaller than that for the Box–Muller transform.

For testing purposes, it is desirable to be able to reproduce earlier simulations but, on the other hand, the stream of random numbers should not be identical for every simulation. Thus, it is important to be able to change the starting seed of the RNG and to make sure that different seeds result in similar results. It is also important to realize that, with the introduction of the stochastic Landau–Lifshitz equation, the ASD simulations now simulate stochastic processes where a unique outcome of the simulations is not guaranteed. Thus, it is imperative to ensure proper statistics of the results when simulating the spin dynamics. What can be considered proper statistics varies from case to case, depending on system size and on which physical observables are of interest. A more detailed discussion of observables and their statistics follows in Section 7.5.

7.5   Extraction of observables

As the main purpose of performing ASD simulations is to estimate one or more physical properties, the extraction and analysis of these observables naturally constitute an important part of an ASD implementation. In Section 7.5.1, we provide a description of the most common observables and how they can efficiently be sampled.

7.5.1   Trajectory-based observables

The majority of observables extracted from an ASD simulation are based on the individual magnetic moment trajectories $\mathbf{m}_i(t)$, where $i = 1, 2, \ldots, N$, which constitute the raw output that comes from integrating the equation of motion. From these trajectories, many other useful observables can be extracted as well. The ferromagnetic order parameter is straightforwardly calculated as the average of the magnetic moments but, in more general terms, the static order parameter for a given magnetic ordering can be expressed, in terms of the staggered magnetization $\mathbf{M}_s$, as

---

[3] Also, other choices of distributions for the stochastic field are possible; see e.g. Milstein et al., 2002.

$$\mathbf{M_s} = \frac{1}{N} \sum_i^N e^{i\mathbf{Q}\cdot\mathbf{r}_i} \mathbf{m}_i, \tag{7.27}$$

where $\mathbf{Q}$ is the ordering wavevector. For collinear antiferromagnets or ferrimagnets, the expression in Eqn (7.27) can typically be simplified by expressing the staggered magnetization as a given combination of the magnetization of each sublattice (see e.g Sergienko and Dagotto, 2006). For general, non-collinear magnetic orders, the calculation of the order parameter is closely related to the Fourier transform of the spatial correlation function, which will be discussed further in this chapter.

As a complement to regular magnetic order parameters or individual magnetic moment trajectories, more intricate observables can be most useful. In the context of frustrated and chemically disordered magnets, Kawamura (1992) defined the local scalar chirality as

$$\chi_i^\mu = \mathbf{m}_{i+\hat{\mathbf{e}}_\mu} \cdot \left(\mathbf{m}_i \times \mathbf{m}_{i-\hat{\mathbf{e}}_\mu}\right), \tag{7.28}$$

where $\mu \in \{x, y, z\}$, and $\hat{\mathbf{e}}_\mu$ is a unit vector along one of the principal axes of a simple cubic lattice. We note that, if calculated for the unit vector $-\hat{\mathbf{e}}_\mu$, the chirality will change sign. The summation of the three spins in Eqn (7.28) can be generalized to other crystal systems and structures. For spin-spiral orderings, the local vector chirality can be defined as

$$\mathbf{c}_i^v = \sum_j \mathbf{m}_i \times \mathbf{m}_j, \tag{7.29}$$

where the index j runs over the moments that are nearest neighbours to $\mathbf{m}_i$. From the local scalar or vector chirality, chiral order parameters can be obtained as averages:

$$\chi^\mu = \frac{1}{N} \sum_i \chi_i^\mu, \tag{7.30}$$

$$\mathbf{C}^v = \frac{1}{N} \sum_i \mathbf{c}_i^v. \tag{7.31}$$

Another property that can be of interest for particular systems is the spin-driven electronic polarization $\mathbf{P}$ that can emerge in insulating spin-spiral ordered systems (Tokura et al., 2014). A model to capture the polarization due to spin currents between pairs of tilted magnetic moments was proposed by Katsura et al. (2005) and reads

$$\mathbf{P} = P_0 \sum_i \sum_j \hat{\mathbf{r}}_{ij} \times \left(\mathbf{m}_i \times \mathbf{m}_j\right), \tag{7.32}$$

where $\hat{\mathbf{r}}_{ij}$ is the unit vector parallel to the direction of the bond between the two moments $\mathbf{m}_i$ and $\mathbf{m}_j$; P_0 is a proportionality constant whose value can be assigned by comparison with Berry phase calculations of the electronic polarization, as done, for example, by Giovannetti et al. (2011) and Hellsvik et al. (2014). On an exchange of site indices i and j, the vector chirality as defined in Eqns (7.29) and (7.31) will change sign, whereas the polarization expressed by Eqn (7.32) is invariant.

In case the topology of the magnetic structure is of interest, topological numbers such as the Pontryagin, or skyrmion, number Q can be calculated from the gradient of the magnetization. For a two-dimensional system, the topological number is given by

$$Q = \frac{1}{4\pi} \int \int dx\, dy\, \mathbf{m}(\mathbf{r}) \cdot [\partial_x \mathbf{m}(\mathbf{r}) \times \partial_y \mathbf{m}(\mathbf{r})], \tag{7.33}$$

where ∂_x and ∂_y are spatial derivatives. In the strict sense, Q is only defined for a continuous magnetization density $\mathbf{m}(\mathbf{r})$ but can be estimated also for discrete spin systems such as those treated by ASD. Then the gradient needs to be calculated on a discrete mesh, and Eqn (7.33) transforms from an integral over the magnetization to a summation over the atomic moments.

We close this section by mentioning that inspiration for other observables to be used in ASD simulations can be provided from review papers on spin liquids (Balents, 2010) and spin ice (Bramwell and Gingras, 2001).

7.5.2 Correlation functions

A very useful set of observables is the set of correlation functions, which come in a range of different variations. As the name indicates, they have in common that they measure the correlation between magnetic moments over different distances and/or times. Here, the focus will be on pair correlation functions, and the first example is the equal-time, connected correlation function:

$$C_c^{\alpha\beta}(\mathbf{r}) = \frac{1}{N} \sum_{\substack{i,j \text{ where} \\ \mathbf{r}_i - \mathbf{r}_j = \mathbf{r}}} \langle m_i^\alpha m_j^\beta \rangle - \langle m_i^\alpha \rangle \langle m_j^\beta \rangle, \tag{7.34}$$

where $\alpha, \beta \in \{x, y, z\}$ denote Cartesian components of the magnetic moment vectors, and the brackets $\langle \ldots \rangle$ denote ensemble, or thermal, averages. The term 'connected' in this context means that the rightmost term in Eqn (7.34) is present, that is, the square of the expectation value of the measured observable will be subtracted from the correlation function (Newman and Barkema, 1999). Omitting this term instead gives the disconnected correlation function. The pair correlation function can, among other things, be used to obtain the correlation length which then is a measure of how long-range the magnetic order is in a given system. For a system in non-equilibrium, the pair correlation function will depend on time, $C_c^{\alpha\beta}(\mathbf{r}) = C_c^{\alpha\beta}(\mathbf{r}, t)$, and can be used to follow the growth in time of spatial correlations, for instance, when a system is quenched from a high temperature paramagnetic state to a temperature below the critical temperature (Hellsvik et al., 2008).

Measuring the correlation over time instead of over spatial coordinates gives the time-displaced autocorrelation

$$C_0(t + t_{\mathrm{w}}, t_{\mathrm{w}}) = \langle \mathbf{m}_i(t_{\mathrm{w}} + t) \cdot \mathbf{m}_i(t_{\mathrm{w}}) \rangle. \tag{7.35}$$

With the autocorrelation function, it is possible to get information on how a system relaxes over time. For a system in thermal equilibrium, the autocorrelation function is time translationally invariant: $C_0(t + t_{\mathrm{w}}, t_{\mathrm{w}}) = C_0(t)$. If the system is brought out of equilibrium at time $t_{\mathrm{w}} = 0$ by, for example, a change in the temperature of its environment, its autocorrelation function will depend on the waiting time t_{w}. It can thus also be used as a measure of how well-equilibrated a system is, which is important in order to ensure that the system in question has been properly thermalized during an ASD simulation. Sampling of the two-time autocorrelation function has been a central component in numerical simulations to study the aging dynamics of glassy systems such as Ising spin glasses (Belletti et al., 2008) and Heisenberg spin glasses (Berthier and Young, 2004). Due to the very circumstance that these systems are slowly relaxing, with important features evolving over a logarithmic rather than linear time scales, numerical simulations are very expensive. In work by Belletti et al. (2008) and Berthier and Young (2004), Monte Carlo schemes were used to evolve the system after a quench from a high temperature to a temperature below the spin glass freezing point. By order of magnitude, one Monte Carlo sweep, which performs on average one attempt to flip or rotate each spin, corresponds to 100 fs of physical time. Thus, for the case that a time step $h = 0.01–0.10$ fs is used in ASD simulations of the physical evolution of a Heisenberg spin glass (Skubic et al., 2009; Banerjee et al., 2011; Pal et al., 2012), the computational effort is some three or four orders of magnitude larger than when mimicking the dynamics with Monte Carlo trial moves. Figure 7.3 shows the autocorrelation sampled in ASD simulations for a logarithmically distributed set of waiting times $\{t_{\mathrm{w}}\}$ for the frustrated and chemically disordered diluted magnetic semiconductor $Zn_{1-x}Mn_xSe$, $x = 0.25$, and averaged over 20 random configurations of the Mn atoms. The simulations were performed following a quenching protocol in which the initial magnetization configuration was random, corresponding to an infinite temperature, and at $t_{\mathrm{w}} = 0$ the temperature was set to $T = 10$ K, which is below the spin glass freezing temperature for this material. The system relaxes over many orders of magnitude in time, as is expected for a spin glass system (Berthier and Young, 2004; Skubic et al., 2009).

Both spatial and temporal fluctuations can be studied simultaneously in the time-dependent correlation function $C(\mathbf{r}, t)$, which is similar to the correlation functions in form:

$$C^{\alpha\beta}(\mathbf{r}, t) = \frac{1}{N} \sum_{\substack{i,j \text{ where} \\ \mathbf{r}_i - \mathbf{r}_j = \mathbf{r}}} \langle m_i^{\alpha}(t) m_j^{\beta}(0) \rangle - \langle m_i^{\alpha}(t) \rangle \langle m_j^{\beta}(0) \rangle. \tag{7.36}$$

The correlation function defined in Eqn (7.36) can thus describe how the magnetic order evolves both in space and over time. The perhaps most valuable application of

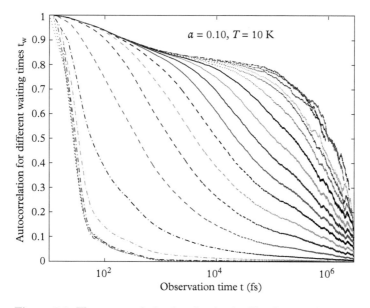

Figure 7.3 *The autocorrelation function for the diluted magnetic semiconductor $Zn_{1-x}Mn_xSe$, $x = 0.25$. The simulation follows a quenching protocol where the initial magnetization configuration of the Mn atoms is random; T, temperature. The different curves correspond to different waiting times t_W. The exchange parameters were calculated via the coherent potential approximation, as for $Ga_{1-x}Mn_xAs$ in Kudrnovský J., Turek, I., Drchal, V., Máca, F., Weinberger, P., and Bruno, P., (2004), Exchange interactions in III-V and group-IV diluted magnetic semiconductors. Phys. Rev. B, 69, 115208, and kindly provided by Dr Josef Kudrnovský.*

$C(\mathbf{r}, t)$ is, however, obtained by a Fourier transform over space and time, to give the dynamic structure factor

$$S^{\alpha\beta}(\mathbf{q}, \omega) = \frac{1}{\sqrt{2\pi N}} \sum_{\mathbf{r}} e^{i\mathbf{q}\cdot\mathbf{r}} \int_{-\infty}^{\infty} e^{i\omega t} C^{\alpha\beta}(\mathbf{r}, t) \, dt, \qquad (7.37)$$

which is closely related to what is measured by inelastic neutron or electron scattering on magnetic materials. More details on these scattering studies will be given in Chapter 9 but, in brief, scattering experiments are naturally analysed in terms of the differential cross section (Squires, 2012), which for many materials is proportional to the dynamical structure function. This means that, by simulating the dynamical structure factor, the relation between the wavevectors $\mathbf{q}$ and the frequency ω for the magnetic excitations, that is, magnons, in the material can be obtained.

In the definitions above, both the dynamical structure factor $S(\mathbf{q}, \omega)$ and the pair correlation function $C_c(\mathbf{r})$ have been defined as 3×3 matrices. In materials with a collinear ordering, typically only the diagonal elements of the correlation functions are of interest. As an example, consider a ferromagnetic material with the quantization axis along the z-axis. In order to relate a simulated dynamical structure factor with inelastic scattering experiments, one needs to keep in mind that these experiments only measure fluctuations perpendicular to the quantization axis. Thus, in the current example, it is the $S^{xx}(\mathbf{q}, \omega)$ and $S^{yy}(\mathbf{q}, \omega)$ elements that would correspond to spin wave excitations. In materials with a non-collinear ordering, the quantization axis is not fixed in the material, so non-diagonal elements of the dynamical structure factor are needed to ensure a good correspondence between simulations and experiments. In the work by Babkevich et al. (2012) a lucid discussion can be found on how spherical neutron polarimetry can be used to investigate the relative population of domains in the high temperature chiral phase of the type II multiferroic CuO. Experimentally measured elements of the polarization matrix (as used in Eqns (1) and (4) in Babkevich et al., 2012) connect to the Fourier transform of the equal-time pair correlation function, which can be sampled in an ASD or Monte Carlo simulation.

7.5.3   Correlation functions and Fourier transforms

The sampling and calculation of most observables usually takes much less time than calculating the effective fields and evolving the equations of motion does. However, the spatial correlation functions introduced in Section 7.5.2 contain products of pairs of atoms and, if large-enough distances $|\mathbf{r}|$ are considered, then effectively all pairs in the simulation box need to be summed over. This implies a quadratic scaling, for example, $\mathcal{O}(n^2)$, of the number of magnetic moments in the system on the cost of calculating the spatial correlation functions. Since all other time-critical operations in the ASD method can be made to scale linearly with system size, the calculation of the correlation functions can actually be the dominating part for large simulations, if implemented directly. Luckily, this can be avoided by, instead, sampling the Fourier transform of the correlation function directly (Newman and Barkema, 1999), since

$$S_c^{\alpha\beta}(\mathbf{q}) = \sum_{\mathbf{r}} e^{i\mathbf{q}\cdot\mathbf{r}} C_c^{\alpha\beta}(\mathbf{r}) \tag{7.38}$$

$$= \frac{1}{N} \sum_{\mathbf{r}} \sum_{\substack{i,j \text{ where} \\ \mathbf{r}_i - \mathbf{r}_j = \mathbf{r}}} e^{i\mathbf{q}\cdot\mathbf{r}} \left[\langle m_i^\alpha m_j^\beta \rangle - \langle m_i^\alpha \rangle \langle m_j^\beta \rangle \right] \tag{7.39}$$

$$= \frac{1}{N} \langle \sum_i e^{-i\mathbf{q}\cdot\mathbf{r}_i} \left(m_i^\alpha - \bar{m}^\alpha \right) \sum_j e^{i\mathbf{q}\cdot\mathbf{r}_j} \left(m_j^\beta - \bar{m}^\beta \right) \rangle \tag{7.40}$$

$$= \frac{1}{N} \langle s^\alpha(\mathbf{q}) s^\beta(\mathbf{q}) \rangle, \tag{7.41}$$

where $s^\alpha(\mathbf{q})$ now is the Fourier transform of $M_i^\alpha = m_i - \bar{m}^\alpha$, where $\bar{m}^\alpha$ is the average of the α component of the magnetization. That means that, instead of performing the summation over N^2 pairs of magnetic moments at each sampling time step, it is instead enough to calculate the Fourier transform of the site-resolved magnetic moments. If the lattice where the moments sit allows for it, then FFT methods, which scale like $N \log N$, can be employed to sample $S_c(\mathbf{q})$ and, after all averaging has been performed, a backwards Fourier transform can be used if $C_c(\mathbf{r})$ is sought after. If it is the dynamical structure factor that is being measured, then there is not even a need for backtransforming the correlation function to real space, since it is the Fourier space representation that is needed.[4]

7.5.4 Sampling, averaging, and post-processing

As mentioned several times in this chapter, at finite temperature, the ASD method is stochastic, and thus proper sampling and statistics are needed in order to extract meaningful data from the simulations. This is also true for correlation functions in general and the dynamical structure factor in particular. The connection between the dynamical structure factor and the corresponding magnon spectra is made by identifying the maximum values in the correlation function so that, for each reciprocal wavevector $\mathbf{q}$, there is at least one frequency ω which has a high intensity, and that implies that a magnon with that particular wavevector is associated with the frequency. Therefore, it is necessary that the peaks in the dynamical structure factor that defines the magnon spectra are clearly visible and that no spurious peaks occur. Better statistics can always be obtained by performing an average over a larger system and/or a larger number of replicas but there are also other algorithms that can be applied in order to improve the quality of the sampled correlation functions.

Since the way that the correlations functions are most efficiently sampled is by means of Fourier transforms, the fact that the samplings are on a finite interval causes spurious peaks in the correlation functions. This can partly be remedied by the use of so-called window functions, which can help smooth out the resulting spectra (Shastry, 1984). The sampled correlation function can also be improved via post-processing, which has been shown to give a dramatic improvement in visualizing the resulting spectra. Noisy data can be smoothened by convolution with a smearing function, which is often a Gaussian function with suitable support. However, smearing also tends to diffuse the appearance of the proper maximum peaks in the spectra. One successful algorithm for achieving smooth spectra while retaining the original peaks was devised by Bergman et al. (2010). The algorithm consists of the combination of a Gaussian convolution and an edge-enhancing operation that is performed as follows. For each measured value of the wavevector $\mathbf{q}_i$, the frequency dimension of the structure factor $S(\mathbf{q}_i, \omega)$ is convoluted with a Gaussian with a suitable width. The smeared function $\tilde{S}(\mathbf{q}_i, \omega)$ is then normalized and, in order to enhance the peaks in the spectra, the normalized function is fed into a power function with an exponent $n > 1$ in order to yield $S_{\text{out}}(\mathbf{q}_i, \omega) = \tilde{S}(\mathbf{q}_i, \omega)^n$.

---

[4] The calculation of the autocorrelation function can also be optimized by calculating the Fourier transforms of the time-dependent magnetization. More details on how that is done can be found in other textbooks, such as the famous book by Newman and Barkema (1999).

The optimal value of the exponent can be iterated according to what best fits the need of the post-processing. The resulting curves can then be combined to provide the full $S_{out}(\mathbf{q}_i, \omega)$ spectra. It is also possible to perform a similar algorithm along the ω axis but, when tested, this was not found to improve the final results significantly.

7.5.5 Thermodynamic observables

While the strength with the ASD method is that it is capable of capturing dynamical processes, the method can in fact also be used to simulate processes such as phase transitions. In that sense, the ASD method has the same capabilities as most Monte Carlo methods. In order to ensure efficiency in simulating equilibrium states with ASD, it is advisable to increase the damping parameter used in the LLG equation.[5] While changing the damping parameter will obviously change the description of the dynamical processes, the nature of the equilibrium state is not affected by an overly large damping. When characterizing a phase transition, a simple estimate of the critical temperature can be obtained by studying the vanishing of the order parameter. However, due to the amount of fluctuations present at high temperatures and the finite size of the simulation box, the decrease of the order parameter is typically not abrupt enough to allow for an accurate determination of the critical temperature. In order to better accommodate for finite size scaling effects, other observables are preferable. A straightforward method to correct for finite size scaling is to use the cumulant crossing method (Binder, 1981), in which the reduced fourth-order cumulant U_4, or Binder cumulant, is introduced as

$$U_4 = 1 - \frac{\langle m^4 \rangle}{3 \langle m^2 \rangle^2}. \tag{7.42}$$

This approach can be used to estimate the phase transition temperature of systems with second-order phase transitions. For the Heisenberg model, in the thermodynamic limit, the value of U_4 goes to $\frac{4}{9}$ for $T > T_c$ where T_c is the critical temperature for the phase transition. On the other hand U_4 goes to $\frac{2}{3}$ for $T < T_c$ as the lattice size increases. For large-enough systems, the curves for U_4 for different lattice sizes cross at a 'fixed point' with value U^* and, from this crossing, an improved estimate of the critical temperature can be obtained (Binder, 1981; Stanley, 1999).

Other measures of interest for studying phase transitions include the zero field susceptibility, and the specific heat of the magnetic system. It can be shown (D. Landau and Binder, 2005) that the isothermal susceptibility $\chi = (\partial \langle m \rangle / \partial E)_T$ is related to the fluctuations of the systems magnetization m so that, at zero wavevector,

$$k_B T \chi = \langle m^2 \rangle - \langle m \rangle^2 = \sum_i \sum_j [\langle \mathbf{m}_i \cdot \mathbf{m}_j \rangle - \langle \mathbf{m}_i \rangle \cdot \langle \mathbf{m}_j \rangle] \tag{7.43}$$

---

[5] For systems with inherently slow dynamics, such as the spin glass systems discussed in Section 7.5.2, the efficiency of using ASD simulations for sampling phase transitions is very low, compared to the more efficient Monte Carlo simulations.

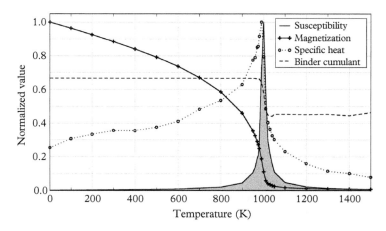

Figure 7.4 *Temperature dependence of the magnetization, magnetic susceptibility, specific heat, and the Binder cumulant for body-centred cubic Fe, as obtained by Monte Carlo simulations performed with the UppASD code. The simulations were performed for a 40 × 40 × 40 unit cell system by using Heisenberg exchange interactions calculated from multiple scattering theory.*

with indices i, j running over all atoms of the simulation cell. If the system considered is not a ferromagnet, the staggered magnetization should enter the expression for the zero field susceptibility instead. On an equal footing, the specific heat relates to the fluctuations of the system's energy, E, as

$$k_B T^2 C_V = \langle E^2 \rangle - \langle E \rangle^2, \tag{7.44}$$

where C_v is volumetric heat capacity, and both the susceptibility and the specific heat diverge at the critical temperature for system sizes approaching the thermodynamic limit. However, even for relatively small systems, these observables should have maximum values close to the critical temperature. Using the specific heat for identifying the phase transition has the advantage that it works even if the order parameter is not known. As an example of the aforementioned thermodynamic measurables, their behaviour as a function of temperature in bcc Fe is shown in Fig. 7.4, where it can be seen that the drop in magnetization coincides with the peaks in the susceptibility and specific heat. It can also be noted that the calculated T_c agrees well with the experimental value.

7.6 Aspects of efficiency and optimization

An atomistic simulation framework enables studies with very high resolution but comes with the drawback that, for most systems of interest, a very large number of atomistic spins need to be simulated. Thus, it is important to ensure that the implementation is fast and efficient. Since what constitutes the time-critical part of an ASD program can vary with the type of system considered, care needs to be taken with most, if not all, algorithms that enter the simulation framework. As has been noted earlier, choosing efficient algorithms can reduce the running time by orders of magnitude. A prime example of this is that a good solver like the semi-implicit midpoint solver SIB (Mentink et al., 2010) or the Depondt–Mertens solver (Depondt and Mertens 2009), allows for time steps which may be 10–100 times larger than what is possible with a simple Euler solver. All major parts of the ASD methodology can be made to scale linearly with the system size. This is indeed a good thing when it comes to efficiency considerations but it also means that the inclusion of algorithms that scale quadratically with the system size can become major bottlenecks, even if those algorithms are not used many times during each simulation. Notable examples in this category were mentioned in Section 7.5 and include the construction of a neighbour map that is used for mapping exchange interactions, and the sampling of pair correlation functions. A particular case is the calculation of the dynamical structure factor where using the $\mathcal{O}(N^2)$ algorithm can, for large systems, demand a running time which is three to four orders of magnitude longer than that needed when the Fourier-transformed correlation function is sampled instead.

Besides through choosing efficient algorithms, significant impact on the performance of an ASD method can be obtained by writing efficient code.[6] In scientific computing applications, it is often advisable to use trusted and efficient external libraries to as large an extent as possible. This was mentioned in Section 7.4.10 in connection with RNGS but also holds more generally. One example that is applicable to the ASD methodology as described here is that the Fourier transforms of the correlation functions can be done using established FFT libraries such as FFTW (Frigo and Johnson, 2005). On the other hand, the ASD framework typically does not explicitly make use of large systems of linear equations or eigenvalue problems, and thus external libraries for linear algebra such as Basic Linear Algebra Subprograms (BLAS) and LAPACK (Linear Algebra Package), are typically not needed. An exception is when implicit solvers are used. For instance, the implementation by D'Aquino et al. (2005) of the IMP solver used a special quasi-Newton iterative technique to solve large systems of globally coupled non-linear equations. Furthermore, sparse linear algebra packages have been put to use in connection with a particular choice of parallelization scheme (Evans et al., 2014).

In addition to using external libraries and efficient algorithms, the obvious possibility for speeding up the ASD method is to use parallelization techniques. In UppASD, several different types of parallelization have been implemented and tested, and a selection of these schemes will be presented in Section 7.7.

---

[6] How to optimize code is outside the scope of this book but, for the interested reader, there is plenty of literature on the subject, and a good starting point for this purpose is the book by Goedecker and Hoisie (2001).

7.7 Parallelization approaches

Concerning parallelization of computationally heavy methods, there have historically been two dominant approaches: memory-sharing parallelization, and memory-distributed parallelization. Nowadays, memory-shared parallelization is often implemented using OpenMP (Open Multi-Processing) programming interface, while the dominant interface for memory-distributed parallelization is the Message Passing Interface (MPI). Thus, the following discussion will use the names of these interfaces when discussing parallelization techniques, but the methods can alternatively be implemented using other interfaces. A recent trend in parallel computing is to use not only large arrays of central processing units (CPUs) but also the computational power available in more specific types of hardware, such as GPUs. For this purpose, other programming platforms and interfaces are typically needed, such as OpenCL (Stone et al., 2010) and CUDA (Nickolls et al., 2008).

7.7.1 Shared memory parallelization

Parallelization for hardware assets that share memory, by means of OpenMP or another interface, is typically easy in the sense that it can be implemented on top of a serial algorithm without introducing too many changes into the original code. Since modern day CPUs have several cores available, an OpenMP parallelization can also be thought of as getting increased performance almost 'for free' since it makes it possible to use an increased amount of the hardware that already exists in the computer. On the other hand, the shared memory approach can be said to be hard because it is often difficult to obtain good scaling properties for the algorithms when the number of active cores increase. The decreased scaling with increasing number of cores is mainly due to the fact that OpenMP parallelization involves forking and joining of computational tasks, and that these operations come with a non-negligible amount of overhead cost. To reduce this overhead, it is advised to use regions that are as large and computationally heavy as possible for each parallelization initialization.

Since the efficiency of the parallelization grows with the amount of needed computation, it is easier to parallelize tasks that take a long time to perform. From this point of view, it is a drawback for the ASD framework that a simulation typically consists of a very large number of time steps but the computational cost for each step is quite limited. Therefore, excellent scaling by means of a shared memory approach cannot be expected here. Despite this, in the UppASD framework, there are still a few algorithms that lend themselves quite well to be OpenMP parallelized. The calculation of the effective field is typically the most costly part of each time step and, as can be seen in the equations in Section 7.2, the effective field acting on each moment can actually be calculated independently of the other moments during that time step. Thus, the effective field calculation can indeed be parallelized using OpenMP. This is done in UppASD, and the scaling with respect to the number of processors is quite good for systems where a large number of neighbours are included in the calculation of the exchange contributions to the effective field. Several measurement/sampling algorithms can also be successfully parallelized using a shared memory approach, in particular the correlation functions described in Section 7.5.2.

A more complicated scenario is found when parallelizing the solvers of the equations of motion. In principle, the solvers themselves are trivially parallelizable over the number of magnetic moments but, for finite temperatures, the generation of the thermal stochastic field can pose a problem. The issue here is that it is important that the stochastic field is uncorrelated and, from the parallelization point of view, that means that the RNG needs to be thread-safe, that is, that multiple copies of the RNG can work in parallel. Most RNGs, in fact, do not work well in parallel and therefore the computational cost of generating random numbers for the thermal field may form a bottleneck if the rest of the solver is parallelized. One possible solution for this problem is to use a different seed for the RNG on each core. However, this approach has to be used with great care as, for most RNGs, it is very difficult to predict how close the sequences generated by different seeds can end up being. If two or more seeds do generate overlapping streams of random numbers, the resulting stochastic field is not to be trusted, since the elements in it are consequently no uncorrelated.

7.7.2   Distributed memory parallelization

Distributed memory parallelization approaches often need a significant reformulation of both algorithms and computer code in order to achieve acceptable performance gains and scaling. While using such an approach might take longer than introducing primitive shared memory algorithms would, it often brings improved performance. In distributed memory parallelization, at least in the implementations done in UppASD, there is no forking or joining of processes; instead, multiple instances of the same program run in parallel to each other and only share selected amounts of data at particular times during the simulation. In fact, this approach is very suitable for the ASD framework, and distributed memory parallelization can actually be used in several ways, with close to ideal speed-up.

7.7.2.1   *Ensemble parallelization*

For the ASD methodology, there is the possibility of employing what is called trivial parallelism, which is when an algorithm is naturally parallel already as it is formulated. This parallelization is based on the fact that, at finite temperatures, the LLG equations are stochastic, wherefore a proper amount of statistics needs to be gathered for each simulation. For many situations, the increased statistics can be obtained by simply increasing the system size or simulation time, but a general and more elegant way to improve the statistics is to employ ensemble averaging. That means that the same simulation is run in several instances, with the only difference between the different simulations being the sequence of random numbers that contribute to the stochastic fields. This provides an excellent way to parallelize the ASD method. Ensemble parallelization has been included in UppASD and effectively works the same way as running several simulations side by side with each other by means of scripting, but here data are exchanged between the different ensembles when the observables are sampled. This enables statistical averaging to be performed inside the code, and no particular post-processing is needed on the output data. Due to the limited coupling between the different parallel instances using this parallelization scheme, the scaling with respect to number of processors is excellent.

7.7.2.2 *Spatial decomposition*

Ensemble averaging provides an efficient method of parallelization but there are scenarios where this approach is not applicable. It could, for instance, be the case when a very large system, hypothetically, even a simulation of a complete spintronic device, is considered. Then, the number of moments contained in the simulation box may be too large to even fit memory-wise on a single CPU, or the computational cost would prohibit achieving long-enough simulation times. In this case, another approach for parallelization can be pursued by dividing the simulated system into smaller parts where each processor now takes care of the time evolution of only its own small simulation box. This is called spatial decomposition and is also an obvious and common way to partition large numerical simulations in general (see Fig. 7.5).

When using spatial decomposition, the decomposed sub-boxes cannot be treated as if they are isolated from each other, as the time evolution of each considered moment depends on the local effective field acting on that moment and, as seen in Section 7.2, the effective field depends on the surrounding moments. That means that, while moments placed in the middle of a sub-box may very well only interact with other moments within the box, moments closer to the boundaries of the box do, however, interact with moments in neighbouring boxes, and thus information about the moments needs to be exchanged between neighbouring boxes at every time step. The ASD methodology has, in this perspective, the drawback that the calculation of the effective field is performed quite fast, wherefore the time spent passing information between boxes can take

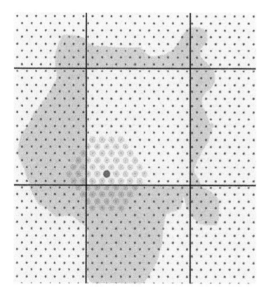

Figure 7.5 *Schematic view of how spatial decomposition can look. The boxes denote different processors that keep the information about the atomic moments, shown as dots that are situated within the box. Since any particular moment interacts with other moments, as is illustrated by the light grey filled circles, each processor also needs information about moments that resides on other processors. The configuration of these non-local moments, shown as the dark grey area, is copied to neighbouring processors before every time the effective fields are calculated.*

a considerable amount of time, compared to the time spent doing actual calculations. This will, of course, lead to degraded parallelization performance. On the other hand, ASD has two benefits in this regard: First, that the moment sits on a fixed lattice and, second, that it has a constant and often short-range, at least compared with the size of the simulation box, set of neighbour interactions. That means that the size and layout of regions where information needs to be passed between neighbouring boxes are known a priori and do not need to be recalculated during the course of the simulation. As a result, the program can spend time optimizing the parallel topology and the message-passing regions at the beginning of the simulations. The short-rangedness of the interactions also brings the advantage that, as long as each sub-box is significantly larger than the exchange interaction range for the contained moments, there are indeed moments in the box that do not need any information from neighbouring boxes. This can be used to 'hide' the communication between boxes, so that those contributions to the effective fields that can be calculated without non-local information can be performed while this information is being passed in the background. By the time the local contributions have been calculated, the data from neighbouring boxes have hopefully arrived, so that the non-local contributions can be calculated as well. A decomposition method of the kind described here has been implemented in UppASD and has been shown to scale well up to several thousands of processors. This is illustrated in Fig. 7.6, where the size of each decomposed sub-box is kept fixed but, as the number of processors used in the

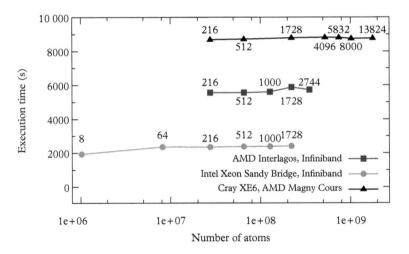

Figure 7.6 *Weak scaling behaviour for the Message Passing Interface implementation of UppASD FeCo in a B2 structure, using material-specific parameters from density functional theory. The size of each of the decomposed sub-boxes is kept fixed, and the execution time (in seconds) for a test simulation as function of the total system size is displayed. The number close to each data point corresponds to the number of processes used in the simulation.*

simulation increases, the total system size also increases. Since the computing time is more or less constant for each of the three different tested computers, it can be concluded that, for the chosen size of decomposition, excellent weak scaling behaviour is found and it is possible to simulate systems consisting of more than 10^9 moments, if a sufficient number of processes are used.

7.7.2.3 Hybrid parallelization schemes

Even better performance can be obtained if several parallelization schemes are combined in an effective fashion. That has also been done in the UppASD implementation where shared memory parallelization, which makes it possible to take advantage of both spatial decomposition and ensemble averaging, is combined with the feature that several processors can share so that each ensemble or sub-box, shared memory parallelization is in turn used within these boxes/ensembles.

7.7.3 GPU parallelization

Currently, much of the increased computational power in supercomputers comes not only from increasing numbers of parallel CPUs but also by dedicated, or hybrid, massively parallel accelerators based on GPUs. Thus, it is desirable to be able to harness the massively parallel properties of GPUs, in addition to using the efficient CPU parallelization schemes described in Section 7.7.2. When it comes to GPU programming, GPU resources can either be employed by writing specific kernels for the tasks that should

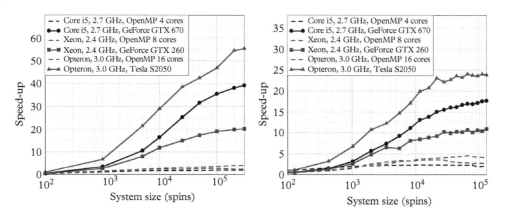

Figure 7.7 *Scaling behaviour for the CUDA implementation of UppASD. The left-hand panel shows a limiting case where very few neighbours contribute to the effective field, while the right-hand panel shows a simulation where most of the time is spent calculating the effective field. The speed-up is relative to the single-core CPU performance of each of the tested architectures. Legend denotes different computer architectures.*

be done, or by using provided libraries and interfaces. Since the GPU architectures are excellent for liner algebra and transform operations, there exist several highly efficient GPU-accelerated versions for the BLAS, LAPACK, and FFT packages. As discussed in Section 7.6, ASD makes typically little use of large linear algebra operations but FFTs can be used for the calculation of correlation functions. Substantial development in this area is to be expected, and it is not yet clear which programming standard will prevail. Currently, the difficulty in achieving optimal performance when parallelizing for GPUs lies in the fact that there is a significant amount of latency when sending data between the GPU and the CPU; however, hand, the instruction set for contemporary GPUs is not as versatile as that for CPUs, with the consequence that is not trivial to make the full program run on GPUs only. As a result, typical GPU implementations use the GPU as an additional resource, while the main program is run on CPU architecture. Once again, there is in this aspect a drawback with the ASD methodology that is, that every iteration is quite fast and thus there is a need to push data between the GPU and the CPU more often than would be desired. Nevertheless, with UppASD, it is possible to use GPU parallelization using the CUDA framework (Nickolls et al., 2008). Then most of the simulation is actually performed by the GPU, where the calculation of the effective fields and the evolution of the equations of motions are performed by customized computing kernels, while the cuRAND library is used to provide the random numbers for the stochastic field. When data need to be measured, then that is done by making the current configurations available to the CPU, so that the regular measurement and sampling routines that are programmed for the CPU can be used. This gives a good balance with respect to the efficiency portability between the different parallelization approaches. In Fig. 7.7 we show two examples of the GPU implementation of UppASD: the left panel shows the case for short-range interactions, and the right panel shows the case for long-range interactions. The speed-up indicated in the figure is relative to the single-core CPU performance of each of the tested architectures. For short-range interactions, the most recent GPU yields a speed-up of ~ 50, while somewhat lower performance is found for the case of long-range interactions. Nevertheless, GPU performance is much better than the comparable OpenMP performance, for all tested architectures.

Part 3

Applications

In order to establish an understanding of material properties, measurements of the ground state and excited states are necessary, to give information about the interactions in the material and the properties that result from these interactions. Examples of such properties include magnetic ordering, heat and electric transport, superconductivity, and structural properties. Typical methods for determining bulk crystal structures are X-ray diffraction and neutron scattering while, for surfaces, many other high-resolution methods exist. One method in particular is low energy electron diffraction, which gives information about the atomic arrangement of a surface. Magnetic properties and elementary spin excitations are frequently probed via magnetometry, the magneto-optical Kerr effect, X-ray magnetic circular dichroism (XMCD), ferromagnetic resonance, or neutron scattering in both elastic and inelastic modes. Elastic neutron scattering gives information about the magnetic ground-state structure, whether the material is ferromagnetic, antiferromagnetic, or has a more complicated, non-collinear arrangement of the moments. Inelastic scattering probes both energy and momentum transfer and can therefore be used to map out the excitation spectra of the material, for example, magnon dispersion. However, for low dimensional magnetic systems such as ultrathin magnetic film and nanostructures, neutron scattering experiments are not as useful as they are for bulk, due to the fact that rather large samples are required. For that reason, measurements of magnetic excitations were for a long time limited to bulk materials, until it was shown that spin polarized electron energy loss spectroscopy (SPEELS) can detect surface and interface magnons (Vollmer et al., 2003; G. Tang and Nolting, 2007; Prokop et al., 2009; Zakeri, 2014). A review has recently been published that describes how measured magnon spectra of surface and interface systems compare to theoretical results from atomistic spin dynamics (Etz et al., 2015). Moreover, recent experimental developments in pump-probe techniques (Beaurepaire et al., 1996), using ultra fast high intensity lasers (possibly in combination with XMCD), allow for magnetization measurements in the sub-picosecond regime.

The development of the theoretical side of atomistic spin dynamics simulations, as outlined in Chapters 1–7, has hence its counterpart in an equally exciting and rapid development on the experimental side. This part of the book covers some of these experimental developments and how they describe the dynamical aspects of magnetic materials. Whenever possible, we will compare measured results to the theory described in Chapters 1–7 of this book.

8

Ferromagnetic Resonance

In Parts 1 and 2 we covered the theoretical aspects of magnetism and magnetization dynamics, as well as practical aspects of implementation of the stochastic Landau–Lifshitz equation in efficient software. In this chapter, we focus on the most natural and frequently used experimental method for studying magnetization dynamics, namely ferromagnetic resonance (FMR). This experimental technique has evolved into a powerful method for studies of magnetization dynamics in materials. It is, by far, the most common method for extracting damping parameters in materials and is also a reliable technique for estimating the precession frequencies of magnetic systems, leading to detection of the magnetic g-factor, magnetic anisotropy, and saturation magnetism. The method was first experimentally described by Griffiths (1946) and the theoretical formalism was developed by Kittel (1948) and Suhl (1955) in 1948 and 1955, respectively, and comprehensively summarized by Vonsovskii (1966). The main idea behind FMR is that the interplay between static fields and time-dependent fields affects the magnetic moments in the sample. As already pointed out in Chapter 4, when a static external magnetic field **B** is applied to a ferromagnet, it causes the magnetization **M** to align itself to the field direction. If a weak, time-dependent microwave field **b** is applied in addition, the perpendicular components to the magnetization will exert a torque that can cause magnetization precession, so that the energy of the microwave field is absorbed. When the frequency of the microwave field matches the natural frequency of the system, the microwave absorption is maximized; this is the principle underlying FMR. The absorption spectra when varying the frequency is observed as a Lorentzian peak and, as is described in this chapter, from the linewidth of the peak, it is possible to extract material properties like Gilbert damping, the effective Landé g-factor, and anisotropy fields.

8.1 Experimental set-up and demagnetization field

A typical experimental set-up is schematically illustrated in Fig. 8.1. The sample holder is positioned between electromagnets that provide a static magnetic field across the sample. In order to access anisotropy fields, it is useful to have a sample holder that can be rotated. The microwave source directs the microwaves along a waveguide to the sample holder, and the detector monitors the absorption. FMR experiments are performed

Atomistic Spin Dynamics. Olle Eriksson, Anders Bergman, Lars Bergqvist, Johan Hellsvik. First Edition.

Figure 8.1 *Schematic illustration of a typical set-up for ferromagnetic resonance experiments. The electromagnets generate a static magnetic field (B) along the magnetic moment while an alternating microwave field (b) is produced perpendicular to the moment, causing precession. The frequency of the microwave field is varied until resonance is found; M, magnetization.*

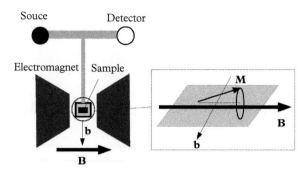

either under constant microwave frequency while varying the strength of the static the field (field-swept meausurement, giving under field linewidth) or a constant field while varying the microwave frequency (frequency-swept measurement, giving the frequency linewidth). Each of the methods complement the other and has its pros and cons, which depend on the material property of interest, as is very well described in more detail by Farle et al. (2013).

A demagnetization field $\mathbf{B_d}$, described in Section 7.2.2 and which gives rise to shape anisotropy, plays an important role in FMR experiments, since the samples in those experiments are macroscopically sized. The field is often written as (Chikazumi, 1997)

$$\mathbf{B_d} = -\mu_0 \widetilde{N} \cdot \mathbf{M}, \tag{8.1}$$

where $\widetilde{N}$ is the demagnetization tensor, and μ_0 is the magnetic constant. Note that, here, $\mathbf{M}$ is the total magnetization of the sample and not the individual moment size. For an ellipsoidal sample with principle axes coinciding with the coordinate axis, the demagnetization tensor is diagonal:

$$\widetilde{N} = \begin{pmatrix} N_x & 0 & 0 \\ 0 & N_y & 0 \\ 0 & 0 & N_z \end{pmatrix}. \tag{8.2}$$

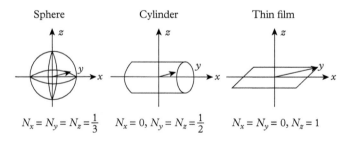

Figure 8.2 *Demagnetization factors for sphere, cylinder, and thin film geometries.*

Typical values for the demagnetization factors N_x, N_y, and N_z for common geometries are illustrated in Fig. 8.2. For practical reasons, most experimental work is performed in thin film geometry and, in this geometry, the only contribution to the demagnetization occurs for magnetization components perpendicular to the film plane, that is, in the z-direction.

8.2 Kittel equations

We now want to establish the resonance condition of a ferromagnet in various set-up types. The theory was developed by Kittel (1948), and an important remark is that the magnetization in this work is treated as a classical macro spin, to simplify the analysis, ignoring atomistic and quantum effects. This approximation may be seen as an averaging over many atomic magnetic moments, such that one replaces $\sum_i \mathbf{m}_i(t)$ with an average property, $\mathbf{M(t)}$. The derivation here is loosely based on the derivation found in Craik (1998) and Wei (2015).

8.2.1 Damping and anisotropy ignored

The most simple set-up which still establishes the FMR resonance condition is to first ignore the damping and anisotropy of the system. The different fields acting on the ferromagnet are then the exchange field $\mathbf{B}_{ex}$, the external static field $\mathbf{B}$, the external microwave field $\mathbf{b}$, and the demagnetization field $\mathbf{B}_d$. Furthermore, if the external static field is large and there is microwave field, then the magnetization is completely aligned with the field, with saturation magnetization M_s. In this situation, the exchange field is often disregarded, since it does not contribute to the torque, and all atomic moments corotate. Without loss of generality, $\mathbf{B}$ is considered to point along the z-axis of the system, that is, $\mathbf{B} = B\hat{z}$, where $\hat{z}$ is the coordinate unit vector. The microwave field is time dependent and written as $\mathbf{b}(t) = be^{i\omega t}(\hat{x}, \hat{y}, \hat{z})$, where the frequency ω is in the gigahertz region characteristic of microwaves. With $\mathbf{b}$ assumed small in comparison with $\mathbf{B}$, the magnetization stays close to its original size and orientation, that is, the saturation magnetization M_s, but with added, small, time-dependent components such that

$$\mathbf{M(t)} = \mathbf{M} + \tilde{\mathbf{m}}(t) \approx M_s\hat{z} + \tilde{\mathbf{m}}(t) = M_s\hat{z} + \mathbf{m}e^{i\omega t}, \tag{8.3}$$

with $\mathbf{m} = m_x\hat{x} + m_y\hat{y} + m_z\hat{z}$. The magnetization corresponds to a uniform mode where all the moments are precessing in phase. The demagnetization field, Eqn (8.1), together with the demagnetization tensor, Eqn (8.2), and the magnetization gives

$$\mathbf{B}_d = -\mu_0\begin{pmatrix} N_x & 0 & 0 \\ 0 & N_y & 0 \\ 0 & 0 & N_z \end{pmatrix}\begin{pmatrix} m_x \\ m_y \\ M_s + m_z \end{pmatrix} = -\mu_0\begin{pmatrix} N_x m_x \\ N_y m_y \\ N_z(M_s + m_z) \end{pmatrix}. \tag{8.4}$$

A solution to the Landau-Lifshitz equation (without damping) is sought, and when replacing individual atomic moments with an average, $\mathbf{M}(t)$, this equation takes the form

$$\frac{d\mathbf{M}(t)}{dt} = -\gamma\mathbf{M}(t) \times \mathbf{B}_{\text{tot}}$$

$$= -\gamma\,(\mathbf{M}_s + \mathbf{m}) \times (\mathbf{B} + \mathbf{b} + \mathbf{B}_d)$$

$$= -\gamma\,(\mathbf{M}_s \times \mathbf{B} + \mathbf{M}_s \times \mathbf{b} + \mathbf{m} \times \mathbf{B} + \mathbf{m} \times \mathbf{b} + \mathbf{M}_s \times \mathbf{B}_d + \mathbf{m} \times \mathbf{B}_d)\,. \quad (8.5)$$

Each term of the vector products gives

$$\mathbf{M}_s \times \mathbf{B} = 0, \quad (8.6)$$

$$\mathbf{M}_s \times \mathbf{b} = -(M_s b_y)\hat{x} + (M_s b_x)\hat{y}, \quad (8.7)$$

$$\mathbf{m} \times \mathbf{B} = (m_y B)\hat{x} - (m_x B)\hat{y}, \quad (8.8)$$

$$\mathbf{m} \times \mathbf{b} = 0, \quad (8.9)$$

$$\mathbf{M}_s \times \mathbf{B}_d = (\mu_0 N_y m_y M_s)\hat{x} - (\mu_0 N_x m_x M_s)\hat{y}, \quad (8.10)$$

$$\mathbf{m} \times \mathbf{B}_d \approx -(\mu_0 N_z m_y M_s)\hat{x} + (\mu_0 N_z m_x M_s)\hat{y}, \quad (8.11)$$

where we have neglected the small components in the last line. The first and fourth contributions are 0 because the saturation magnetization is parallel to the external field, and the time-dependent magnetization is parallel to the microwave field. Moreover, only transversal magnetization components, m_x and m_y, change with time, a result which follows from the original assumption of a dominating static magnetic field. Collecting all terms and combining Eqns (8.5), (8.6) and (8.3), the macroscopic form of the Landau–Lifshitz equation now takes the form

$$\begin{bmatrix} \frac{dM_x(t)}{dt} \\ \frac{dM_y(t)}{dt} \end{bmatrix} = \begin{bmatrix} i\omega m_x \\ i\omega m_y \end{bmatrix} = \begin{bmatrix} \gamma M_s b_y - \gamma m_y B - \mu_0\gamma(N_y - N_z)m_y M_s \\ -\gamma M_s b_x + \gamma m_x B - \mu_0\gamma(N_z - N_x)m_x M_s \end{bmatrix}. \quad (8.12)$$

As $b \ll B$, the terms involving the microwave field can be disregarded. A non-trivial solution to Eqn (8.12) is given by the secular equation

$$\begin{vmatrix} i\omega & \gamma\left(B + \mu_0(N_y - N_z)M_s\right) \\ -\gamma\left(B - \mu_0(N_z - N_x)M_s\right) & i\omega \end{vmatrix} = 0. \quad (8.13)$$

Rearranging terms, the FMR resonance condition in the field B is finally obtained:

$$\omega_0 = \gamma\sqrt{[B + \mu_0(N_x - N_z)M_s]\,[B + \mu_0(N_y - N_z)M_s]}. \quad (8.14)$$

The resonance frequency for a sphere where all the components of $\tilde{N}$ are equal becomes

$$\omega_0 = \gamma B \quad (8.15)$$

and, for a thin film with its surface normal parallel to the z-axis, one gets

$$\omega_0 = \gamma(B - \mu_0 M_s). \tag{8.16}$$

8.2.2 Including anisotropy

Anisotropy, such as uniaxial anisotropy, gives rise to an additional field, which is referred to as the anisotropy field and which often is written in a form that is similar to that of the demagnetization field:

$$\mathbf{B}_k = -\mu_0 \tilde{K} \cdot \mathbf{M}, \tag{8.17}$$

where $\tilde{K}$ is the anisotropy tensor, which has properties similar to those of the demagnetization tensor, being diagonal for ellipsoidal samples, with its principle axis along the coordinate axis. This tensor is related to the anisotropy constants, which are described in Chapter 3, and may therefore be obtained from DFT calculations. Deriving the resonance condition by performing a treatment similar to that used in Section 8.2.1, after some algebra, one ends up with the following expression for the resonance frequency:

$$\omega_0 = \gamma\sqrt{\left\{ B + \mu_0 \left[(N_x - N_z) + (K_x - K_z) \right] M_s \right\} \left\{ B + \mu_0 \left[(N_y - N_z) + (K_y - K_z) \right] M_s \right\}}. \tag{8.18}$$

If the saturation magnetization is pointing along the easy axis, taken here as the z-axis for simplicity, the anisotropy field $\mathbf{B}_k = \left(\frac{2K_1}{M_s}\right) \hat{z}$, where K_1 is the uniaxial anisotropy constant. The only surviving element in the anisotropy tensor is then $K_z = -\frac{2K_1}{\mu_0 M_s^2}$. For a sphere, the resonance frequency with uniaxial anisotropy then becomes

$$\omega_0 = \gamma \left[B + \left(\frac{2K_1}{M_s} \right) \right]. \tag{8.19}$$

Even without an external field, resonance occurs if the sample is magnetically saturated along the easy axis. By varying the direction of the external field, it is possible using FMR to extract M_s and K_1 (and higher-order anisotropy constants), as well as the gyromagnetic ratio γ, from which the Landé g-factor can be estimated.

8.2.3 Full treatment including damping

In the analysis in Section 8.2.3, damping was neglected; thus, in the resulting FMR spectra, there was with zero linewidth. Since all real materials have damping, this is not a very realistic assumption and therefore damping needs to be factored into the analysis. As will be clear in this section, the FMR technique is perhaps the most powerful experimental tool for estimating damping, by analysing the lineshape of the resonance peak in the spectrum. In this section, the resonance condition for the general case, including damping, will be derived. However, in order to simplify the analysis, a thin film geometry will be assumed from the outset. Moreover, it is also assumed, as in the analysis in Section 8.2.3, that the external magnetic field and the anisotropy field are pointing along the surface normal of the film, that is, we assume that the material exhibits perpendicular magnetic anisotropy. The microwave field is directed in the film plane, perpendicular to the external field. The total effective field then becomes

$$\mathbf{B}_{tot} = b\hat{x} + (B + B_k + B_d)\hat{z} = b\hat{x} + (B + B_k - \mu_0 M_s)\hat{z}. \tag{8.20}$$

The solution to the macroscopic Landau–Lifshitz–Gilbert equation with damping is sought, and the resulting equation takes the following form:

$$\frac{d\mathbf{M}(t)}{dt} = -\gamma \mathbf{M}(t) \times \mathbf{B}_{tot} + \frac{\alpha}{M_s} \mathbf{M}(t) \times \frac{d\mathbf{M}(t)}{dt}. \tag{8.21}$$

The form for time-dependent magnetization is given by Eqn (8.3), and combining it with Eqn (8.20) gives the precession term as

$$-\gamma \mathbf{M}(t) \times \mathbf{B}_{tot} = -\gamma \begin{vmatrix} \hat{x} & \hat{y} & \hat{z} \\ m_x & m_y & (m_z + M_s) \\ b & 0 & (B + B_k - \mu_0 M_s) \end{vmatrix}$$

$$= -\gamma \begin{bmatrix} m_y(B + B_k - \mu_0 M_s) \\ (m_z + M_s)b - m_x(B + B_k - \mu_0 M_s) \\ -m_y b \end{bmatrix}$$

$$\approx -\gamma \begin{bmatrix} m_y(B + B_k - \mu_0 M_s) \\ M_s b - m_x(B + B_k - \mu_0 M_s) \\ 0 \end{bmatrix}. \tag{8.22}$$

The last line is approximated using the fact that $b << B$ and m_x, m_y, and $m_z << M_s$, so that higher-order (small) terms can therefore be neglected. The precession therefore only takes place in the xy-plane, that is, in the film plane. Now, the damping term needs to be taken care of and it becomes

$$\frac{\alpha}{M_s} \mathbf{M}(t) \times \frac{d\mathbf{M}(t)}{dt} = \frac{\alpha}{M_s} \begin{vmatrix} \hat{x} & \hat{y} & \hat{z} \\ m_x & m_y & (m_z + M_s) \\ \frac{dm_x}{dt} & \frac{dm_y}{dt} & \frac{dm_z}{dt} \end{vmatrix}$$

$$= \frac{\alpha}{M_s} \begin{bmatrix} m_y\frac{dm_z}{dt} - (m_z + M_s)\frac{dm_y}{dt} \\ (m_z + M_s)\frac{dm_x}{dt} - m_x\frac{dm_z}{dt} \\ m_x\frac{dm_y}{dt} - m_y\frac{dm_x}{dt} \end{bmatrix}$$

$$\approx \alpha \begin{bmatrix} -\frac{dm_y}{dt} \\ \frac{dm_x}{dt} \\ 0 \end{bmatrix}. \tag{8.23}$$

Collecting all terms and reshuffling, the equation of motion for the transversal magnetization components becomes

$$\frac{dm_x}{dt} + \gamma m_y \left(B + B_k - \mu_0 M_s\right) + \alpha \frac{dm_y}{dt} = 0, \tag{8.24}$$

$$\frac{dm_y}{dt} - \gamma m_x \left(B + B_k - \mu_0 M_s\right) - \alpha \frac{dm_x}{dt} = -\gamma M_s b. \tag{8.25}$$

Performing the time derivatives of the magnetization the equation of motion gives the form

$$i\omega m_x + \gamma m_y \left(B + B_k - \mu_0 M_s\right) + i\alpha\omega m_y = 0, \tag{8.26}$$

$$-i\omega m_y + \gamma m_x \left(B + B_k - \mu_0 M_s\right) + i\alpha\omega m_x = \gamma\mu_0 M_s h, \tag{8.27}$$

and, in matrix form,

$$\begin{pmatrix} i\omega & \gamma\left(B + B_k - \mu_0 M_s\right) + i\alpha\omega \\ \gamma\left(B + B_k - \mu_0 M_s\right) + i\alpha\omega & -i\omega \end{pmatrix} \begin{pmatrix} m_x \\ m_y \end{pmatrix} = \begin{pmatrix} 0 \\ \gamma\mu_0 M_s h \end{pmatrix}. \tag{8.28}$$

In order to tidy up the notation and make everything more transparent, the following variables are introduced:

$$\omega_B = \gamma(B + B_k), \tag{8.29}$$

$$\omega_M = \gamma\mu_0 M_s. \tag{8.30}$$

Consequently, Eqn (8.28) takes the form

$$\begin{pmatrix} i\omega & (\omega_B - \omega_M) + i\alpha\omega \\ (\omega_B - \omega_M) + i\alpha\omega & -i\omega \end{pmatrix} \begin{pmatrix} m_x \\ m_y \end{pmatrix} = \begin{pmatrix} 0 \\ \omega_M h \end{pmatrix} \tag{8.31}$$

or, similarly,

$$\frac{1}{\omega_M} \begin{pmatrix} (\omega_B - \omega_M) + i\alpha\omega & -i\omega \\ i\omega & (\omega_B - \omega_M) + i\alpha\omega \end{pmatrix} \begin{pmatrix} m_x \\ m_y \end{pmatrix} = \begin{pmatrix} h \\ 0 \end{pmatrix}. \tag{8.32}$$

We next write $\mathbf{h} = (h, 0)$, and $\mathbf{m} = (m_x, m_y)$. A susceptibility tensor $\tilde{\chi}$ is then introduced that measures the magnetic response such that

$$\mathbf{m} = \tilde{\chi}\mathbf{h}, \tag{8.33}$$

in compact form and by elements

$$\begin{pmatrix} m_x \\ m_y \end{pmatrix} = \begin{pmatrix} \chi_{xx} & \chi_{xy} \\ \chi_{yx} & \chi_{yy} \end{pmatrix} \begin{pmatrix} h \\ 0 \end{pmatrix}. \tag{8.34}$$

We next invert the matrix equation, Eqn (8.32). The susceptibility tensor is, after some algebra, finally obtained:

$$\tilde{\chi} = \frac{\omega_M}{\left[(\omega_B - \omega_M)^2 - \omega^2(1 + \alpha^2)\right] + i\left[2\alpha\omega(\omega_B - \omega_M)\right]} \begin{bmatrix} (\omega_B - \omega_M) + i\alpha\omega & i\omega \\ -i\omega & (\omega_B - \omega_M) + i\alpha\omega \end{bmatrix}.$$
(8.35)

Of most interest is the response of the transversal components, that is the diagonal part of the susceptibility tensor, for instance, χ_{xx}. The susceptibility tensor is a complex quantity and, in order to extract more information from it, dividing it into real and imaginary parts is helpful:

$$\mathrm{Re}(\chi_{xx}) = \frac{\omega_M(\omega_B - \omega_M)\left[(\omega_B - \omega_M)^2 - \omega^2(1 - \alpha^2)\right]}{\left[(\omega_B - \omega_M)^2 - \omega^2(1 + \alpha^2)\right]^2 + 4\alpha^2\omega^2(\omega_B - \omega_M)^2},$$
(8.36)

$$\mathrm{Im}(\chi_{xx}) = -\frac{\alpha\omega\omega_M\left[(\omega_B - \omega_M)^2 + \omega^2(1 + \alpha^2)\right]}{\left[(\omega_B - \omega_M)^2 - \omega^2(1 + \alpha^2)\right]^2 + 4\alpha^2\omega^2(\omega_B - \omega_M)^2}.$$
(8.37)

The simulated frequency dependence of the real and imaginary parts of the susceptibility is displayed in Fig. 8.3, for typical parameters. The imaginary part has its maximum at the resonance frequency where the real part is 0 and, from this condition, the resonance

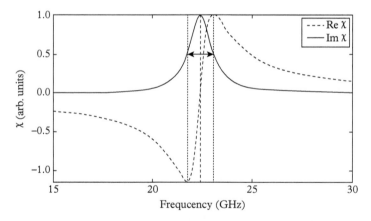

Figure 8.3 *Simulated ferromagnetic resonance spectra in a thin film geometry with an external magnetic field perpendicular to the film, from the real and the imaginary parts of the susceptibility χ, using the following parameters: $\mu_0 M_s = 0.9$ T; $B = 0.1$ T; $B_k = 0.0$ T; and $\alpha = 0.03$. The resonance frequency is marked with a dashed line, and the corresponding frequency linewidth is marked with dotted lines and the double arrow; α, damping factor; μ_0, magnetic constant; B, magnetic field; B_k, anisotropy field, M_s, saturation magnetization.*

frequency ω_0 is obtained as

$$\omega_0^2 = \frac{(\omega_B - \omega_M)^2}{1 - \alpha^2}. \tag{8.38}$$

As an example, setting damping to 0, neglecting anisotropy, so $B_k = 0$, and using the definitions in Eqns (8.29) and (8.30), the resonance frequency is $\omega_0 = \gamma(B - \mu_0 M_s)$, so that the results of Eqn (8.16) are recovered in the case when the external field is aligned perpendicularly to the film.

Around the resonance, the imaginary part has a Lorentzian form where the full width half maximum is proportional to the damping α. More precisely, the angular frequency linewidth $\Delta\omega$ or the frequency linewidth Δf of the susceptibility becomes

$$\alpha = \frac{\Delta\omega}{2(\omega_B - \omega_M)} = \frac{\Delta\omega}{2\gamma(B + B_k - \mu_0 M_s)} = \frac{2\pi\Delta f}{2\gamma(B + B_k - \mu_0 M_s)}. \tag{8.39}$$

Using the parameters from Fig. 8.3, the frequency linewidth is calculated as 1.4 GHz. The linewidth is directly proportional to the external field and the damping. Thus, the damping can be easily extracted by varying the external field, and a linear fit of the linewidth then gives the damping parameter. In fact, the experimental damping parameters shown in Figs 6.5 and 6.7 were detected from FMR measurements.

If the geometry is changed such that the external magnetic field is instead aligned in the plane of the film, along an easy axis, the resonance condition will change. As in Eqn (8.3), the magnetization will consist of the saturation magnetization M_s, which now is pointing in, for example, the $\hat{x}$-direction, with an added, small, time-dependent component from the microwave field:

$$\mathbf{M(t)} = \mathbf{M} + \tilde{\mathbf{m}}(t) \approx M_s\hat{x} + \tilde{\mathbf{m}}(t) = M_s\hat{x} + \mathbf{m}e^{i\omega t}. \tag{8.40}$$

The total field will consist of the in-plane external field and the anisotropy field, both of which lie along the $\hat{x}$ direction, an in-plane microwave field that is applied such that it is in the $\hat{y}$ direction, and a perpendicular demagnetization field, yielding

$$\mathbf{B}_{\text{tot}} = (B + B_k)\hat{x} + b\hat{y} + B_d\hat{z} = (B + B_k)\hat{x} + b\hat{y} - \mu_0 m_z\hat{z}. \tag{8.41}$$

Using an analysis similar to that used for the perpendicular field case, that is, solving the Landau–Lifshitz–Gilbert equation and calculating the susceptibility tensor, the final result becomes

$$\text{Re}(\chi_{xx}) = \frac{\omega_M\left\{\omega_B(\omega_B + \omega_M)^2 - \omega^2\left[\omega_B(1 - \alpha^2) + \omega_M(1 + \alpha^2)\right]\right\}}{\left[\omega_B(\omega_B + \omega_M)^2 - \omega^2(1 + \alpha^2)\right]^2 + \alpha^2\omega^2(2\omega_B + \omega_M)^2} \tag{8.42}$$

$$\text{Im}(\chi_{xx}) = -\frac{\alpha\omega\omega_M\left[(\omega_B + \omega_M)^2 + \omega^2(1 + \alpha^2)\right]}{\left[\omega_B(\omega_B + \omega_M)^2 - \omega^2(1 + \alpha^2)\right]^2 + \alpha^2\omega^2(2\omega_B + \omega_M)^2}. \tag{8.43}$$

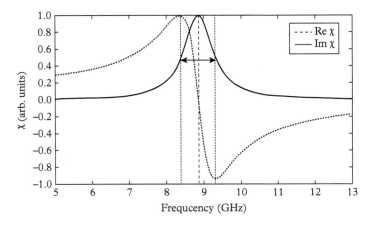

Figure 8.4 *Simulated ferromagnetic resonance spectra in a thin film geometry with an external magnetic field in plane along the easy axis of the film, from the real and the imaginary parts of the susceptibility χ, using the following parameters: $\mu_0 M_s = 0.9\,T$; $B = 0.1\,T$; $B_k = 0.0\,T$; and α= 0.03. The resonance frequency is marked with a dashed line, and the corresponding frequency linewidth is marked with dotted lines and the double arrow.*

In Fig. 8.4, the real and imaginary parts of the susceptibility tensor are shown as functions of frequency, in the case of an in-plane magnetic field. It is immediately noted that the resonance frequency in this case is lower than that for a perpendicular magnetic field (Fig. 8.3). The demagnetization field has a minimum when the magnetization is oriented in plane. If the applied external field is also oriented in plane, then there is no barrier to overcome, as in the case of a perpendicularly applied field, so the resonance becomes lower.

The conditions for resonance in the case of an in-plane magnetic field are extracted from the root of the real part of the susceptibility tensor:

$$\omega_0^2 = \frac{\omega_B(\omega_B + \omega_M)^2}{\omega_B(1 - \alpha^2) + \omega_M(1 + \alpha^2)}. \tag{8.44}$$

Setting the damping to zero and using the definitions of ω_B and ω_M as in Eqns (8.29) and (8.30), the resonance condition takes the form

$$\omega_0 = \sqrt{\omega_B(\omega_B + \omega_M)} = \gamma\sqrt{(B + B_k)(B + B_k + \mu_0 M_s)}, \tag{8.45}$$

which is normally denoted the Kittel equation in the literature. If surface anisotropy and interface anisotropy are taken into account, in a way that is relevant for thin samples with perpendicular anisotropy, M_s in Eqn (8.45) is replaced by the effective magnetization

$M_{\text{eff}} = M_s - \frac{2K_s}{M_s d}$, where K_s is the surface anisotropy constant, and d is the film thickness. The frequency linewidth then becomes

$$\Delta f = \frac{\alpha}{2\pi}(2\omega_B + \omega_M) = \frac{\alpha\gamma}{2\pi}[2(B + B_k) + \mu_0 M_s], \qquad (8.46)$$

from which the damping parameter can be extracted.

As mentioned in Section 8.1, experiments are either performed using fixed field or fixed frequency. So far, a fixed magnetic field and varying frequency has been assumed. However, from the susceptibility tensor and Eqn (8.45), it is clear that the resonance condition is fulfilled also from a fixed frequency and varying field, a set-up that is more common in practice. The dependence of the resonance frequency on the applied magnetic field, for various values of saturation magnetization and anisotropy field, is displayed in Fig. 8.5. If an anisotropy field is present, there is a resonance even in zero applied field, in which case the moments precess in the anisotropy field.

The field linewidth takes the form

$$\Delta B = \frac{2\alpha\omega_0}{\gamma} = \frac{4\pi\alpha f_0}{\gamma}, \qquad (8.47)$$

which is displayed in Fig. 8.6 for fixed M_s but different values of α and B_k. The linewidth is linear with respect to the external field in a large interval and, from the slope, the damping parameter can be extracted.

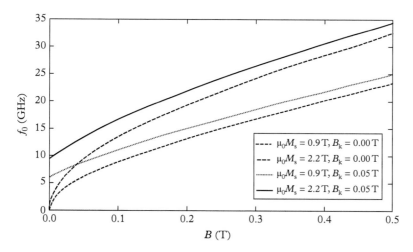

Figure 8.5 *Simulated ferromagnetic resonance frequency (f_0) in thin film geometry as a function of external magnetic field (B) in plane along the easy axis of the film for different values of $\mu_0 M_s$ and anisotropy field B_k; μ_0, magnetic constant; M_s, magnetization saturation.*

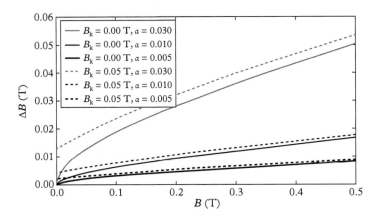

Figure 8.6 *Simulated ferromagnetic resonance field linewidth at the resonance frequency in thin film geometry as a function of external magnetic field in plane along the easy axis of the film for different values for the anisotropy field B_k and damping α. The saturation magnetization was set to $\mu_0 M_s = 0.9$ T.*

The whole analysis in this chapter is based on the assumption of uniform magnetization precession, where all moments are in phase with each other. However, in real materials and experiments, one very often observes deviations from the theoretical predictions, mostly originating from non-uniform precession and sample inhomogenities which causes frequency shift and additional linewidth broadening. The measured frequency linewidth is therefore often expressed in the form

$$\Delta f_{\text{exp}} = \Delta f + \Delta f_{\text{extrinsic}} \tag{8.48}$$

or

$$\Delta B_{\text{exp}} = \Delta B + \Delta B_{\text{extrinsic}} \tag{8.49}$$

for the field linewidth. All additional effects not taken care of in the FMR analysis are put in the extrinsic term. An important process for causing these extrinsic effects is the so-called two-magnon scattering (Arias and Mills, 1999, 2000; Mills and Arias, 2006; Lenz et al., 2006; Zakeri et al., 2007), which has its origin in local variations of the local effective fields in inhomogeneous samples, as these give rise to finite wavelength magnons in addition to the uniform mode. The uniform mode may then interact with these magnons and decay. Mills and Arias (Arias and Mills, 1999, 2000; Mills and Arias, 2006) developed a model for estimating the linewidth contribution of the two-magnon scattering:

$$\Delta B_{2M}(\omega) = \gamma \sin^{-1} \sqrt{\frac{\sqrt{\omega^2 + (\omega_0/2)^2} - \omega_0/2}{\sqrt{\omega^2 + (\omega_0/2)^2} + \omega_0/2}}, \tag{8.50}$$

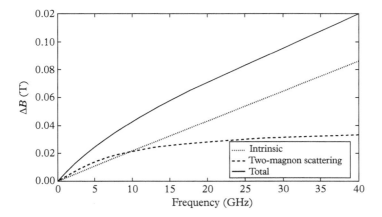

Figure 8.7 *Schematic illustration of the frequency dependence of the field linewidth in a ferromagnetic resonance experiment with an in-plane field. The two-magnon scattering strength γ was put equal to the damping α.*

where γ is the strength of the two-magnon scattering. It displays a steep slope at low frequencies and saturation at higher frequencies, as illustrated in Fig. 8.7. It is clearly an advantage to measure over a large frequency range (or a larger field range) to obtain a linear behaviour, but doing so is associated with many experimental difficulties.

In addition to extrinsic effects, in reality the microwaves have a non-uniform distribution within the sample, due to there being a finite skin depth which microwaves can penetrate. If the sample is thicker than the skin depth, the magnetic excitation is not uniform, and the FMR spectrum may consist of several peaks, the uniform mode and peaks associated with finite wavelength magnons (see Chapter 9).

8.3 The Smit–Suhl equation

A disadvantage of the FMR modelling described in Section 8.2 based on the approach by Kittel is the assumption of specific geometries, so the resonance condition needs to be recalculated for each specific condition, such as a specific set-up of the field direction. In a more general approach, developed by Smit, Beljers, and Suhl (Smit and Beljers, 1955; Suhl, 1955), no assumption of the geometry is made. The approach is, instead, based on the variational properties of the total energy. Using spherical coordinates, a general magnetization direction can be written as

$$\mathbf{m} = M_{\mathrm{s}}(\cos\phi\sin\theta\hat{x} + \sin\phi\sin\theta\hat{y} + \cos\theta\hat{z}), \tag{8.51}$$

where ϕ and θ are the spherical angles, as illustrated in Fig. 8.8. The total magnetic energy ϵ_{tot} may be expressed to be the contribution of the sum

$$\epsilon_{\mathrm{tot}} = \epsilon_{\mathrm{zee}} + \epsilon_{\mathrm{ani}} + \epsilon_{\mathrm{dem}} + \epsilon_{\mathrm{exch}}, \tag{8.52}$$

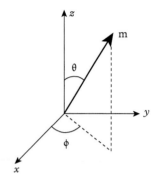

Figure 8.8 *General magnetization direction* **m**, *with the corresponding spherical angles* θ *and* ϕ.

where ϵ_{zee} is the Zeeman energy from the external magnetic field, ϵ_{ani} is the anisotropy energy, ϵ_{dem} is the demagnetization energy, and ϵ_{exch} is the exchange energy. If uniform precession is assumed, the exchange energy can be neglected. As in the analysis in Section 8.2, the resonance condition is obtained by solving the macroscopic Landau–Lifshitz–Gilbert equation and looking for periodic solutions (Smit and Beljers, 1955; Suhl, 1955). After some algebra, the end result is the following:

$$\left(\frac{\omega}{\gamma}\right)^2 = \frac{1}{m^2 \sin^2 \theta} \left[\frac{\partial^2 F}{\partial \theta^2} \frac{\partial^2 F}{\partial \phi^2} - \left(\frac{\partial^2 F}{\partial \theta \, \partial \phi} \right)^2 \right], \tag{8.53}$$

evaluated for the equilibrium angles of the magnetization. Similarly, the frequency linewidth has the general expression (Vonsovskii, 1966)

$$\Delta f = \frac{\alpha \gamma}{2\pi M_s} \left[\frac{\partial^2 F}{\partial \theta^2} + \frac{1}{\sin^2 \theta} \frac{\partial^2 F}{\partial \phi^2} \right]. \tag{8.54}$$

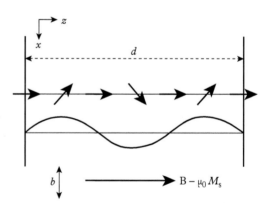

Figure 8.9 *Spin wave resonance in a thin film with thickness d. The surface moments are pinned to the surface, and the microwave field excites spin waves with an odd number of half wavelengths (three, in this case); b, microwave field; B, magnetic field; μ_0, magnetic constant; M_s, saturation magnetization. Arrows tilted out of the z-direction are thought to have a precession on a cone aligned along the z-direction.*

8.4   Spin wave resonance

In thin films, it is possible to excite finite wavelength spin waves in addition to the uniform precession, due to confinement effects through which the anisotropy field felt by the surface moments is different from that felt by the moments in the film. The surface moments may be pinned to the surface, due to the surface anisotropy, but, since the rest of the moments may be precessing, spin waves with an odd number of half wavelengths are excited within the film (see Fig. 8.9). If the thickness of the film is d, then $n(\frac{\lambda}{2}) = d$, where λ is the wavelength. At the same time, $q = \frac{2\pi}{\lambda}$, so that $q = \frac{n\pi}{d}$. With small wavevectors, the spin wave dispersion is $\sim Dq^2 = D(\frac{n\pi}{t})^2$, where D is the spin wave stiffness.

The condition for resonance follows the analysis for thin film with perpendicular field with added spin wave frequencies contributions

$$\omega_0 = \gamma(B + B_k - \mu_0 M_s) + Dq^2 = \gamma(B + B_k - \mu_0 M_s) + D\left(\frac{n\pi}{d}\right)^2. \qquad (8.55)$$

By varying the field strength and measure the resonance field against n^2, the spin wave stiffness D can be determined from a linear fit.

9

Magnons

In this chapter, we give several examples of how the multiscale approach for atomistic spin dynamics, as described in Parts 1 and 2, performs for describing magnon excitations of solids. Due to the recent experimental advancements in detecting such excitations for surfaces and multilayers, we focus here primarily on spin wave excitations of two-dimensional systems. The discussion can easily be generalized to bulk magnets and, in fact, some examples of bulk properties are given in this chapter as well. Magnons can be categorized as dipolar and exchange magnons, where the latter are in the range of terahertz frequency and are the main focus of this chapter. These are the magnons that are best suited to test the validity of the stochastic Landau–Lifshitz (SLL) equation, and the theoretical evaluation of relevant parameters from DFT. We note that, recently reviews on the experimental and theoretical aspects of the magnon excitations of thin films have been published (Zakeri, 2014; Etz et al., 2015) and some of the material presented in this chapter overlaps with these works. Before we continue, we note that studies of static or dynamic properties of magnetic thin films or nanostructures, require attention also to the substrate these films are grown on. The substrate should be preferably non-magnetic, with small or no coupling to the magnetic system, if the purpose of the study is to investigate the magnetic film alone. Another requirement is that the material should be easy to grow and stable from a structural and chemical point of view (Gradmann, 1993; Vaz et al., 2008). Cu fulfils these criteria, so it is an ideal substrate, in particular for Co, since both metals can be found in the fcc structure, with only a small lattice mismatch with each other. For this reason, in this chapter we compare results from different levels of theory and experiment, primarily for films of Co on Cu, but we also discuss Fe on Cu, Fe on W, and Fe on Ir. As this chapter shows, the atomistic spin dynamics approach reproduces measurements of magnon excitations with good accuracy, a result which shows that the calculated parameters are accurate, as well as the basic theory of the atomistic spin dynamics simulation, that is, the SLL equation.

9.1 Spin excitations in solids

A fundamental concept of condensed matter physics and properties of materials is the concept of quasiparticles (L. Landau and Lifshitz, 1980; Giulinani and Vignale, 2005).

Atomistic Spin Dynamics. Olle Eriksson, Anders Bergman, Lars Bergqvist, Johan Hellsvik. First Edition.
© Olle Eriksson, Anders Bergman, Lars Bergqvist, Johan Hellsvik 2017. First published in 2017 by Oxford University Press.

The behaviour of a complicated, many-body system often boils down and is understood from simpler models with weakly interacting or even non-interacting quasi-particles. These quasiparticles are sometimes fundamental quanta of collective excitations. Phonons represent the quanta for collective lattice vibrations, while magnons represent the quanta for spin excitations, to name the most common ones. These excitations behave as bosonic particles and therefore follow Bose–Einstein statistics.

To introduce some of the basic concepts, we start with the simple example of a ferromagnetic spin chain, as shown in Fig. 9.1. The spins interact with each other through the exchange interactions and, for simplicity, let us restrict the interactions only to nearest neighbours. Moreover, we assume that the Heisenberg Hamiltonian, introduced in Chapter 2, describes the magnetic excitation spectrum.

The only excitations that are eigenstates to the Heisenberg Hamiltonian are collective excitations; a single spin flip is not an eigenstate and is therefore not allowed. Instead, the excitation associated with the flipping of a spin is 'smeared' out over the whole spin chain, which gives rise to spin excitations in the form of a wave, creating spin waves (see the left-hand side of Fig. 9.1). The mathematical form of this wave is described below, but we note first that, according to Fig. 9.1, this wave has, in addition to an excitation energy, a characteristic wavelength, or a wavevector of reciprocal space. The relationship between excitation energy and wavevector is referred to as dispersion, and a method to evaluate it is the topic of Section 9.3.

The magnon energy dispersion relation may be obtained by evaluating the torque on each magnetic moment and solving the equation of motion for the spins using the SLL equation, Eqn (4.1). Here, we rewrite it in a form where damping is neglected and we work with spin **S** instead of moment $\mathbf{m} = -g\mathbf{S}$, where g is the Landé g-factor, and where

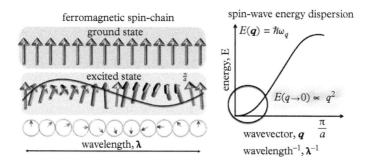

Figure 9.1 *Ferromagnetic spin chain in the ground state and an excited state. A spin flip is 'smeared' over the whole system, creating a spin wave with wavelength λ and wavevector $\mathbf{q}$. The characteristic energy dispersion $E(\mathbf{q})$ is represented on the right-hand side; a, lattice constant; $\hbar$, Plank's constant divided by 2π; ω_q, frequency. Figure adapted from Etz, C., Bergqvist, L., Bergman, A., Taroni, A., and Eriksson, O., (2014), Atomistic spin dynamics and surface magnons, Psi-k Highlight, 123, 1–34.*

we have made an explicit expression of the effective magnetic field in terms of the $\mathcal{J}_{ij}$ Heisenberg exchange parameter:

$$\frac{d\mathbf{S}_i}{dt} = \gamma \sum_j \mathcal{J}_{ij} (\mathbf{S}_i \times \mathbf{S}_j). \tag{9.1}$$

Introducing the quantum mechanical spin raising operator $S^+ = S_x + iS_y$, one can from Eqn (9.1) obtain

$$i\frac{d\mathbf{S}_i^+}{dt} = 2\mathcal{J}S \sum_j (S_i^+ - S_j^+). \tag{9.2}$$

Assuming a wave-like solution in the form $S_i^+ = A_i \exp[i(\mathbf{q} \cdot \mathbf{R}_j - \omega t)]$, that is, a magnon with wavevector $\mathbf{q}$ and frequency ω, where A_i is the amplitude of the excitation at position $\mathbf{R}_i$ (spin wave amplitude), one may evaluate the magnon energy dispersion relation from Eqn (9.2). The result is

$$E(\mathbf{q}) = \hbar\omega = 2z\mathcal{J}S \left[1 - \frac{1}{z} \sum_i \cos(\mathbf{q} \cdot \mathbf{R}_i) \right], \tag{9.3}$$

where z is the number of nearest neighbours (=2 for the spin chain). At small wavevectors, $\mathbf{q} \to 0$, the energy dispersion is proportional to q^2, which follows from Eqn (9.3) after a Taylor expansion is performed on the right-hand side. For antiferromagnetic spin chains, a similar analysis shows that the dispersion relation is instead linear at small wavevectors.

The magnons described above, that originate from the exchange interactions, are referred to as exchange magnons and are the shortest wavelength spin waves that can occur in materials and hence possess the highest frequencies (which can exceed 1 THz). However, there is a large variety of magnons over a considerable interval of momentum and frequency, as illustrated in Fig. 9.2. On the opposite length scale of exchange magnons, one finds magnons originating from coherent precession of moments, a precession that originates from long-range electrostatic dipolar interactions and magnetic anisotropy. These dipolar spin waves have wavelengths of the order of micrometres, and frequencies in the range of gigahertz. In an intermediate regime, a mixture of dipolar and exchange spin waves are present, as Fig. 9.2 shows. In addition to spin wave excitations, a metallic magnet also has an additional type of excitation, namely Stoner excitations, which arise from excitations of the dispersive electronic structure (Stoner, 1936; Slater, 1937; Stoner, 1938; van Kranendonk and Van Vleck, 1958; Gunnarsson, 1976; Gokhale and Mills, 1994; Mohn, 2006; Kübler, 2009). As discussed in Chapter 2, ferromagnets experience an energy separation of majority (spin-up) and minority (spin-down) electron bands, called exchange splitting. A possible kind of excitation for such an electronic structure is when a spin-up electron leaves its occupied state below the Fermi level while a spin-down electron enters a previously unoccupied

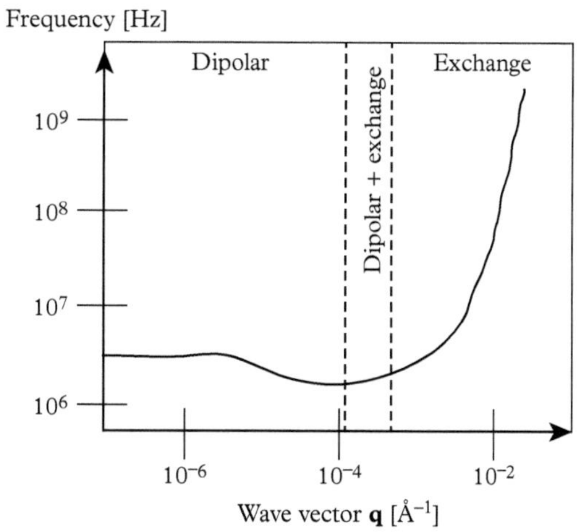

Frequency [Hz]

Figure 9.2 *Schematic classification of spin waves in a magnetic material, with corresponding ranges for the wavevector* **q** *and for frequency. Figure adapted from Zakeri, K., and Kirschner, J., (2013), 'Probing Magnons by Spin-Polarized Electrons' in S. O. Demokritov and A. N. Slavin, eds,* Magnonics: From Fundamentals to Applications, *Topics in Applied Physics, vol. 125, Springer, Berlin, Heidelberg, pp. 83–99.*

spin-down state. This creates a hole in the majority electron band, and an extra electron in the minority band, so that an electron–hole pair is formed. This is known as a Stoner excitation. These electron–hole pairs carry a total spin of 1. In addition, they also have a dispersion, and the wavevector of the Stoner excitation can take any value in the Brillouin zone (BZ; Mohn, 2003). Stoner excitations are responsible for longitudinal fluctuations of the magnetic moment. In addition, magnons that enter regions where Stoner excitations are allowed (Stoner continuum) become damped and the corresponding magnon lifetime decreases. This is a mechanism known as Landau damping (L. Landau, 1946). In Fig. 9.3, typical Stoner excitations for both a bulk and a low dimensional metallic ferromagnet are illustrated. The Stoner gap, Δ, is defined as the lowest energy of Stoner excitations. In bulk materials, only a few Stoner excitations are available at low wavevectors, and they have relatively large energy, as Fig. 9.3 shows. For this reason, magnons possess a long lifetime in this regime of wavevectors, where Landau damping is low. In metallic ultrathin ferromagnets, a wavevector perpendicular to the surface normal is not conserved and this opens up an additional path for Stoner excitations, which are more pronounced in low dimensional magnetic systems.

Most of the discussion of this chapter concerns thin film and surface magnetism, and we make therefore a short comment on the magnetism of two-dimensional systems. Strictly speaking, magnons should not exist for a single monolayer of a ferromagnetic thin film, if the interactions are symmetric. This is a consequence of the Mermin–Wagner theorem (Mermin and Wagner, 1966). However, in real systems with overlayers that are grown on top of a substrate, there is always some symmetry-breaking interaction that gives rise to magnetic order, even at finite temperatures. All the materials discussed in this chapter are in this category, and hence both experiment and theory, as we will describe in this chapter, have indeed found magnetic excitations of these two-dimensional materials.

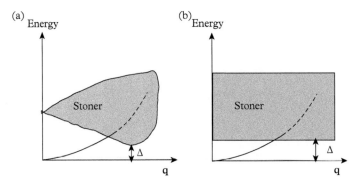

Figure 9.3 *Schematic spin wave excitation spectra with wavevector* q *for (a) bulk and (b) a low dimensional magnetic system. The Stoner continuum, separated by the Stoner gap Δ, is marked in grey where the spin waves experience Landau damping and a finite lifetime. Figure adapted from Zakeri, K., and Kirschner, J., (2013), 'Probing Magnons by Spin-Polarized Electrons' in S. O. Demokritov and A. N. Slavin, eds,* Magnonics: From Fundamentals to Applications, *Topics in Applied Physics, vol. 125, Springer, Berlin, Heidelberg, pp. 83–99.*

9.2   Experimental methods

Scattering experiments are the most common methods for probing excitations in solids. In short, they are all based by sending a beam of particles (neutrons, electrons, photons) towards the solid, and its response and the reflected beam are studied. The particles may collide with the system of interest, with conservation of energy but not momentum (elastic scattering) or with transfer of energy and momentum (inelastic scattering). The elastic scattering yields information of the underlying crystalline or magnetic structure, while inelastic scattering yields information of excitations. The loss or gain of energy or momentum of the incident particles is related to the annihilation or creation of quasiparticles, which are characteristic of the excitation of interest. For magnetic excitations in particular, inelastic scattering using neutrons, electrons, and light is dominating. These methods all probe different length scales. Neutron scattering is very common for bulk materials due to its long penetration depth. However, the relatively weak interaction between the neutron and the materials requires a large volume of the material to be investigated. Electron scattering is more popular for thin films, due to the high surface sensitivity that is possible in such experiments. If the magnons originate from dipolar interactions, it is possible to use light as a probe, via Brillouin light scattering. In Fig. 9.4, we show a schematic representation of the energy and momentum values that are possible to obtain using various experimental methods.

Magnon dispersion curves for bulk system have been measured by neutron scattering for a variety of bulk magnetic systems in the past decades; for instance, bulk Co was measured in 1960 (Sinclair and Brockhouse, 1960). Moreover, the importance of neutron scattering experiments performed by B. N. Brockhouse and C. G. Shull was

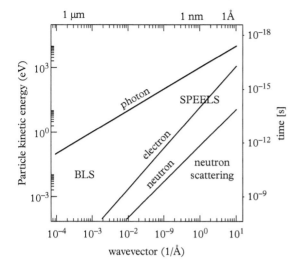

Figure 9.4 *Energy and momentum transfer range achieved in inelastic scattering experiments when using photons, electrons, or neutrons; BLS, Brillouin light scattering; SPEELS, spin polarized electron energy loss spectroscopy. Figure adapted from Etz, C., Bergqvist, L., Bergman, A., Taroni, A., and Eriksson, O., (2014), Atomistic spin dynamics and surface magnons, Psi-k Highlight, 123, 1–34; that figure, in turn, was adapted from Sandratskii, L. M., (1991), Symmetry analysis of electronic states for crystals with spiral magnetic order. I. General properties, J. Phys. Condens. Matter, 3, 8565–86.*

recognized when the two scientists were awarded the Nobel Prize in Physics in 1994. However, measuring magnon dispersion in magnetic thin films is by far a more challenging task and was first accomplished by Vollmer et al. (2003) for an eight-monolayer film of Co on top of a Cu(100) surface. The experimental tool that was used is called spin polarized electron energy loss spectroscopy (SPEELS), which we describe in the next paragraph. Subsequent works showed that it is possible to reduce the thicknesses down to a single monolayer while still obtaining a good signal (Etzkorn et al., 2004; Vollmer et al., 2004). Co films on Cu(100) are well behaved; the interactions are completely dominated by exchange, and the films are grown with good epitaxy with little interdiffusion and/or interface roughness. The magnetic ground state is ferromagnetic for all thicknesses of Co, making it a nicely behaved system to test the methodology and comparisons with experiments and theoretical calculations.

SPEELS has, in recent years, become the preferred experimental method for studying spin excitation in low dimensional magnetic materials. In this chapter, we give an elementary introduction to this method; for more details, see the excellent review articles and books by Zakeri (2014); Zakeri and Kirschner (2013); Zakeri, et al., 2013; and Ibach et al. (1996). In Fig. 9.5, a schematic figure of a typical experimental set-up is illustrated. An incoming monochromatic electron beam is focused on the sample. The electrons in the incident beam are either parallel (spin up) or antiparallel (spin down) to the majority spins of the sample. The incident electrons will interact with the sample electrons, and the reflected (scattered) electron beam is collected and its energy measured for a given geometry. Assuming that the energy and momentum of the incident and scattered electron beam are ϵ_i, $\mathbf{k}_i$, and ϵ_f, $\mathbf{k}_f$, respectively, the spin wave energy ϵ with propagation vector $\mathbf{q}$ of the sample are given by the expressions (Zakeri, 2014)

$$\epsilon = \epsilon_f - \epsilon_i \tag{9.4}$$

$$\mathbf{q} = -\Delta K_{\parallel} = \mathbf{k}_i \sin\theta - \mathbf{k}_f \sin(\theta_0 - \theta), \tag{9.5}$$

Figure 9.5 *The scattering geometry used in a spin polarized electron energy loss spectroscopy experiment, showing the magnetization* **M** *and spin wave energy E, with the propagation vector* **q** *in plane. An incoming monochromatic electron beam with known energy, momentum, and incident angle θ to the sample normal is focused on the sample, and the reflected (scattered) beam is measured for both spin orientations; the energy and momentum of the incident and scattered electron beam are* E_i, $\mathbf{k}_i$, *and* E_f, $\mathbf{k}_f$, *respectively. Figure adapted from Zakeri, Kh., (2014), Elementary spin excitations in ultrathin itinerant magnets,* Phys. Rep., 545, 47–93.

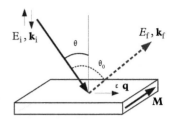

where θ is the angle of the incident beam to the sample normal, and θ_0 is the angle between reflected and incident beam. Total angular momentum has to be conserved in the experiment and thus magnon excitations can only be probed with incident electrons that are spin down with respect to the sample magnetization. A typical SPEELS spectrum will therefore have a peak in the intensity in the minority spin channel, due to magnon excitation, as function of energy loss, while the majority channel is more or less flat. Since this is a scattering experiment, a natural way to analyse it is via the differential cross section and the dynamical structure function, as we will describe in Section 9.5.

9.3   Adiabatic magnon spectra and the frozen magnon method

When theoretically treating magnons, it is common to make use of the adiabatic approximation (see Chapters 2 and 4). In this approach, the atomic moments are regarded as rigid spins, and their internal motion is assumed to be described by a Heisenberg-like Hamiltonian. The adiabatic approximation is appropriate for the class of materials with well-defined local magnetic moments. In energetic terms, the exchange splitting between the majority and minority spin channels should be large in comparison to the magnon energy. For the common transition metals Fe, Co, and Ni, the *d*-bandwidth is experimentally found to be 3.1 eV, 3.8 eV, and 3.4 eV, respectively (Eastman et al., 1980). The corresponding measured exchange splitting is 1.5 eV, 1.1 eV, and 0.3 eV, respectively. Since the magnon energies are of the order of 0.3 eV or less, it is clear that the adiabatic approximation works reasonably well for Fe and Co but should be used with caution for Ni. If the adiabatic approximation is not applicable, spin-flip Stoner excitations, causing longitudinal fluctuation of the local moment, become important even at low temperatures and/or for magnons with long wavelengths.

In the frozen magnon method (Sandratskii, 1991 a, b; Rosengaard and Johansson, 1997; Halilov et al., 1998; Brown et al., 1999; Niu et al., 1999; Pajda et al., 2000; Pajda et al., 2001; Grotheer et al., 2001; Udvardi et al., 2003; Sandratskii and Bruno, 2006; Kübler, 2009; Ležaić et al., 2013), the total energy of spin waves is calculated directly using density functional theory. Practically, this is done by assuming a magnetic configuration of the spin-spiral type, where the magnetic moment $\mathbf{m}_i = m(\cos(\mathbf{q} \cdot \mathbf{R}_i) \sin \theta, \sin(\mathbf{q} \cdot \mathbf{R}_i) \sin \theta, \cos \theta)$, where $\mathbf{q}$ is the wavevector of the spin spiral (the propagation vector), θ is the azimuthal angle between the magnetic moment and the spin-spiral propagation vector, and $\mathbf{R}_i$ is the position vector of the magnetic moment at site i. Using the Heisenberg Hamiltonian and comparing the total energy of the spin spiral $E(\mathbf{q}, \theta)$ with a ferromagnetic reference state, one arrives, after some algebraic manipulations, at the following expression for the excitation energy:

$$\Delta E(\mathbf{q}, \theta) = \sum_{j \neq 0} \mathcal{J}_{0j} [1 - \exp(i\mathbf{q} \cdot \mathbf{R}_{0j})] \sin^2 \theta, \tag{9.6}$$

where $\mathbf{R}_{0j} = \mathbf{R}_0 - \mathbf{R}_j$, and $\Delta E(\mathbf{q}, \theta) = E(\mathbf{q}, \theta) - E(0, \theta)$. This expression is conveniently reorganized into

$$\Delta E(\mathbf{q}, \theta) = (\mathcal{J}(0) - \mathcal{J}(\mathbf{q})) \sin^2 \theta, \tag{9.7}$$

where $\mathcal{J}(\mathbf{q})$ is the Fourier transform of the Heisenberg exchange parameters $\mathcal{J}_{ij}$, which was introduced in Chapter 2, that is,

$$\mathcal{J}(\mathbf{q}) = \sum_{j \neq 0} \mathcal{J}_{0j} \exp(i\mathbf{q} \cdot \mathbf{R}_{0j}). \tag{9.8}$$

Reciprocally, the exchange interactions in a real space representation may be obtained by the inverse Fourier transform of $\mathcal{J}(\mathbf{q})$, that is,

$$\mathcal{J}_{0j} = \frac{1}{N} \sum_{\mathbf{q}} \mathcal{J}(\mathbf{q}) \exp(-i\mathbf{q} \cdot \mathbf{R}_{0j}). \tag{9.9}$$

Hence, it is possible to evaluate the exchange parameters in a q-space representation by using Eqn (9.7) in combination with Eqn (9.8). It is also possible to obtain a real space representation of the exchange interaction, either directly from Eqn (2.23) or from Eqn (9.9). Independently of the approach, the exchange parameters should ideally be the same, since they are the result of the same basic electronic structure. A comparison of exchange parameters is hence of interest. In Fig. 9.6, we compare for body-centred cubic (bcc) Fe, exchange interactions calculated via the frozen magnon method to those obtained via Eqn (2.23). The two different approaches give more or less the same set of exchange parameters and the small differences are attributed to different implementation and/or numerical aspects of the calculations.

Once the Fourier transform of the exchange interactions, $\mathcal{J}(\mathbf{q})$, or the spin-spiral total energy differences are obtained, the adiabatic magnon spectrum can be obtained. Before generalizing the expressions to an arbitrary number of atoms, for simplicity, we first give the expressions in the case of a single atom in the primitive cell:

$$\omega(\mathbf{q}) = \frac{4}{M} [\mathcal{J}(0) - \mathcal{J}(\mathbf{q})], \tag{9.10}$$

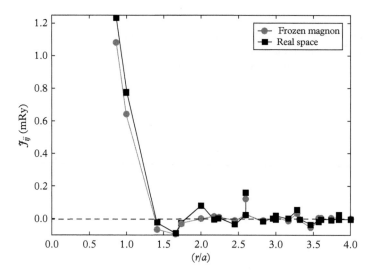

Figure 9.6 *Exchange interactions $\mathcal{J}_{ij}$, of body-centred cubic Fe as a function of distance (given in terms of position r divided by the lattice constant a); interactions obtained via the frozen magnon method by using Eqn (9.9) (curve labelled 'Frozen magnon') are compared to those obtained via Eqn (2.23) (curve labelled 'Real space'). Units in milli-Rydbergs.*

and

$$\omega(\mathbf{q}) = \lim_{\theta \to 0} \frac{4}{M} \frac{\Delta E(\mathbf{q}, \theta)}{\sin^2 \theta}. \tag{9.11}$$

Note that this is the same expression, although with slightly different notation, as the magnon dispersion expression for the ferromagnetic chain, Eqn (9.3). Hence, Eqns (9.7) and (9.11) can be used in density functional theory (DFT)-based theories to evaluate the total energy and magnon dispersion of, in principle, any material. The generalization to an arbitrary number of N atoms in the unit cell is straightforward; two additional indices α, β, denoting atom type, need to be introduced such that Eqn (9.8) changes into

$$\mathcal{J}^{\alpha\beta}(\mathbf{q}) = \sum_{j \neq 0} \mathcal{J}^{\alpha\beta}_{0j} \exp(i\mathbf{q} \cdot \mathbf{R}_{0j}), \tag{9.12}$$

where $\mathcal{J}^{\alpha\beta}_{0j}$ is the exchange interaction between atom types α and β. The spin wave energies will then be given by the eigenvalues of a general $N \times N$ matrix (Anderson, 1963; Halilov et al., 1998), here expressed in block form:

$$4 \begin{bmatrix} \frac{1}{M_\alpha} \sum_\gamma^N \left(\mathcal{J}^{\alpha\gamma}(0) - \mathcal{J}^{\alpha\alpha}(\mathbf{q}) \right) & -\frac{\mathcal{J}^{\alpha\beta}(\mathbf{q})}{M_\alpha} \\ -\frac{\mathcal{J}^{\alpha\beta}(\mathbf{q})^*}{M_\beta} & \frac{1}{M_\beta} \sum_\gamma^N \left(\mathcal{J}^{\gamma\beta}(0) - \mathcal{J}^{\beta\beta}(\mathbf{q}) \right) \end{bmatrix}. \tag{9.13}$$

We leave it to the reader to rewrite the matrix in the case when spin-spiral energies, Eqn (9.7), are used as input instead of the Fourier transform of the exchange inter- actions. It is worth noting that evaluating the total energy for spin-spiral configurations, $E(\mathbf{q}, \theta)$, in the frozen magnon method can be done in two different ways: either by self- consistent calculations or by employing the force theorem (Mackintosh and Andersen, 1975; Dederichs et al., 1984). Self-consistent calculations have the advantage that the size of the magnetic moments is allowed to relax at the expense of being much heavier than those in calculations based on the force theorem as, in the latter, the potential and size of the moments are 'frozen', and only the sum of the occupied eigenvalues is needed to obtain the excitation energy.

Direct calculations of spin-spiral configurations, and from these the evaluation of magnon dispersions, have been extensively employed for bulk systems (Sandratskii et al., 2007; Kübler, 2009; Etz et al., 2015). However, significantly fewer studies on low dimensional systems, such as magnetic thin films, have been published. In Fig. 9.7 we give an example of a calculation or a single magnetic monolayer of Cr and for a single magnetic monolayer of Mn, both on top of a Cu(111) substrate (from Kurz et al., 2001). The magnetic moment of a Cr monolayer is very dependent on the wavevector, unlike that of Mn; this result suggests that Cr is not well described by a standard Heisenberg Hamiltonian. By scanning the wavevectors in the two-dimensional BZ and calculating the total energy, the magnetic ground-state configuration can be determined and it was found to be a non-collinear state for both Cr and Mn. Figure 9.7 shows that the ground state for Cr is a spin-spiral state with a pitch vector coinciding with the $\bar{\mathrm{K}}$ point of the BZ. For Mn on Cu (111), the pitch vector of the stable configuration coincides with the $\bar{\mathrm{M}}$ point. Similar studies have also been performed on other magnetic thin

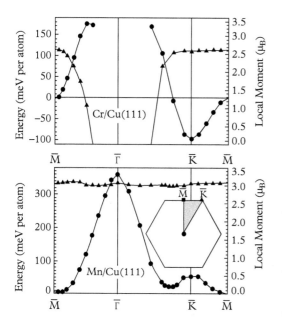

Figure 9.7 *Total energies (circles) and mag- netic moments (triangles) for spin-spiral states along high-symmetry lines in the two- dimensional Brillouin zone (inset), obtained via the frozen magnon method, for (upper panel) a Cr monolayer and (lower panel) an Mn monolayer, both on top of a Cu(111) sub- strate; $\bar{\mathrm{M}}$, $\bar{\Gamma}$, and $\bar{\mathrm{K}}$ are high-symmetry points of the Brillouin zone. Reprinted figure with permission from Kurz, Ph., Bihlmayer, G., Hirai, K., and Blügel, S., Phys. Rev. Lett., 86, 1106, 2001. Copyright 2001 by the American Physical Society.*

film systems, which show complex, non-collinear magnetic structures (Kubetzka et al., 2005; Bode et al., 2007; Kudrnovský et al., 2009; Deák et al., 2011; Heinze et al., 2011; Zimmermann et al., 2014).

A more common approach for obtaining adiabatic magnon dispersion is to first calculate exchange interactions in real space, from Eqn (2.23), and from their Fourier transforms, used in Eqn (9.13), obtain the dispersion relationship. As a demonstration of this method, we consider a system of eight monolayers of Co on top of Cu(100). The thin film system of eight monolayers of Co on top of Cu(001) has historical importance, since it was the first system where magnon dispersion was successfully measured using the SPEELS technique. According to Eqn (9.13), eight eigenvalues for each wavevector, **q**, are expected and these are displayed in Fig. 9.8, together with experimental data from Vollmer et al. (2003). It may be noted that experiments only pick up the acoustic branch and that the dispersion relation for this magnon agrees rather well with the calculated acoustic branch, although there is room for improvement. Since the experimental data in Fig. 9.8 were taken at room temperature, whereas theory in Fig. 9.8 does not have temperature effects at all, it is likely that this is a major reason for the disagreement. As we will see in Section 9.5, finite temperature effects improve on the comparison between experiment and theory. The reason for why experiments do not detect any of the optical modes in Fig. 9.8 is a matter of debate and will be discussed in more detail in Section 9.5 as well. We note, however, that, in any experiment, Landau damping due to Stoner excitations will destroy the magnon dispersion at large **q** values, so that any excitation with a **q**-vector close to the BZ boundary will be difficult to detect. However, excitations close to the centre of the BZ should be detectable.

More information about magnon excitation can be obtained from the eigenvectors that correspond to the eigenvalues of Eqn (9.13). As an example, the contribution from

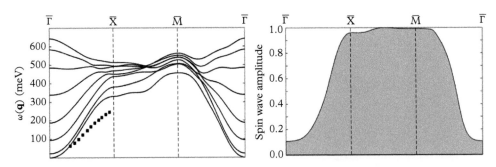

Figure 9.8 *(Left) Calculated adiabatic magnon spectra ω(**q**) for Co₈/Cu(001) (lines), and magnons from experimental spin polarized electron energy loss spectroscopy data (squares) from Vollmer, R., Etzkorn, M., Anil Kumar, P. S., Ibach, H., and Kirschner, J., (2003), Spin-polarized electron energy loss spectroscopy of high energy, large wave vector spin waves in ultrathin fcc Co films on Cu(001), Phys. Rev. Lett., 91, 147201. (Right) Contribution from the interface Co layer to the spin wave amplitude of the acoustic branch; $\bar{\Gamma}$, $\bar{X}$, and $\bar{M}$ are high-symmetry points of the Brillouin zone.*

each layer to the amplitude of each of the spin wave branches is obtained by taking the square of the corresponding eigenvector coefficient. In the right panel of Fig. 9.8, we show the contribution of the interface layer, that is, the Co layer adjacent to Cu, to the lowest lying acoustic branch. At the Γ-point, all layers contribute equally due to the uniform precession. However, for wavevectors away from the Γ-point, the interface layer has a significantly larger contribution and, in particular, along the X–M edge line, where it completely dominates. In addition, the eigenvectors give information about the magnetic state and the internal orientation between the moments for a given wavevector. As an example, the eight possible magnetic configurations corresponding to the eight magnon branches at the Γ-point are visualized in Fig. 9.9. In this figure, the in-plane (m_x) component of magnetic moment is displayed for each atom and magnon branch. In the lowest lying branch, which is the acoustic branch, denoted by '1', all moments have similar in-plane magnetization, that is, a ferromagnetic state. If one moves into the second branch, that is, the first optical branch, which is denoted by '2', then the directions of the interface atom and surface atom have opposite direction of their in-plane magnetization, that is, they are antiferromagnetically coupled and, overall, the eight atoms form a spiral state. If one continues higher in energy, more and more layers become antiferromagnetic coupled and, in the highest lying branch, which is denoted by '8', each layer is antiferromagnetically coupled to the adjacent layers.

A second example of magnons in reduced dimensions is the measured excitation spectrum of two Fe layers on top of W(001), as illustrated in Fig. 9.10. The bottom

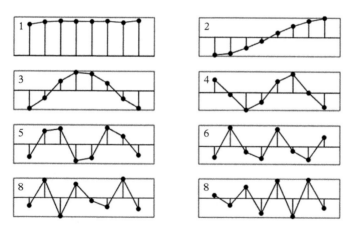

Figure 9.9 *Magnetic configurations for Co$_8$/Cu(001) at the Γ-point for each of the magnon branches, which are denoted 1–8, where 1 is the lowest lying acoustic branch, and 8 is the highest lying optical branch. In each figure, the leftmost atom is the interface atom, and the rightmost atom is the surface atom. In each panel, the in-plane magnetization is plotted for each of the eight Co atoms, which are represented by the black circles.*

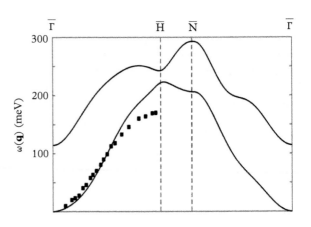

Figure 9.10 *Adiabatic magnon spectra* $\omega(\mathbf{q})$ *for Fe2/W(110), together with experimental spin polarized electron energy loss spectroscopy data from Tang, W. X., Zhang, Y., Tudosa, I., Prokop, J., Etzkorn, M., and Kirschner, J., (2007). Large wave vector spin waves and dispersion in two monolayer Fe on W(110),* Phys. Rev. Lett., **99**, *087202.* $\bar{\Gamma}$, $\bar{H}$, *and* $\bar{N}$ *denote high-symmetry points in the body-centred cubic (110) two-dimensional Brillouin zone.*

Fe layer is denoted the interface layer, while the upper Fe layer is denoted the surface layer. In contrast to the $Co_8/Cu(001)$ system, where there is almost no induced magnetic moment on Cu, the W substrate hybridizes much more strongly and obtains a finite induced magnetic moment. However, here, this induced moment will be neglected, except from its influence on the Heisenberg exchange parameters, so that we consider in the spin Hamiltonian only two Fe layers. According to Eqn (9.13), two eigenvalues for each wavevector $\mathbf{q}$ are expected, and they are displayed in Fig. 9.10, together with experimental data from W. Tang et al. (2007). Experiments only report magnon dispersion for the acoustic branch, something we will return to in Sections 9.4 and 9.5. For now, we note that the agreement between experiment and theory is seen to be very good, especially for long wavelength excitations. Overall, comparing experimental and theoretical magnon excitations, as done in Figs 9.8 and 9.10, is the most natural way to validate the accuracy of the calculated Heisenberg exchange parameters. Although the comparison is done in reciprocal space, it gives indirect information about the real space exchange parameters and, as Figs 9.8 and 9.10 show, theory does a rather good job at evaluating such parameters. We will discuss this further, in connection to finite temperature effects, in Section 9.5.

As in the analysis of the $Co_8/Cu(001)$ system, additional information is provided by the eigenvectors. More specifically, the eigenvectors give the internal magnetic states of the layers, and the contribution from each of the layers to the spin wave amplitude, as shown in Fig. 9.11. In the left panel of the figure, the contribution from the interface layer is displayed, while the contribution from the surface layer is displayed in the right panel. At the $\bar{\Gamma}$-point, each layer contributes equally to the two branches, and the two possible magnetic states correspond to a ferromagnetic state and an antiferromagnetic state (not shown), respectively. The most pronounced differences between the layers can be found between the $\bar{H}$–$\bar{N}$ line and the $\bar{N}$–$\bar{\Gamma}$ line. Along the $\bar{H}$–$\bar{N}$ line, the higher-lying branch represents a pure interface mode, while the lower branch only has a contribution from the surface layer, as is also the case for most of the $\bar{N}$–$\bar{\Gamma}$ line.

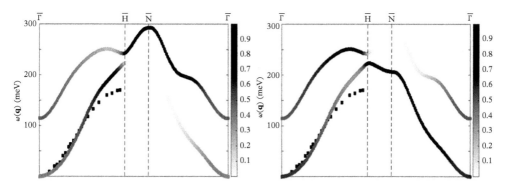

Figure 9.11 *Contribution to the spin wave amplitude of an Fe bilayer on W(110) from the interface (left) and from the surface Fe layer (right). Predicted values are compared to experimental spin polarized electron energy loss spectroscopy data (marked in squares) from Tang, W. X., Zhang, Y., Tudosa, I., Prokop, J., Etzkorn, M., and Kirschner, J., (2007), Large wave vector spin waves and dispersion in two monolayer Fe on W(110), Phys. Rev. Lett., 99, 087202; ω, frequency; q, wavevector; Γ̄, H̄, and N̄ are high-symmetry points of the Brillouin zone.*

9.4 Dynamical magnetic susceptibility

An important quantity for the description of magnetic excitations is the dynamical magnetic susceptibility $\chi(\omega, \mathbf{q})$ (Edwards, 1962, 1967; Callaway, 1968; Edwards and Rahman, 1978; Cooke et al., 1980; Edwards and Muniz, 1985; H. Tang et al., 1998; Savrasov, 1998; Aryasetiawan and Karlsson, 1999; Buczek et al., 2009; Lounis et al., 2010; Şaşıoğlu et al., 2010). Since it is both frequency and wavevector dependent, both collective magnon and Stoner excitations are, in principle, properly included in the description. The poles of the dynamic susceptibility are associated with the magnon excitations, where the real part gives the energy of the magnon while the imaginary part is associated with the lifetime of the excitation. The (transverse) dynamical susceptibility may be written in the following form:

$$\chi(\omega, \mathbf{q}) = \frac{A(\mathbf{q})}{\omega - \epsilon(\mathbf{q})/\hbar + i\Delta(\mathbf{q})}, \tag{9.14}$$

where $A(\mathbf{q})$ has a direct relation to the amplitude of the magnon wave function, $\epsilon(\mathbf{q})$ is the magnon energy, and $\Delta(\mathbf{q})$ is the inverse lifetime. In order to calculate $\chi(\omega, \mathbf{q})$, it is necessary to solve a Dyson equation, here written as

$$\chi(\omega, \mathbf{q}) = \frac{\chi_0(\omega, \mathbf{q})}{1 - U\chi_0(\omega, \mathbf{q})}, \tag{9.15}$$

where $\chi_0(\omega, \mathbf{q})$ is the so-called Kohn–Sham–Lindhardt susceptibility for non-interacting electrons (Kübler, 2009), obtained from the eigenvalues and eigenvectors of the Kohn–Sham equation, and U is the exchange–correlation kernel describing the electron correlations in the system. This kernel is often assumed to be local in space and time, using the so-called adiabatic version of the local density approximation, which was discussed in Chapter 1. Maybe the most complete method for computing $\chi(\omega, \mathbf{q})$ with the least amount of approximations is based on time-dependent DFT within the linear response, at $T = 0$ K, as this approach provides a parameter-free description of the excitations. Due to the computational expense of these calculations, the formalism was applied only to bulk systems until recently, when algorithmic and computer advances have allowed calculations on low dimensional systems. As a typical example of what can be achieved with this method (Buczek et al., 2011), the calculated spin wave dispersions for a single Fe monolayer (in the (100) or (110) orientation), both when the layer is free standing and when it is on top of a Cu (100) or W(110) substrate, are displayed in Fig. 9.12. From the results, it is clear that the strength of Stoner excitations via the Landau damping, as revealed by the broadening of the spectra, is strongly dependent on the substrate and its hybridization with the magnetic thin film. Moreover, the Stoner excitation becomes important for high energies and/or large wavevector transfer, in agreement

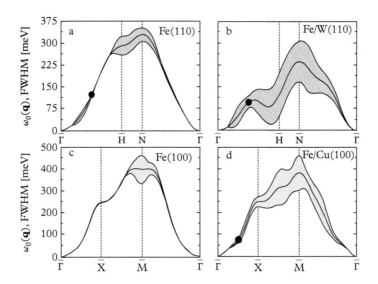

Figure 9.12 *Spin waves of Fe monolayer thin films. Results for (a) a free-standing film in the (110) surface orientation, (b) an Fe monolayer on top of a W(110) substrate, (c) a free-standing Fe film in the (100) surface orientation, and (d) an Fe monolayer on top of a Cu(100) substrate. The thick line denotes the dispersion relation, and the shaded area indicates the broadening due to Landau damping (full width half maximum (FWHM) of the corresponding spectral density); ω_0, resonance frequency; $\mathbf{q}$, wavevector; $\bar{\Gamma}$, $\bar{H}$, and $\bar{N}$ are high-symmetry points of the Brillouin zone. Reprinted figure with permission from Buczek P., Ernst, A., and Sandratskii, L. M., Phys. Rev. B, 84, 174418, 2001. Copyright 2001 by the American Physical Society.*

with the observation that the SPEELS signal becomes significantly more diffuse in this region (Prokop et al., 2009).

Alternative approaches to calculating dynamical magnetic susceptibility are provided by empirical tight binding theory and many-body perturbation theory. In particular, empirical tight binding has had great success in predicting spin wave dispersions and lifetimes for low dimensional systems such as magnetic thin films, clusters, and adatoms. As an example, we mention that, in order to describe the SPEELS data successfully for the eight-monolayer Co/Cu(001) system, Costa et al. (2004a,b) employed empirical tight binding theory to calculate the dynamical magnetic susceptibility. Along the $\overline{\Gamma}$–$\overline{X}$ direction, their calculated magnon modes are in good agreement with the experimentally determined ones, and theory was found to be able to capture the broadening of peaks for the 'acoustic' mode and the absence of the standing spin wave modes, the 'optical' branches. This study indicates that longitudinal fluctuations as well as the Landau damping process, through which the spin waves decay into the Stoner continuum, are present in the Co/Cu(001) system.

9.5   Surface magnons from atomistic spin dynamics

The atomistic spin dynamics method allows us to address dynamical properties in general in a spin system, at any temperature, which is a clear advantage. The spin system could, in this approach, also be perturbed by external stimuli, like an external magnetic field or an electrical current. The combination of first-principles calculations and the atomistic spin dynamics approach allows for finite temperature studies of magnetic materials (Chen and Landau, 1994; Tao et al., 2005; Skubic et al., 2008; Evans et al., 2014) at the expense of introducing an adiabatic approximation. It is, however, also possible to account for Stoner excitations in the atomistic spin dynamics method, for example, as described by Ma and Dudarev (2012), who added an extra Landau-like term to the Hamiltonian. An alternative method was suggested by Chimata et al. (2012), who introduced temperature effects from the electron subsystem, via the Fermi–Dirac distribution function, into DFT calculations for atomic magnetic moments and exchange parameters. A third approach was suggested by Szilva et al. (2013); in their work, in addition to temperature entering the Fermi–Dirac distribution function, magnetic moments and exchange interactions were evaluated from DFT calculations for a non-collinear magnetic configuration, which was identified to be representative of the magnetic moment configuration at a given temperature.

In this chapter, we do not consider temperature effects on this level. Instead, we focus on how temperature effects, as they enter the stochastic field of the SLL equation, which was described in Chapter 5, influence magnetism and, in particular, magnon excitations. In practice, magnons are obtained from atomistic spin dynamics simulations for each temperature by first equilibrating the atomic spin system. After this, one performs a sampling step, where relevant properties are evaluated and connected to observables, as outlined in Chapter 7. As we shall see in this section, it is the spin correlation functions and the dynamical structure function that are relevant to obtaining the magnon dispersion. The approach used in atomistic spin dynamics simulations to evaluate information that can be used to compare to experimental data is in principle not different from the approach used in molecular dynamics simulations.

Magnons are typically measured using scattering experiments, and hence the relevant quantity for interpretation is the differential cross section. This function is normally evaluated using Fermi's golden rule and, for scattering in a magnetic system, the relevant expression is (Squires, 1997)

$$\frac{d^2\sigma}{d\Omega\, dE} \propto |\langle k'\sigma'\lambda'|V_m|k\sigma\lambda\rangle|^2\, \delta(E_\lambda - E_{\lambda'} + \hbar\omega). \tag{9.16}$$

Here, V_m is the magnetic scattering potential, k and σ are the wavevector and spin state, respectively, of the particle that scatters, which could be a neutron or, as in SPEELS experiments, an electron, and λ specifies the state available for the excitation process. After some algebra and the assumption made in Chapter 2, that is that one can replace the electron density with an integrated quantity, the atomic magnetic moment, as shown in Fig. 2.7 and the text around this figure, one arrives at (Squires, 1997)

$$\frac{d^2\sigma}{d\Omega\, dE} \propto S(\mathbf{q}, \omega), \tag{9.17}$$

that is, the differential cross section is proportional to the dynamical structure function. Hence, any scattering experiment is well suited for analysis in terms of $S(\mathbf{q}, \omega)$ (see also Eqn (7.37) for implementation in atomistic spin dynamics simulations). The peak value of the dynamical structure function along particular directions in the reciprocal space determines the magnon energies that are available for detection in a scattering experiment (Skubic et al., 2008; Tao et al., 2005; Chen and Landau, 1994; Bergman et al., 2010), and the full width half maximum (FWHM) value is related to the lifetime of the magnon. Several examples of how this theory compares to experiments follow below.

9.5.1   Thin films of Co on Cu substrates

We discuss first results for Co$_n$/Cu(001) and Co$_n$/Cu(111) films, for different thicknesses n of Co, and at different temperatures, as shown in in Figs 9.13, 9.14, and 9.15. The magnetic excitations in these systems are investigated in two different limits. First, we consider a low temperature ($T = 1$ K) with a very low damping constant $\alpha = 3 \times 10^{-4}$ (see Eqn (4.1)). The value of the damping parameter was chosen to provide a very weak coupling between the temperature bath and the spin system. Second, we considered simulations performed at elevated temperatures (typically 300 K) and, for these simulations, we used a higher value of the damping parameter, namely $\alpha = 0.05$, to ensure a connection to the temperature bath. In principle, the Gilbert damping should be calculated from first principles, as outlined in Chapter 6, but here we want to illustrate the explicit dependence of damping and temperature on the magnon spectra and have for this reason, selected the values for damping. Since most experiments of surface magnons

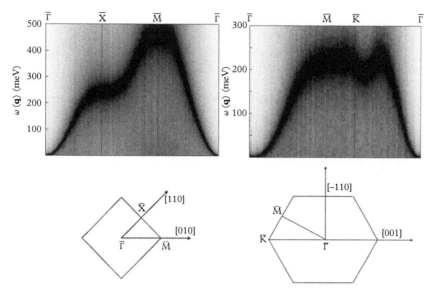

Figure 9.13 *Spin wave dispersion spectra* ω(**q**) *obtained from atomistic spin dynamics simulations of one layer of Co/Cu(001) and one layer of Co/Cu(111). Both sets of data were obtained at T = 200 K and by using a realistic value of the damping constant,* α = 0.05. *(Top left). Spectra for one layer of Co/Cu(001). (Bottom left) The corresponding surface Brillouin zone for the [001] orientation, containing points* Γ̄, X̄, *and* M̄. *(Top right) Spectra for one layer of Co/Cu(111). (Bottom right) The corresponding surface Brillouin zone for the [111] orientation, containing points* Γ̄, X̄, *and* M̄. *Reprinted figure with permission from Bergqvist L., Taroni, A., Bergman, A., Etz, C., and Eriksson, O., Phys. Rev. B, 87, 144401, 2013. Copyright 2013 by the American Physical Society.*

were taken at elevated temperatures, the results from the final temperature simulations are the ones best suited for a comparison to the experiment data.

Figure 9.13 displays the spin wave spectra for a one monolayer-thick Co layer on a Cu substrate, both when the layer is in the [001] orientation (upper left) and when it is in the [111] orientation (upper right). The low temperature simulated spectra for these systems are almost identical to their calculated adiabatic magnon spectra (comparison not shown), which is to be expected since the atomistic spin dynamics simulations and the adiabatic evaluation of magnons make use of the same exchange interactions. In fact, it is quite rewarding that the two approaches come to the same result at $T = 0$ K. Next, we consider finite temperature effects. The calculated critical temperature of the one-monolayer Co/Cu(111) system is around 255 K, while the one-monolayer Co/Cu(001) has a Curie temperature of 370 K. In order to make a comparison between the two surface orientations, the temperature in the simulations was set to $T = 200$ K, to ensure that the simulation conditions for both systems would be well below the ordering temperature. Figure 9.13 shows that the surface magnons depend critically on the orientation of the substrate, as the magnon dispersion for the [111]-oriented substrate is much softer than that for the [001]-oriented surface. This is due both to the geometry the range and nature of the Heisenberg exchange interactions. The [111] surface is close packed, with six nearest neighbours in hexagonal symmetry, while the [001] more surface is open, with only four nearest neighbours. The strength of the nearest-neighbour interaction is much stronger (~1.89 mRy) for the [001] orientation and, consequently, a stiffer magnon dispersion is observed, compared to that for the [111] orientation, which is characterized by a weaker nearest-neighbour exchange (~1.06 mRy).

Figure 9.13 also shows that, even if Landau damping is ignored in the simulations, there is still one source of damping present, namely the increased transversal fluctuations, which is caused by the finite temperature fluctuations of the atomic spins. This is the main reason why the magnon spectra in the figure have finite widths. Since the simulations done for the Co/Cu(111) system are performed at a temperature that is closer to the Curie temperature, compared to those done for the Co/Cu(001) system, thermal effects are more important, and hence the width of the magnon curve for the Co/Cu(111) system is broader than that for Co/Cu(001).

For the system consisting of two Co monolayers on Cu(001), two branches of the magnon spectra are expected and, indeed, observed in both the adiabatic spectra (not shown) and in the atomistic spin dynamics simulations (see Fig. 9.14). One mode corresponds to the acoustic branch, and the second, which is at a higher energy, corresponds to the optical branch. We start by analysing the acoustic branch, for which it may be observed that, even if the microscopic interactions in the Hamiltonian are the same in both the left and the right panels of Fig. 9.14, the spin wave excitation dispersions are different for different temperatures. According to the previous discussion, higher temperatures cause increased broadening but, in addition to this, temperature has a direct effect on excitation energy. This is due to the fact that a magnon is a dynamical

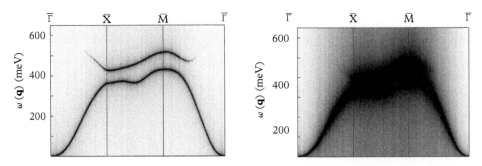

Figure 9.14 *Spin wave dispersion spectra $\omega(\mathbf{q})$ obtained from atomistic spin dynamics simulations of two-monolayer Co/Cu(001); $\bar{\Gamma}$, $\bar{X}$, and $\bar{M}$ are high-symmetry points of the Brillouin zone. (Left) Spectra obtained at $T = 1\ K$, and a small damping constant $\alpha = 3 \times 10^{-4}$. (Right) Spectra obtained at $T = 300\ K$, and a damping constant $\alpha = 0.05$. Reprinted figure with permission from Bergqvist L., Taroni, A., Bergman, A., Etz, C., and Eriksson, O., Phys. Rev. B, 87, 144401, 2013. Copyright 2013 by the American Physical Society.*

object and so it, as well as its dispersion, will be affected by temperature. In Fig. 9.14 this can clearly be seen; for example, the magnon energy of the acoustic branch at the $\overline{X}$-point is clearly lower than that for the simulations done at the higher temperature. Hence, a rather general phenomenon can be seen in Fig. 9.14; elevated temperatures tend to soften the magnon dispersion.

In the standard adiabatic magnon dispersion approach without eigenvector analysis, as performed in Section 9.3, both the optical and the acoustic branches have the same intensity. The dynamic structure factor naturally gives information about the relative intensities of all branches, as they would appear in a scattering experiment. As observed in Fig. 9.14, even at low temperatures and very small damping (1 K and 3×10^{-4}, respectively), the optical mode vanishes close to the zone centre: near the $\overline{\Gamma}$-point, for both the $\overline{\Gamma}$–$\overline{X}$ and the $\overline{\Gamma}$–$\overline{M}$ directions. The reason for the suppression and vanishing of intensity of the optical branch close to the $\overline{\Gamma}$-point was analysed by Taroni et al. (2011) and can be understood from the dynamical structure factor, Eqn (7.37). In the limit of $q \to 0$, this function can, for the optical branch, be seen to be a sum of alternating positive and negative contributions from the spin correlation function, Eqn (7.36), which results in a vanishing dynamical structure factor. The acoustic branch, in contrast, is simply a sum of positive contributions and hence yields a significant contribution to $S(\mathbf{q}, \omega)$.

A complementary way to analyse the absence of optical magnons close to the zone centre is to analyse the susceptibility, which, for $Co_2/Cu(001)$, can be written in the following form (Jensen and Mackintosh, 1991):

$$\bar{\bar{\chi}}(\mathbf{q} + \tau, \omega) = \frac{1}{2}(1 + \cos \phi)\bar{\bar{\chi}}_{Ac}(\mathbf{q}, \omega) + \frac{1}{2}(1 - \cos \phi)\bar{\bar{\chi}}_{Op}(\mathbf{q}, \omega). \qquad (9.18)$$

Here, $\bar{\bar{\chi}}_{Ac}$ and $\bar{\bar{\chi}}_{Op}$ are the susceptibilities that originate from the acoustic and optical branches, respectively; $\mathbf{q}$ is a reciprocal vector within the first BZ; $\tau = [hkl] = h\mathbf{b}_1 + k\mathbf{b}_2 + l\mathbf{b}_3$, and $\phi = \tau \cdot \rho$, where ρ is a real space vector that connects two sublattices. From Eqn (9.18), it can be seen that the total susceptibility can be decomposed into an acoustic and an optical contribution. By changing the momentum transfer, by varying τ, the relative intensity of the two branches will change. When wavevectors in the primitive BZ are probed, the acoustic term dominates and this information is then detected in an experiment. However, when wavevectors outside the first BZ are probed, it is possible to find situations where the optical term dominates, at the expense of the acoustic response. Since, as described in Section 9.4, the susceptibility is linked to magnon excitation, it is possible to identify both acoustic and optical magnon excitations by carefully choosing an experiment with momentum transfer both inside and outside the first BZ. A practical example of how this works is discussed below. For now, we note that, if there are more than two atoms in the unit cell, this analysis becomes more involved but the principle is the same. Based on first-principles calculations of the electronic structure Costa et al. (2004a,b) have shown that the optical modes of thin film magnets vanish if the magnon spectra are simulated by using a dynamical susceptibility, much in the same way as discussed for Eqn (9.18).

In Section 9.3, the adiabatic magnon spectra for eight-monolayer Co on Cu(001) were shown and compared to the SPEELS experimental data from Vollmer et al. (2003). In Fig. 9.15 the spin wave excitation spectra from atomistic spin dynamics simulations, as revealed by the dynamical structure factor, is displayed for this structure at a temperature of 300 K. As in the bilayer case, the most striking difference between the dynamical treatment and the static adiabatic magnon spectra approach is the clear suppression of the intensity of the optical modes close to the zone centre. The lowest lying acoustic branch is, on the other hand, little affected by dynamics and temperature; however, the dispersion at room temperature is slightly softer (by ∼25 meV) than that at zero temperature (cf Figs 9.8 and 9.15). Qualitatively, theory is in good agreement with SPEELS electron energy loss spectroscopy and (EELS) data (Vollmer et al., 2003), although the calculated values are still a bit higher than the experimental observed values, which most likely is due to an overestimation of the exchange interactions, as they are obtained from a collinear magnetic arrangement, in contrast to the experimental

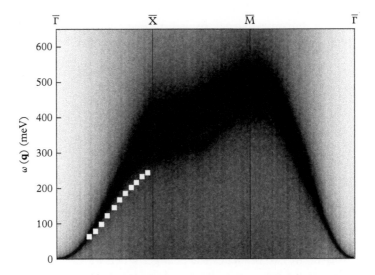

Figure 9.15 *Spin wave dispersion spectra $\omega(\mathbf{q})$ obtained from atomistic spin dynamics simulations of eight-monolayer Co/Cu(001) at T = 300 K and with a damping constant $\alpha = 0.05$; $\bar{\Gamma}$, $\bar{X}$, and $\bar{M}$ are high-symmetry points of the Brillouin zone. Experimental values are given as white squares (data from Vollmer, R., Etzkorn, M., Anil Kumar, P. S., Ibach, H., and Kirschner, J., (2003), Spin-polarized electron energy loss spectroscopy of high energy, large wave vector spin waves in ultrathin fcc Co films on Cu(001), Phys. Rev. Lett., 91, 147201). Reprinted figure with permission from Etz C., Bergqvist, L., Bergman, A., Taroni, A., and Eriksson, O., J. Phys. Condens. Matter, 27, 243202, 2015. Copyright 2015 by the IOP publishing.*

situation, where a certain degree of non-collinearity is present. The approach by Szilva et al. (2013) should, in principle, take this into account.

9.5.2   A comparison of approaches

Recently, efforts have been made for the experimental determination of layer-resolved exchange interactions. Theoretical studies are able to predict the size and type of inter- and intra-layer exchange coupling in almost any type of systems, provided that the exchange and correlation functional (see Chapter 1) used in the calculation of the electronic structure are accurate. In an experimental work by Ibach and co-workers (Rajeswari et al., 2014), the EELS technique was used to probe Co layers of different thicknesses (from four to eight monolayers) deposited on a Cu(001) substrate, and a comparison between the measured dispersion curves and theoretical predictions (Costa et al., 2004b; Bergqvist et al., 2013) was performed. Obtaining layer-resolved exchange interactions from such experiments, or from any experiment, for that matter, remains a challenge. However, work by Rajeswari et al. (2014) set a benchmark for existing theoretical models used to calculate exchange interactions and obtaining magnon dispersion. In their study, EELS data were compared with data obtained via three different theoretical methods: (i) the strategy employed by Costa et al. (2004b), (ii) the atomistic spin dynamics approach using ab initio exchange interactions (Bergqvist et al., 2013), and (iii) a nearest-neighbour Heisenberg model, with constant exchange coupling. The main conclusion from Rajeswari et al. (2014) is that the best fit to the experimental data is provided by the nearest-neighbour Heisenberg model with a constant exchange parameter of 15 meV. However, one cannot conclude that this represents a good description of the magnetic interactions of the Co/Cu(001) system, since there is no metal known to have significant nearest-neighbour exchange interaction but vanishing long-range interactions. Indeed, first-principles calculations of the exchange interactions of the Co/Cu(001) system show that there are significant contributions from long-range interactions (Bergqvist et al., 2013). In fact, the experiments by Rajeswari et al. (2014) are equally well reproduced by using the first-principles-calculated, layer-resolved exchange parameters from the work by Bergqvist et al. (2013) when these parameters are scaled to 85 % of their value. The necessity of downscaling the calculated exchange interactions by 15 % in order to give a very good fit of the measured data is most likely due to the fact that the exchange parameters are evaluated for a collinear configuration, instead of the disordered non-collinear finite temperature configuration, as discussed by Szilva et al. (2013).

9.5.3   Fe on Cu(001)

In contrast to the Co overlayers on Cu(100), Fe layers on Cu(100) represent a rather complex system, due to the fact that, at low temperatures and ambient pressure, Fe does not naturally stabilize in the fcc lattice, as discussed in Chapter 3, and therefore it is difficult to grow thicker fcc Fe layers with good quality. Moreover, fcc Fe shows a rather complicated magnetic phase diagram in which many magnetic configurations with similar energies exist, depending on the volume. For instance, it has been argued (van Schilfgaarde et al., 1999) that non-collinear magnetic structure is the main driving force for the Invar behaviour in Fe-rich FeNi alloys. Ultrathin Fe layers on Cu(100) have been extensively studied in the past (Li et al., 1990; Hjortstam et al., 1996; Lorenz and Hafner, 1996; Schmitz et al., 1999). It is generally accepted that thicknesses of one to three monolayers exhibit a ferromagnetic configuration. However, when the fcc Fe films become more than three monolayers thick, more complex magnetic structures may occur. For instance, Sandratskii (2010) proposed that, when fcc Fe films on Cu(001) are more than three monolayers thick, they show the following magnetic configuration: $\downarrow\uparrow\uparrow$, where each arrow represents the direction of the magnetization within the top three Fe monolayers in the system.

From first-principles calculations, the ferromagnetic ground state for a three-monolayer thin film can be obtained. The calculated magnon dispersion, from atomistic spin dynamics simulations, of this system is displayed in Fig. 9.16. Note that the figure contains information in the limits of low damping and low temperature as well as high damping and room temperature. At low temperature and damping, the different branches are distinct from each other while, at higher damping and room temperature, the branches have a significant broadening and hence overlap. The value of the spin stiffness constant in Fig. 9.16 is lower than that for a single monolayer (shown in Fig. 9.12). This is atypical; often thinner layers display weaker dispersion than thicker layers or bulk systems do. This interesting result indicates that there is a softness of the magnon dispersion spectra of $Fe_3/Cu(001)$, caused by the approaching magnetic instability for even thicker layers. There are no experimental data available for this system, one of the

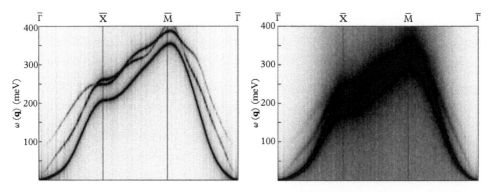

Figure 9.16 *Spin wave dispersion spectra $\omega(\mathbf{q})$ obtained from atomistic spin dynamics simulations of three-monolayer Fe/Cu(001); $\bar{\Gamma}$, $\bar{X}$, and $\bar{M}$ are high-symmetry points of the Brillouin zone. (Left) Spectra at T = 1 K, and a small damping constant $\alpha = 3 \times 10^{-4}$. (Right) Spectra at T = 300 K, and a damping constant $\alpha = 0.05$. Reprinted figure with permission from Bergqvist, L., Taroni, A., Bergman, A., Etz, C., and Eriksson, O., Phys. Rev. B, 87, 144401, 2013. Copyright 2010 by the American Physical Society.*

reasons being the difficulty of growing thick fcc Fe layers on Cu, but an experimental study of this system would be highly interesting.

9.5.4   Fe on Ir(001)

Fe on an Ir(001) substrate is an interesting system in many ways. Ir is a heavy element, with a large spin–orbit interaction that manifests itself in several intriguing magnetic properties, as will be discussed in Section 9.6. For instance, a single monolayer of Fe on Ir(001) displays a complex magnetic structure in the form of a spin spiral. This magnetic state is found both in theoretical calculations and experiment (Kubetzka et al., 2005; Martin et al., 2007; Kudrnovský et al., 2009; Deák et al., 2011). On the Ir(111) substrate, a single monolayer of Fe exhibits a so-called nano-skyrmion lattice (Heinze et al., 2011). Alternatively, this state could be referred to as a multiple-q state. Note that skyrmionic states of magnets are discussed in a separate chapter of this book, Chapter 10, and are not discussed further in this section. Thicker Fe layers on Ir (001) show ferromagnetic order. Using the SPEELS method, an experimental magnon dispersion was obtained at room temperature for six layers of Fe on Ir(001) (Zakeri, Chuang, et al., 2013). The acoustic magnon branch of the spectra was detected with high intensity, and the data are compared with atomistic spin dynamics simulations in Fig. 9.17. The simulations used first-principles calculations of the exchange interactions and were performed at a temperature of 300 K and with a damping parameter of 0.01.

It is clear that the acoustic mode has the highest intensity and is clearly visible throughout the whole first BZ. Experiment and theory compare reasonably well for this mode (see Fig. 9.17). In the experiments, it was possible to probe energies up to 150 meV, and wavevector transfers $\Delta K_\parallel$ up to 0.5–0.8 Å^{-1}, for both the $\overline{\Gamma}$–$\overline{X}$ and the $\overline{\Gamma}$–$\overline{M}$ directions. For these energy and wavevector transfer ranges, one would expect that first and second low energy modes would be observed in the experiment and, indeed, experimentally, at the $\overline{X}$-point, there was some indication of an additional peak in the spectra at around 200 meV (Zakeri, Chuang, et al., 2013), a result which corresponds well with the theoretical spectra for the second low energy mode. However, it is important to understand

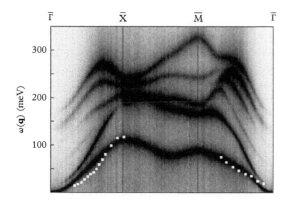

Figure 9.17 *Spin wave dispersion spectra* $\omega(\mathbf{q})$ *from atomistic spin dynamics simulations of six-monolayer Fe on Ir(001) at 300 K. Experimental values obtained by spin polarized electron energy loss spectroscopy (data from Zakeri Kh., Chuang, T.-H., Ernst, A., Sandratskii, L. M., Buczek, P., Qin, H. J., Zhang, Y., and Kirschner, J., (2013), Direct probing of the exchange interaction at buried interfaces,* Nat. Nanotechnol., *8, 853–8) are marked by white squares;* $\overline{\Gamma}$, $\overline{X}$, *and* $\overline{M}$ *are high-symmetry points of the Brillouin zone.*

that the electron beam used in such experiments has a short penetration depth into the film, being a highly surface sensitive experimental tool. This leads to a strongly attenuated signal in the interior of the film, which, in addition to Landau damping, causes a weakened signal.

9.5.5   Fe on W(110)

The SPEELS method was applied to a single magnetic layer of Fe on W(110), by Prokop and co-workers (2009). This is a prototype system for low dimensional magnetism and has over the years been subjected to a number of experimental investigations (Elmers, 1995; Elmers et al., 1996). The experimental study by Prokop et al. (2009) showed that the magnon dispersion energies were considerably softer than those of bulk Fe. It was observed that the softening of a monolayer of Fe on W(110) was in strong contrast to previous theoretical calculations based on the random phase approximation (Muniz and Mills, 2002), although it should be noted that similar calculations (Costa et al., 2004b) agree better with the experimental SPEELS results (Prokop et al., 2009). An effort to resolve the discrepancy between theory and experiment was done using a combination of first-principles calculations and atomistic spin dynamics simulations (Bergman et al., 2010), and these results are discussed briefly here.

First-principles theory demonstrates that the magnetic order of one Fe layer on W(110) is ferromagnetic (Bergman et al., 2010). The calculated exchange interactions are long range, typical for low dimensional magnets. Due to a rather strong antiferromagnetic next-nearest-neighbour interaction, shown in the left-hand side of Fig. 9.18, the ferromagnetic state is only marginally lower in energy compared to more complex, non-collinear states (Bergman et al., 2010). Depending on the cut-off range of the exchange interactions, and on numerical details of the first-principles calculations, the exchange interactions can, in some situations, be found to favour a non-collinear spin-spiral ground-state ordering of the Fe moments (Bergman et al., 2010), a fact which was also noticed in an earlier theoretical investigation (Wu and Freeman, 1992). However, the substrate element, W, is heavy and has a rather strong spin–orbit interaction, which manifests itself in a strong magnetocrystalline anisotropy of around 4 meV/atom. This anisotropy favours an in-plane magnetization direction along the long axis (X. Qian and Hübner, 2001) and favours a collinear state.

Including only exchange interactions and magnetocrystalline anisotropy in the effective spin Hamiltonian, the magnon spectra for one Fe layer on W(110) is noticeably softer than that for bulk bcc Fe. This is due to the geometry of the 110 surface, and the substantial hybridization between Fe and the tungsten substrate. To come closer to the experiments, which were performed at $T = 120$ K, temperature-dependent exchange interaction parameters were evaluated. We have discussed various techniques for calculating finite temperature effects of the exchange parameters in this chapter, and here we introduce yet another approximation for how to do it. A completely disordered arrangement of atomic spins has been suggested as a reasonable approximation, when performing first-principles calculations of the electronic structure, atomic magnetic moments, and interatomic Heisenberg exchange interactions (Staunton et al., 1984;

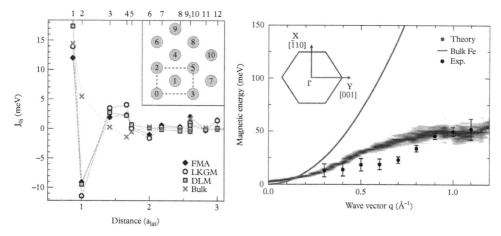

Figure 9.18 *(Left) Calculated exchange interaction parameter $\mathcal{J}_{0j}$ for one-monolayer Fe on W(110) as a function of distance (in units of the lattice constant a_{lat}). The values labelled 'FMA' were obtained using the 'frozen magnon approximation' while the other curves were calculated using Eqn (2.23) for a ferromagnetic solution (curve labelled LKGM), and for a disordered local moment state (curve labelled DLM). Calculated exchange interaction parameters for bulk body-centred cubic Fe are shown for comparison. The inset shows the geometry of the Fe monolayer. (Right) Comparison between experimental and theoretical magnon dispersion curves along the [001] direction for one monolayer Fe on W(110). The dots represent experimentally obtained data, whereas the thick, fuzzy line represents numerically obtained data from the dynamical structure function. For comparison, the experimental spin wave spectrum of bulk body-centred cubic Fe is also displayed; Γ is a high-symmetry point of the Brillouin zone. Reprinted figure with permission from Bergman, A., Skubic, B., Hellsvik, J., Nordström, L., Delin, A., and Eriksson, O., Phys. Rev. B, 83, 224429, 2010. Copyright 2010 by the American Physical Society.*

Gyorffy et al., 1985). A configuration of atoms that are randomly aligned to be either parallel or antiparallel to the z-axis, was then suggested by the standard tools of the electronic structure of alloy theory, that is, coherent potential approximation (Soven, 1967). This approximation is often referred to as the disordered local moment (DLM) approximation (Staunton et al., 1984; Gyorffy et al., 1985) and is relevant for a magnetic material above the ordering temperature. For finite temperatures below the ordering temperature, a partial disorder of atoms having moments parallel or antiparallel to the z-axis have been suggested, in an approach that is referred to as the partial DLM (pDLM) model. In the pDLM model, the net magnetization is reduced and will therefore approximately account for temperature-induced transversal fluctuations. The pDLM concept has, in fact, been employed in related studies of finite temperature magnetism (Korzhavyi et al., 2002; Böttcher et al., 2012), with reasonable success.

By calculation of the average magnetization at $T = 120$ K (which is to be compared to the Curie temperature of $\sim$280 K), a suitable pDLM ratio of 85 % was chosen and assumed to represent an appropriate description of the Fe on W(110) system. Including the temperature-dependent exchange interactions in the atomistic spin dynamics simulations, which were performed at $T = 120$ K, resulted in a substantial softening of the magnon spectra compared to the bulk bcc Fe spectra (see the right panel of Fig. 9.18). The calculated magnon dispersion curve is in rather good agreement with the experiment. Hence, the Fe/W(110) system is one of several examples where a combination of first-principles calculations and atomistic spin dynamics is able to provide not only qualitative agreement with experiments but also a quantitative agreement, and that finite temperature effects are important both for the exchange interactions and for the evaluation of the magnetic excitation spectra.

For two monolayers of Fe on W(110), extensive experimental investigations using SPEELS have been reported. Both spin wave dispersion curves (W. Tang et al., 2007) and estimated magnon lifetimes (Zakeri et al., 2012) have been reported. Lifetimes are further discussed in Section 9.7. Magnon dispersion curves were obtained using the same methodology as described above: atomistic spin dynamics simulations together with first-principles calculations of parameters of a spin Hamiltonian. The magnon excitations are compared to experimental data in Fig. 9.19, along high-symmetry lines within the first BZ. A good agreement between theory and available experimental data is observed along the $\bar{\Gamma}$–$\bar{H}$ line. For this system, finite temperature results from atomistic spin dynamics correspond very well with the adiabatic magnon spectra shown in Fig. 9.10, indicating a rather weak temperature dependence of the exchange interactions. Since there are two Fe layers in this system, two branches are expected in the magnon spectra. For the momentum transfer **q** inside the first BZ, only the acoustic mode gives a high intensity, as was the case in Section 9.5.4, while the optical mode is very weak, as shown in the left-hand side of Fig. 9.19(a). However, according to the discussion for Eqn (9.18), it should be possible to identify modes that are detectable experimentally by considering moment transfers outside the first BZ. This is also explored in the right-hand side of Fig. 9.19, and it may be seen that one then obtains an optical mode with a

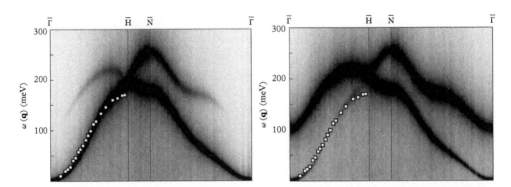

Figure 9.19 *Spin wave dispersion spectra $\omega(\mathbf{q})$ obtained from atomistic spin dynamics simulations of two-monolayer Fe/W(110) at temperature $T = 300$ K, and damping constant $\alpha = 0.01$. Experimental values obtained by spin polarized electron energy loss spectroscopy are marked by white squares; $\bar{\Gamma}$, $\bar{H}$, and $\bar{N}$ are high-symmetry points of the Brillouin zone. (Left) Sampling inside the first Brilloiun zone ($\tau = [00]$). (Right) Sampling shifted by vector $\tau = [10]$. Reprinted figure with permission from Bergqvist, L., Taroni, A., Bergman, A., Etz, C., and Eriksson, O., Phys. Rev. B, 87, 144401, 2013. Copyright 2013 by the American Physical Society.*

very high intensity. This illustrates that scattering experiments of the type discussed in this chapter, that is, ones using SPEELS or EELS, should also be able to detect optical magnon modes, if appropriate momentum transfers are considered.

9.6   Relativistic effects

In recent years, much effort has been focused on relativistic effects and spin–orbit coupling. The ambition is to utilize these interactions to develop novel electronics devices, with improved functionality and properties. As an example, in 2009, Udvardi and Szunyogh predicted that, in magnon spectra, one should be able to observe features that arise from relativistic effects, in particular, the Dzyaloshinskii–Moriya interaction (Eqn (4.33); Udvardi and Szunyogh, 2009). Udvardi and Szunyogh showed that, as a consequence of a finite Dzyaloshinskii–Moriya interaction, the magnon spectra is not symmetric for left- and right-going waves, that is, $\omega(\mathbf{q}) \neq \omega(-\mathbf{q})$ for the same value of $|\mathbf{q}|$. A parameter, $\Delta E = \omega(\mathbf{q}) - \omega(-\mathbf{q})$, was introduced to quantify the asymmetry. The interaction responsible for this asymmetry is the relativistic Dzyaloshinskii–Moriya interaction arising from spin–orbit interactions. A finite value of ΔE, detected experimentally by means of SPEELS, has indeed been found in the magnon spectra of a thin Fe film on W(110) (Zakeri et al., 2010).

The evaluation of the Dzyaloshinskii–Moriya interaction can be seen as a relativistic generalization to the evaluation of the Heisenberg exchange parameters, Eqn (2.19), and thus necessitates a generalized form of the Heisenberg Hamiltonian, where the μ and ν components (μ and $\nu = x, y$, and z) of atomic moments at sites i and j can interact (see Eqn (4.43)). As discussed in Eqn (4.42), this can be decomposed into a non-relativistic Heisenberg contribution and an interaction in the form of a Dzyaloshinskii–Moriya interaction.

The magnon spectra for two monolayers of Fe on W(110) have already been shown in Fig. 9.10 and Fig. 9.19, from a theory that uses only Heisenberg exchange parameters. However, from relativistic first-principles calculations, it is possible to obtain realistic values for the Dzyaloshinskii–Moriya interaction, and using atomistic spin dynamics simulations and information from the dynamical structure factor, the asymmetry was estimated for Fe_2/W(110), for wavevectors within the first BZ, along $\bar{\Gamma}$–$\bar{H}$ and $\bar{H}$–$\bar{\Gamma}$. The results are shown in Fig. 9.20, where they are compared to available experimental data (Zakeri et al., 2010). A qualitatively good agreement is found. However, quantitatively, theory seems to overestimate the spin wave asymmetry by $\sim$4 meV.

In the spin dynamics simulations, the same temperature as that reported for the experimental investigation (room temperature) was used and, in addition, a realistic damping parameter was used. This was done in order for the simulations to stay as close as possible to the experimental conditions. The quantitative difference between theory and experiment, observed in Fig. 9.20, could have several reasons; the most important is likely to be the numerical accuracy needed to calculate the Dzyaloshinskii–Moriya interaction. Since the value of this interaction is very small, there is considerable demand on the methodology used to solve the Kohn–Sham equation, and the evaluation

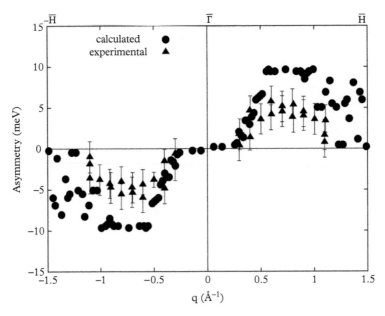

Figure 9.20 *Calculated spin wave asymmetry (see text) for the magnon spectrum of two-monolayer Fe/W(110), using theoretically determined Dzyaloshinskii–Moriya interactions. The experimental values (data from Zakeri, Kh., Zhang, Y., Prokop, J., Chuang, T.-H., Sakr, N., Tang, W. X., and Kirschner, J., (2010), Asymmetric spin-wave dispersion on Fe(110): Direct evidence of the Dzyaloshinskii–Moriya interaction, Phys. Rev. Lett., 104, 137203) have been obtained with M parallel [110]; q, wavevector; H̄ is a Brillouin zone point. Reprinted figure with permission from Bergqvist, L., Taroni, A., Bergman, A., Etz, C., and Eriksson, O., Phys. Rev. B, 87, 144401, 2013. Copyright 2013 by the American Physical Society.*

of Eqn (4.43). Another source of the discrepancy could be the fact that ideal surfaces are considered in the theoretical calculations, without accounting for intermixing at the interface, something which might be relevant in the experiments. In addition, the first-principles calculations were performed for a collinear magnetic structure, whereas experiments certainly have some degree of non-collinearity in them. Nonetheless, it is quite remarkable that relativistic first-principles theory combined with atomistic spin dynamics simulations has sufficient accuracy to give a reasonable estimate of these tiny interactions and the effects they have on magnon dispersions.

9.7 Magnon lifetimes

Magnons, being quasiparticles, as discussed in Section 9.1, in addition to having a specific excitation energy, have an associated lifetime. There are two main mechanisms that determine the magnon lifetime: the Gilbert damping and the Landau damping discussed in Section 9.1. The influence of the first mechanism can be assessed from atomistic spin dynamics simulations, by determining the linewidth due to broadening of the dynamical structure factor, as function of wavevector and temperature. An example of this approach is displayed in Fig. 9.21 for a Co monolayer film on Cu(001), using the same data as in Fig. 9.13. The peak of the structure factor for a given wavevector is fitted with a Gaussian, and the linewidth Δ is then defined as the FWHM of this Gaussian function. The linewidth is seen in Fig. 9.21 to increase with the wavevector, signalling a maximum

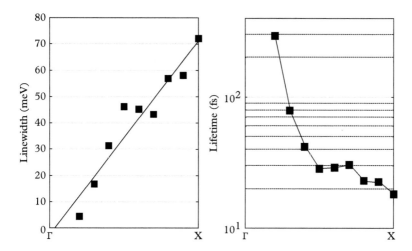

Figure 9.21 *Magnon linewidth (left) and lifetime (right) for one monolayer of Co on Cu(001), as obtained from atomistic spin dynamics simulations at temperature $T = 200$ K, and damping constant $\alpha = 0.05$. Results shown along the Γ–X direction in the Brillouin zone. The straight line on the left-hand side of the figure is a linear fit to the calculated data.*

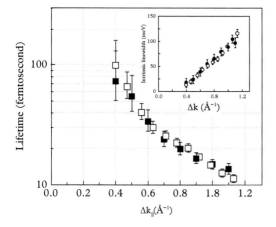

Figure 9.22 *Experimental spin polarized electron energy loss spectroscopy magnon lifetime (inset shows linewidth of the magnon peaks) as a function of wavevector transfer $\Delta k_\parallel$ for two monolayers of Fe on W(110) (solid symbols), and eight monolayers of Co on Cu(001) (open symbols). Reprinted figure from Zhang, Y., Chuang, T.-H., Zakeri, Kh., and Kirschner, J., Phys. Rev. Lett., 109, 087203, 2012. Copyright 2012 by the American Physical Society.*

damping at the zone edge. The linewidth depends less strongly on temperature than on the choice of Gilbert damping parameter α, which is very decisive for the linewidth. The data in Fig. 9.21 were obtained from simulations where the Gilbert damping was set at 0.05. The lifetime τ is inversely proportional to the linewidth, with the relation $\tau = 2\hbar/\Delta$, which is the time it takes for the amplitude of the peak to drop to e^{-1} of its original value. For magnons with wavevectors that are not in the vicinity of the zone centre, the lifetimes in the specific example shown in Fig. 9.21 are of the order of 10–30 fs.

Similarly, in a SPEELS experiment, the magnon lifetime can also be estimated by the broadening of the spectra, but with the important difference that this broadening includes not only the broadening from Gilbert damping but also that from Landau damping, as well as a broadening due to limited instrument resolution. The extrinsic instrumental broadening needs to be removed, and this is normally done by fitting the spectra with both a Gaussian and a Lorentzian function, where the Gaussian contribution stands for the instrument resolution. If this is done properly, one ends up with a spectrum containing only intrinsic contributions to the magnon lifetimes. In Fig. 9.22, intrinsic, experimental magnon lifetimes for two layers of Fe on W(110), and eight monolayers of Co on Cu(001) are displayed. Remarkably, these systems have similar lifetimes despite their very different magnon energies, which differ by a factor of 2. The lifetimes are of the order of a few tens of femtoseconds, values that are surprisingly similar to what is found from the atomistic spin dynamics simulations shown in Fig. 9.21, even if Landau damping is ignored in the latter case.

10

Skyrmions

An important application of spin dynamics is the response of a magnetic material
subjected to external stimuli. In Chapter 9, the response of primarily ferromagnets
to temperature fluctuations, manifested as spin wave excitations and magnons, was
discussed. In this chapter, we are concerned about magnetic materials with more complicated magnetic texture, such as spin spirals, topological magnetic structures, and,
in particular, magnetic skyrmions. Magnetic skyrmions have many appealing and intriguing features that make them interesting for possible applications as well as from a
purely theoretical point of view.

10.1 Background

The vast majority of applications based on magnetic phenomena rely on the existence
of a finite magnetization. Thus, until recently, ferromagnets or ferrimagnets, with a net
magnetization, attracted a larger interest, from an application point of view, compared to
compensated antiferromagnets or spin-spiral materials. With the advent of spintronics,
this paradigm is starting to change and antiferromagnetic materials now play important
roles for providing exchange bias for spin valves and magnetic tunnel junctions. Further, for ferromagnetic systems, more and more focus has been put on domain walls,
since manipulation of these walls has shown promise for pure solid-state-based magnetic storage techniques, such as race track memory (Parkin et al., 2008). Also, materials
with a more exotic magnetic order are becoming increasingly interesting for applications
where notable examples can be found in multiferroics, perhaps predominantly the class
of improper multiferroics, where a magnetic spin-spiral ordering can cause an electric
polarization, and vice versa. Thus, it is not an adventurous guess that complex magnetic
orderings can be expected to play a larger role also when it comes to applications of
magnetic materials.

In this chapter, we will discuss a particular manifestation of non-collinear order,
namely the magnetic skyrmion. Analogously to how domain walls can be seen as a
sole magnetic object even though they consist of many individual magnetic moments,
a magnetic skyrmion can also be seen as a single object which, in many ways, actually
shares physical properties with domain walls or other magnetic solitons. A skyrmion has,

Atomistic Spin Dynamics. Olle Eriksson, Anders Bergman, Lars Bergqvist, Johan Hellsvik. First Edition.

however, several unique properties which make it interesting from many points of view. The concept of a skyrmion as a topological soliton in non-linear field theory was introduced by Tony Skyrme (1961). Since the concept can be applied to a broad scope of continuous field models, skyrmions can be realized in many different physical situations, including high energy physics and particle theory, where the concept was originally conceived, but also in condensed matter physics, where realizations have been predicted and reported in nematic liquid crystals (Wright and Mermin, 1989), Bose–Einstein condensates (Al Khawaja and Stoof, 2001), and magnetic systems (Bogdanov and Yablonskii, 1989; Mühlbauer et al., 2009). Here, we will focus on the latter realization of magnetic skyrmions, which will be denoted simply as skyrmions. We note that, since this is a very intense and active field, a lot of new knowledge will evolve in the years to come, and we give here a basic understanding and description of fundamental concepts of these soliton excitations.

10.2 Magnetism and topology

Deviations from a uniformly magnetized state can be described as topological defects. This picture holds true for domain walls and vortices but also for skyrmions. While we here leave the detailed discussion about topology and magnetism to more devoted textbooks (Munkres, 2000), a brief discussion of topology is still warranted due to the close coupling between skyrmions and topology. Two systems can be said to have the same topology if a continuous deformation can transform one of the systems into another without passing an infinitely large energy barrier. A standard example is that an elastic ball can be deformed into a rod or a cube but, in order to transform it to a torus the surface of the ball must be ruptured. Thus, a cube and a sphere share the same topology, while the torus has another topology. In magnetic systems, the considered deformations correspond to rotations of the spin structure between two magnetic states, which means that a ferromagnetic structure and a single-wavevector spin spiral share the same (trivial) topology, while a vortex in an easy-plane situation has a different topology.

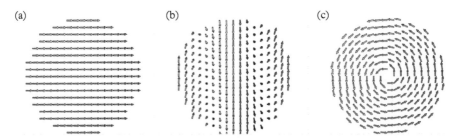

Figure 10.1 *Illustration of planar magnetic structures with and without finite winding numbers. (a) shows a ferromagnetic configuration, and (b) shows a spiral configuration, both having zero winding number, while (c) shows a ring-like structure, or planar vortex, which has a finite winding number.*

In order to classify topological defects, it is practical to introduce the 'winding number' w, which is a measure of how many times the magnetization rotates, or winds, along a circular path circumferencing the defect. In Fig. 10.1, examples of easy-plane magnetic structures are given and, by following the circular paths given in the figure, it can be seen that the ferromagnet and the cycloidal spin spiral have the same winding number, $w = 0$, while the vortex has a finite winding number of $w = 1$. Mathematically, the winding number can be obtained (for two-dimensional systems) from an integral over the whole system as

$$w = \frac{1}{4\pi} \int \int dx \, dy \, \mathbf{M} \cdot \left(\partial_x \mathbf{M} \times \partial_y \mathbf{M} \right), \tag{10.1}$$

where $\mathbf{M}$ is the continuous magnetization. The winding number calculated according to Eqn (10.1) is also called the topological charge or, when discussing skyrmions, the skyrmion number.

10.3 Magnetic skyrmions

After having defined the topological winding number, we now turn our attention to the main subject of this chapter, the magnetic skyrmion. A skyrmion can be described as a topological defect where the magnetization for a system, $\mathbf{M}(\mathbf{r})$, where $\mathbf{r}$ has its origin at the defect, performs a 180° rotation as $|\mathbf{r}|$ goes from 0 to ∞, that is, the magnetization at the centre of the skyrmion points in the opposite direction from the magnetization far away from the skyrmion. How the actual rotation is performed affects the shape and structure of the skyrmion but does not change its topology. In Fig. 10.2 two different rotations are manifested, resulting in a spiral skyrmion when the rotational plane of the magnetization is perpendicular to the radius, as in Fig. 10.2(a), or a hedgehog skyrmion when the rotational plan is parallel to the radius, as in Fig. 10.2(b). These skyrmions can be compared with the vortex displayed in Fig. 10.2(c) where the performed rotation is not 180° but instead only 90° since, far from the defect centre, the magnetization lies in plane. In this depiction, the vortex has an out-of-plane component in the centre, while

(a)   (b)   (c)

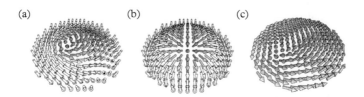

Figure 10.2 *Topological defects in two-dimensional systems: (a) a spiral skyrmion, (b) a hedgehog (or cycloidal) skyrmion, and (c) a vortex with finite out-of-plane polarization.*

vortices often are considered in situations where the magnetization only has finite components in the x- and y-directions. The vortex displayed in Fig. 10.2(c) thus does not have the same topology as the skyrmions in Fig. 10.2(a) and (b); instead, the topological charge of the vortex is in fact half of that of a skyrmion, that is, $w_{vortex} = \pm 1/2$, while $w_{skyrmion} = \pm 1$.

Since a skyrmion has an integer topological charge, it does not have the same topology as a ferromagnet. Given the previous discussion about topology and energy barriers (Section 10.2), it is thus tempting to view the skyrmion as being topologically stable, meaning that an infinite amount of energy needs to be added in order to break the skyrmion structure. There is, however, cause for a certain amount of caution, since this definition is based on the mathematical definition of topology. In any physical representation of a topological system, the energy barriers involved are always finite and thus the concept of topological stability is best handled with a bit of caution when discussing condensed matter realizations of skyrmions. In addition, the topological protection and, in fact, skyrmions as well are formally only defined for continuous fields. The introduction of localized magnetic moments thus also breaks the assumption of continuity, a fact which weakens the stability claims further. That being said, even in discrete lattices and real-world realizations, skyrmions are indeed stabilized by means of their topology, but not against all perturbations.

10.4   Theoretical prediction and experimental identification

The possible existence of skyrmions in Heisenberg ferromagnets was first pointed out by Belavin and Polyakov (1975). They used topological arguments to show that, for each homotopical class, there exists a solution which has a minimum energy, and thus a series of metastable excited states can form for isotropic two-dimensional ferromagnets. The first arguments towards thermodynamically stable skyrmion states came from Bogdanov and co-workers (Bogdanov and Yablonskii, 1989; Bogdanov and Hubert, 1994), who extended the Hamiltonian to include, in addition to the Heisenberg exchange, antisymmetric Dzyaloshinskii–Moriya interactions (DMIs) and a Zeeman term. Their calculations showed that, for intermediate applied magnetic fields, a lattice of close-packed skyrmions can be stabilized and they also suggested candidate materials where these skyrmion lattices could possibly be found. The candidates included the tetragonal rare-earth intermetallics Tb_3Al_2 and Dy_3Al_2 but also materials such as MnSi, FeGe, and others belonging to the $B20$ structural type, which has a cubic unit cell but lacks inversion symmetry. The proposed model needed an applied magnetic field for the skyrmions to be stabilized, but later on Rößler et al. (2006) argued that, by introducing longitudinal fluctuations, as given by Landau theory to the spin model, skyrmion lattices could indeed be stabilized as a ground state, even without applied magnetic fields or other influences. Once again, MnSi was proposed as a possible candidate in which to find skyrmion lattice states.

The phase diagram for MnSi (Ishikawa and Arai, 1984), which has until recently been a matter of debate, is shown in Fig. 10.3(a). The zero temperature ground state is

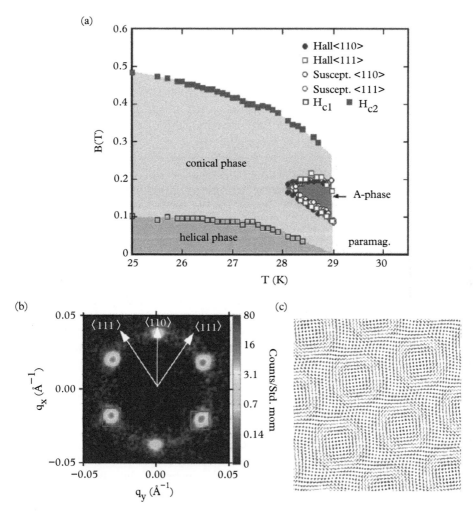

Figure 10.3 *(a) Phase diagram for MnSi. Reprinted figure with permission from Neubauer, A., Pfleiderer, C., Binz, B., Rosch, A., Ritz, R., Niklowitz, P. G., and Böni, P., Phys. Rev. Lett., 102, 186602, 2009. Copyright 2009 by the American Physical Society. (b) Neutron small angle scattering intensities for MnSi. Reprinted figure with permission from Adams, T., Mühlbauer, S., Pfleiderer, C., Jonietz, F., Bauer, A., Neubauer, A., Georgii, R., Böni, P., Keiderling, U., Everschor, K., Garst, M., and Rosch, A., Phys. Rev. Lett., 107, 217206, 2011. Copyright 2011 by the American Physical Society. (c) Real space depiction of a helical triple-q spin spiral, which strongly resembles a skyrmion lattice. Panel (c) is the result of an atomistic spin dynamics simulation.*

a long-wavelength helical spin spiral (shown in Fig. 10.3(c)) which is stabilized by the competition between ferromagnetic Heisenberg exchange and DMIs, which are finite due to the lack of inversion symmetry of the compound. The DMIs are, however, weak in the system, so the wavelength of the spin spiral is long: $\lambda \approx 190$ Å. In addition, there is an even weaker magnetocrystalline anisotropy which pins the spin-spiral structure to the crystallographic <111> direction. For weak applied magnetic fields, the helical spin structure persists but, with increasing fields and temperatures, a phase transition to a conical spin structure takes place. At sufficiently large magnetic fields, there is an additional phase transition to a ferromagnetic state while, at high temperature, there is an expected phase transition to a paramagnetic state. The most interesting area of the phase diagram is, however, found for a range of fields and temperatures between the conical spin state and the paramagnetic state. There exists another phase for which the magnetic structure is expected to be non-trivial, but the precise nature of the phase is not known.

From small angle neutron scattering, it was found by Pfleiderer and co-workers (Mühlbauer et al., 2009) that the so-called A-phase (see Fig. 10.3(a)) had a six fold symmetric intensity pattern in the dynamical structure factor (see Fig. 10.3(b)); this finding was taken as a strong indication that the magnetic structure in the A-phase actually consists of three superposed spin spirals, which are oriented perpendicularly with 120° between them. That spin structure can in turn be interpreted as a two-dimensional lattice of close-packed skyrmions (see Fig. 10.3(c)). This finding was in good agreement with the earlier theoretical predictions, but neutron scattering experiments cannot give decisive information about the actual topology of the structure and thus the question of whether the magnetic structure of the A-phase corresponds to a skyrmion lattice was not fully resolved. Soon afterwards, however, Hall-effect measurements were made (Lee et al., 2009; Neubauer et al., 2009) which strongly supported the claim that the A-phase is indeed a realization of a skyrmion lattice.

Both the theoretical predictions and the experimental findings were done on bulk samples, that is, in a three-dimensional geometry. Up to now in our discussion, the skyrmion has been examined in two-dimensional systems and thus the question of how the skyrmions look along the third dimension may arise. The answer here is that the topological defects that are point defects in two dimensions generalize to line defects in three dimensions, so skyrmions form tube-like structures without any dispersion along the third dimension.

10.5   Dimensionality and stability

As seen in Fig. 10.3(a), the skyrmion phase is only stable in a small window of suitable temperatures and magnetic fields. Thus, the observation of the skyrmion phase in MnSi spurred new efforts into not only finding other materials exhibiting skyrmion states but also finding systems where the area of their stability is larger and preferably the true zero temperature ground state. In accordance with the theoretical predictions, skyrmion lattices have since been found to exist in several other $B20$ structures, including $Fe_{1-x}Co_xSi$, which is noteworthy because it is a doped semiconductor and not a metal like MnSi.

Since inelastic neutron scattering experiments provide information from reciprocal space, only periodic structures are possible to detect and thus it should be seen as an experimental breakthrough when Yu et al. (2010) were able to observe skyrmions in real space via using Lorentz transmission electron microscopy (TEM). As the name suggests, Lorentz TEM makes use of the Lorentz force on electrons to achieve spin contrast. In addition to depicting the skyrmion in real space depiction, Yu et al. also noticed that, in the thin films they studied, the region of the skyrmion phase in the phase diagram was much larger than the tiny window observed in bulk systems. This can be explained from the fact that, if the magnetic sample is thinner than the wavelength of the ground-state, helical spin-spiral structure, the wavevector of this spin spiral has to be confined along the plane and not perpendicular to the surface. Having wavevectors parallel to the surface in turn suppresses the stabilization of conical spin spirals in applied fields perpendicular to the surface. In the bulk MnSi phase diagram discussed in Section 10.4, the conical spin spirals are stabilized over a large area of the phase diagram and, if they are suppressed, as in thin films, then there is a more favourable situation for the skyrmion lattice to be the most stable state. In the considered $Fe_{1-x}Co_xSi$ system, the skyrmion lattice was even stable down to $T = 0$ K, albeit only in the presence of a finite magnetic field.

In order to use skyrmions for possible applications, it is, of course, desirable that the material in question be able to stabilize skyrmions at room temperature. Neither MnSi nor $Fe_{1-x}Co_xSi$ fulfils this condition, since their skyrmions only appear for temperatures up to 40 K. The related compound FeGe does, however, have a much higher critical temperature (280 K) and thus one can expect skyrmions to be stable at relatively high temperatures in this material. This was confirmed in another experiment by Yu et al. (2012), who were able to observe skyrmions in a temperature range of 60–260 K by varying the thickness of the thin film. Another possible way to tune the skyrmion stability region in the phase is by alloying, and it has been shown that, by varying the concentration of $Mn_{1-x}Fe_xSi$, another $B20$ system, it is possible to tune not only the temperature range but also the radius and the chirality of the skyrmions (Shibata et al., 2013).

So far, only materials exhibiting the $B20$ structure have been mentioned in the context of carrying skyrmion states. However, what is needed to stabilize skyrmions is a competition between Heisenberg interactions and DMIs, and thus other crystal structures that lack inversion symmetry could also be expected to host skyrmion lattices in at least part of their phase diagram. One example is the insulating chiral-lattice magnet Cu_2OSeO_3, which has the same space group as the $B20$ ($P2_13$) alloys, albeit with a different atom coordination. The magnetic ground state of Cu_2OSeO_3 has been suggested to be a collinear ferrimagnet with a three-up, one-down type of arrangement (K. Kohn, 1977) and, in thin films, it has been noticed that the ferrimagnetic ordering is helically modulated periodically with a long wavelength resulting from finite but weak DMI. By applying magnetic fields on thin films of Cu_2OSeO_3, it is indeed possible to stabilize a skyrmion lattice (Seki et al., 2012). The significance of finding skyrmions in Cu_2OSeO_3 goes far beyond the notion that just another material exhibits skyrmion lattice states because, unlike $B20$ materials, Cu_2OSeO_3 is an insulating multiferroic and has thus a very different electronic structure than the previously known skyrmion materials. In addition, the multiferroic nature of this material also opens up the possibility of controlling the skyrmions with electric fields through the magneto-electric coupling. This would be very interesting from an applied point of view, since this could open up the possibility of manipulating skyrmions without engendering energy losses from Joule heating.

Since DMIs can only occur in the absence of inversion symmetry, it is possible to obtain finite DMIs even for materials that are centrosymmetric in bulk geometries, by artificially breaking that symmetry through the introduction of surfaces or interfaces to other materials. Superlattices or multilayers have for a long time been known as a possible way to tune various material properties and could thus pose a suitable avenue for the creation of suitable skyrmion-carrying materials, as could suitable deposited surfaces. It can be argued that the intense interest in DMIs and the chiral magnetism of thin film systems was actually started by the observation by Bode et al. (2007) that a single monolayer of Mn deposited on a W(110) surface exhibits a modulation of the seemingly antiferromagnetic order of the Mn spins; this modulation is driven by DMI in the same way that later was found for the thin films of Cu_2OSeO_3 mentioned above and many other systems. Following the arguments that DMI can be enhanced at surfaces, there have been several studies looking for skyrmionic or other chiral magnetic structures in surface systems and, from spin polarized scanning tunnelling microscopy (SP-STM) studies supported by extensive theoretical modelling, Heinze et al. (2011)

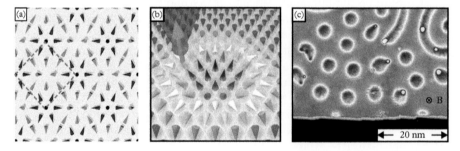

Figure 10.4 *(a) Sketch of the nano-skyrmion lattice for a monolayer of Fe on Ir(111). From Heinze, S., Bergmann, K. von, Menzel, M., Brede, J., Kubetzka, A., Wiesendanger, R., Bihlmayer, G., and Blügel, S., Spontaneous atomic-scale magnetic skyrmion lattice in two dimensions. Reprinted by permission from Macmillan Publishers Ltd: Nature Physics, 7, 713–18, copyright 2011. (b) Schematic picture of a scanning tunnelling microscopy tip creating a skyrmion. Reprinted figure with permission from Romming, N., Kubetzka, A., Hanneken, C., Bergmann, K. von, and Wiesendanger, R., Phys. Rev. Lett., 114, 177203, 2015. Copyright 2015 by the American Physical Society. (c) Spin-polarized scanning tunnelling microscopy image of a PdFe/Ir(111) surface showing both spin-spiral states and skyrmions. Reprinted figure with permission from Romming, N., Kubetzka, A., Hanneken, C., Bergmann, K. von, and Wiesendanger, R., Phys. Rev. Lett., 114, 177203, 2015. Copyright 2015 by the American Physical Society.*

were able to show that the peculiar magnetic order of a monolayer of Fe on Ir(111) has a structure that can be described as a square lattice of skyrmions on an atomic length scale, as can be seen in Fig. 10.4(a). As stressed earlier, the concept of a skyrmion is based on the assumption of continuous fields, and a clearly discrete realization of the spin structure in the work by Heinze et al. (2011) can, for that reason, not have the same sense of topological stability as larger, smoother realizations of magnetic skyrmions. The nano-skyrmion lattice found for Fe/Ir(111) does, however, have another hitherto unique feature: this particular magnetic structure is the true ground state, even at zero temperature and in the absence of external magnetic fields. This stability is due to an additional contribution to the spin model, a four-spin interaction term that arises from electron hopping between four adjacent sites (MacDonald et al., 1988). This additional term can be said to play the role of the stabilizing field, so that no external field is necessary.

10.6 From lattices to individual skyrmions

A very nice extension of the study on Fe/Ir(111) has been performed by adding an additional layer of Pd on top of the original system (Romming et al., 2013). The resulting Pd/Fe/Ir(111) structure has an altered electronic structure resulting in a change of the magnetic interactions compared to the Pd-free surface. As a result, Pd/Fe/Ir(111) behaves like the thin films of the $B20$ systems in that it has a spin-spiral ground state from which a helical skyrmion lattice can be stabilized by applying a suitably strong magnetic field. The period of the spin-spiral state is, however, much smaller than what was found for the $B20$ materials: Romming et al. (2013) reported a period of 6–7 nm. Increasing the field further saturates the system into a ferromagnetic state. By applying an intermediately strong magnetic field where the system is on the border between the skyrmion state and the ferromagnetic state, it is then possible to excite individual skyrmions by applying a spin polarized current from the SP-STM tip through the sample, as illustrated in Fig. 10.4(b) and (c). At low temperatures (4.2–8.0 K), these skyrmions can attach to pinning sites to form defects and are stabilized for several seconds.

Another avenue towards generating individual skyrmions has been demonstrated by Jiang et al. (2015), who used electrical currents and geometrical constrictions to convert striped domains into skyrmions. The type of material, a heavy metal/ferromagnet/insulator (HM/FM/I) stack, that they considered differs from those previously mentioned here in that the ground state of this system is ferromagnetic. However, owing to the emerging DMI at the heavy metal/ferromagnet interface, it is possible to create chiral domains and domain walls in the system. When currents are applied along a plane of a system containing a set of domains, a transversal current occurs, due to the spin Hall effect (Dyakonov and Perel, 1971), which results in a spin accumulation at the HM/FM interface. This spin accumulation, in turn, gives rise to an intrinsic torque, a spin–orbit torque (Emori et al., 2013), that acts on the domain walls in the system. Due to symmetry considerations of the chiral domain walls in the systems, the spin–orbit torque vanishes for domain walls parallel to the applied current and, as a result, the domains are elongated, or stretched, into stripe-shaped domains moving along the current. If a geometrical constriction, such as a wedge or notch, is present in the system, then, from continuity considerations, the current density is increased inside the constriction and, if a stripe domain starts to pass through the constriction, this increased current density cause an acceleration of the domain's head relative to the rest of the domain. If the difference between the current density inside the constriction and that outside it is large enough, the domain will break, so that the head of the domain will relax into a new, circular bubble domain while the rest of the domain will keep its striped shape. This process repeats itself several times as the stripe passes the constriction, until the whole stripe has been converted into bubble domains. In general, there is no guarantee that a circular bubble domain is equivalent to a skyrmion but, in this HM/FM/I system, the interfacial DMIs cause the domains to have chiral Néel walls and thus these bubbles are indeed topologically equivalent to a skyrmion.

10.7   Magnetization dynamics and modelling

In Sections 10.1–10.6, we gave an overview of the experiments that have led from the identification of skyrmion lattices to the possible creation and manipulation of individual skyrmions. A very large part of the research on skyrmions has, however, been of a theoretical or numerical nature. As mentioned in Section 10.5, it is in general enough to have a spin model containing Heisenberg and Dzyaloshinskii–Moriya exchanges together with a Zeeman term in Eqn (4.42), to account for applied external fields (even though additional terms such as anisotropies or many-body exchange interactions can also be introduced). For most systems where skyrmions have been shown to exist, the skyrmions are quite large compared to the atomic length scale, and it becomes relevant to use a micromagnetic description in which the Hamiltonian becomes (Aharoni, 1996)

$$H = \int d\mathbf{r} \left[A(\nabla \mathbf{M})^2 + D\mathbf{M}(\nabla \times \mathbf{M}) - \mathbf{B}\mathbf{M} \right], \tag{10.2}$$

where A is a parameter governing the strength of the exchange energy, D is the DMI vector of a continuum model, and the last term represents the Zeeman energy. The comparison between the micromagnetic parameter D and the atomistic $\mathbf{D}_{ij}$ is not always straightforward, due to the underlying lattice. However, it should in principle be possible to construct parameters of a continuum model, once the atomistic parameters and the underlying lattice are known. Efforts to obtain such coupling are often referred to as multiscale approaches (as are briefly discussed in Section 12.7). However, we remark that one often tries to adapt, for example, A and D to the symmetry of the problem at hand. By analysing how A and D scale with the size of the system, it is possible to combine Eqns (4.42) and (10.2). For skyrmions in a two-dimensional lattice, this can be used to represent the micromagnetic system as a nearest-neighbour Hamiltonian on a square lattice, by introducing a rescaling so that $\mathcal{J}_{ij} = A/d$, $|D_{ij}| = D/d^2$, and $\mathbf{B}^{\text{ext}} = \mathbf{B}/d^3$, where d is the length of a cube where the micromagnetic magnetization density can be considered to be constant. By comparing the calculated phase diagram from this discrete representation of the micromagnetic problem with the experimental phase diagram for thin films of $Fe_{1-x}Co_xSi$, Yu et al. (2010) showed that this simple model can capture the thermodynamics of the real problem surprisingly well.

This coarse-grained nearest-neighbour model has also been used for model studies of the dynamics of skyrmions in *B*20 materials. Many studies of skyrmion dynamics address the question of how skyrmions move in the presence of an applied electric current. In order to model the torques that occur due to the transfer of angular momentum between a spin polarized current and skyrmion or other topological defect which has a finite gradient of the magnetization, the stochastic Landau–Lifshitz–Gilbert equation, Eqn (4.1), can, as discussed in Sec. 7.3, be augmented with additional terms that describe these spin-transfer torques (Slonczewski, 1996; Ralph and Stiles, 2008; Iwasaki et al., 2012). The extra terms that are to be added to the right-hand side of Eqn (4.1) are

$$\frac{p}{2eM}\mathbf{M} - \frac{p\beta}{2eM}\,\mathbf{M} \times [\mathbf{j}(\mathbf{r}) \cdot \nabla]\{\mathbf{M} \times [\mathbf{j}(\mathbf{r}) \cdot \nabla]\mathbf{M}\}. \tag{10.3}$$

The first term describes the adiabatic contributions to the spin-transfer torque, and the second term describes the non-adiabatic contribution, where β determines the strength of this non-adiabatic torque. The parameter p is the spin polarization of the current of the material, and $\mathbf{j}(\mathbf{r})$ is the electrical current density.

Using this model, it has been shown that skyrmions do indeed exhibit current induced dynamics that differ from that of domain walls or spin helices. To exemplify, skyrmions do not seem to be as sensitive as domain walls to pinning on magnetically hard defects (Iwasaki et al., 2012), that is, defects with a large magnetic anisotropy. Non-adiabatic effects do not seem to have as large influence on skyrmions as well, once again contrasted with typical domain wall behaviour. On the other hand, similar simulations has shown that non-magnetic impurities can affect the skyrmion motion if the size of the defect is relatively large compared to the skyrmion radius (see Koumpouras and Bergman, 2016). This latter finding is also in agreement with simulations on finite samples, as a large repulsive effect on skyrmion motion is observed at the edges of such samples (Iwasaki et al., 2013). The surfaces thus act as potential barriers to confine the skyrmions' motion. This could also be seen as a manifestation of the topological protection the skyrmions exhibit, since driving skyrmions out of the sample would break their structure and, as a consequence, the topology of the skyrmion would break as well. And, even though the topological barrier is not infinite, as discussed previously in this section, breaking the topology still comes with a cost of energy. Similar simulations have been performed using a more standardized micromagnetic approach including DMIs (Rohart and Thiaville, 2013) on ferromagnetic systems with a perpendicular magnetic anisotropy and large interfacial DMIs (Fert et al., 2013; Sampaio et al., 2013), as exemplified in Fig. 10.5(a). Landau–Lifshitz–Gilbert simulations on skyrmions has also shown that their dynamics can be described using Thiele's (1973) equations for domain motion and that, due to the rotational nature of the skyrmions, an occurring Magnus force gives a Hall-like contribution to their motion.

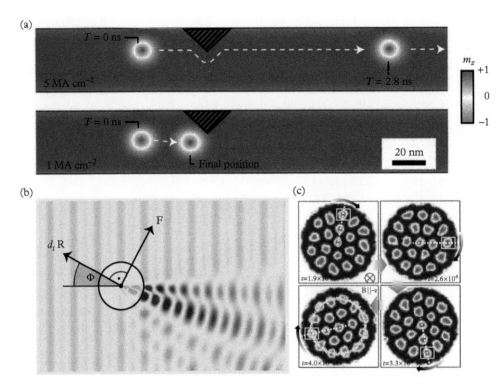

Figure 10.5 *(a) Results from micromagnetic simulations of current-induced skyrmion dynamics. From Fert, A, Cros, V., and Sampaio, J., Skyrmions on the track. Reprinted by permission from Macmillan Publishers Ltd: Nature Nanotechnology, 8, 152–6, copyright 2013. (b) Illustration of magnon–skyrmion scattering. Reprinted figure with permission from Schütte, C., and Garst, M., Phys. Rev. B, 90, 094423, 2014. Copyright 2014 by the American Physical Society. (c) Simulations showing a rotational motion of skyrmions in a temperature gradient on a finite sample. From Mochizuki, M., Yu, X. Z., Seki, S., Kanazawa, N., Koshibae, W., Zang, J., Mostovoy, M., Tokura, Y., and Nagaosa, N., Thermally driven ratchet motion of a skyrmion microcrystal and topological magnon Hall effect. Reprinted by permission from Macmillan Publishers Ltd: Nature Materials, 13, 241–6, copyright 2014.*

Temperature effects are also important to understand in order to be able to have full control over the dynamics of skyrmions. Several studies has included finite temperature simulations and it has been shown that, in agreement with earlier experimental results, temperature gradients combined with very small currents can cause a rotational motion of a lattice of skyrmions (Everschor et al., 2011). Since quantized spin wave excitations in the form of magnons occur in magnetic materials at finite temperatures, studies concerning skyrmion–magnon interaction are also very related to the temperature dependence of skyrmion dynamics. Micromagnetic simulations of magnon–skyrmion scattering, as illustrated in Fig. 10.5(b), have shown that magnons are deflected by skyrmions and that the deflection angle depends strongly on the magnon wavelength

(Iwasaki et al., 2014). The skyrmions are, in turn, deflected by the magnons, which give rise to a topological magnon Hall effect (Schütte and Garst, 2014). This effect was also shown in a related study, where magnons created by a temperature gradient resulted in a rotation of a finite system containing a close-packed lattice of skyrmions (Mochizuki et al., 2014), as can be seen in Fig. 10.5(c).

11

Ultrafast Switching Dynamics

The time-integrated amount of data and stored information is doubled roughly every 18 months and, since the majority of the world's information is stored in magnetic media, the possibility of writing and retrieving information in a magnetic material at ever greater speed and with lower energy consumption has obvious benefits for our society. Hence, the seemingly simple switching of a magnetic unit, a bit, is a crucial process which defines how efficiently information can be stored and retrieved from a magnetic memory. Retrieving, or reading, information from a magnetic medium is performed by sensing the magnetization direction of the bit. This can nowadays be done very efficiently by magnetoresistive spin valves (Baibich et al., 1988; Binasch et al., 1989; Miyazaki and Tezuka, 1995; Moodera et al., 1995) and, while spin-dependent transport is a very interesting topic, it lies outside the scope of this book. Instead, we will in this chapter describe how information can be stored magnetically, with exceptional speed.

11.1 Background

In the present context, storing data means writing information to a magnetic bit by changing the bit's magnetization. From an applied point of view, it is apparent that it is advantageous to be able to switch the magnetization of a bit as fast as possible, since an increased writing speed would lead to a faster operating device. In addition to the pure speed of the magnetization switching, or magnetization reversal, as the process also can be called, there are other aspects that are important for the process, such as minimizing energy losses and heating. As a result, an efficient switching mechanism involves a balance between the switching speed and the amount of external stimuli, be it magnetic fields, electrical currents, or heat. All in all, this balance opens up a lot of questions regarding how and in which kinds of materials magnetization reversal can be controlled efficiently.

Of particular interest here are the concepts of ultrafast magnetism and all-optical control of magnetism, both of which have in recent decades become the basis of intense research. The motivation is natural: the mechanisms behind these phenomena are far from trivial, and the technological implications are huge. Ultrafast magnetization dynamics has naturally become a focus area, not least for practical aspects of information

Atomistic Spin Dynamics. Olle Eriksson, Anders Bergman, Lars Bergqvist, Johan Hellsvik. First Edition.
© Olle Eriksson, Anders Bergman, Lars Bergqvist, Johan Hellsvik 2017. First published in 2017 by Oxford University Press.

technology. The breakthrough experiment in this field is the work by Beaurepaire and co-workers (1996), but it should be noted that even earlier works on fast magnetization dynamics had at that time been published, for example, by Aeschlimann et al. (1991). In this chapter, different aspects of magnetization switching will be discussed, and important and interesting studies pertaining to this topic will be showcased. Since reversal time is a crucial parameter for the dynamical processes in this topic, it is a good starting point for discussing the different time scales present in magnetic phenomena. In Fig. 11.1 we display time scales of observed, dynamical aspects of magnetism. It is quite remarkable that observed magnetic phenomena span a time scale of roughly 30 orders of magnitude. In the figure, the longest time scale is covered by the constant reversal of the Earth's magnetic poles, something which is still a much discussed scientific topic (Glatzmaier and Roberts, 1996). Other magnetic time scales shown in Fig. 11.1 and which are more or less noticeable in every day life are the expected life time of a magnetic storage device, the frequencies used in an AC transformer, and time scales used in NMR imaging of biological tissue and for medical diagnostics. At even higher frequencies, we enter the domain of dipolar and exchange magnons, which were discussed in Chapters 5, 6, 8, and 9. Phenomena with the very shortest time scales, on the order of

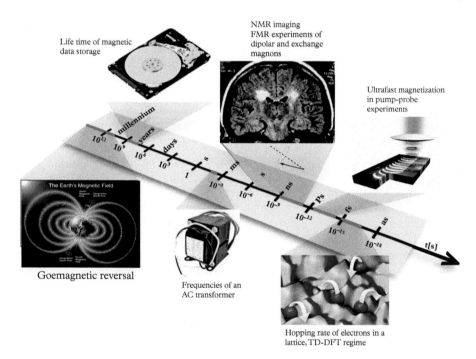

Figure 11.1 *Time scales of different magnetic phenomena, involving geomagnetic reversal, life times ferromagnetic resonance of magnetic hard discs, NMR and ferromagnetic resonance (FMR) experiments, magnon frequencies, and ultrafast magnetization dynamics; TD-DFT, time-dependent density functional theory.*

picoseconds or faster, are referred to as ultrafast and make up the topic of this chapter. Starting from a general discussion about switching processes and concepts important in this context, we will throughout the chapter move towards smaller and faster systems, going from macrospin behaviour down to where the entire concept of localized magnetic moments becomes uncertain. Unless otherwise stated, the discussions in this chapter will be based on magnetic materials which are ferromagnetic and have uniaxial magnetic anisotropy.

11.2   Energy barriers, domains, and domain walls

With the possible exception of the kind of magnetization reversal processes associated with the concept of ultrafast demagnetization, where the adiabatic approximation behind the atomic moments may break down and where electron processes play a more prominent role, most switching mechanisms can be described by the collective, or individual, motion of atomic spins across one or more energy barriers. The barriers are determined by the system's magnetic anisotropy, which can originate from magnetostatic dipole interactions, which give rise to shape anisotropies; spin–orbit effects, resulting in magnetocrystalline anisotropies (see Chapter 3, Section 3.3); or a combination of both. While the magnitude of this energy barrier typically affects the speed of the magnetization reversal process, so that a larger barrier implies a slower switching rate, having a large magnetic anisotropy is important since it keeps the system stable against thermal fluctuations. The energy barrier is proportional both to the magnetic anisotropy and to the volume of the system so, given a particular strength of the magnetic anisotropy energy per atom, there is thus a limit to how small a system can be and still have an energy barrier that is larger than the thermal effects at ambient conditions. Below this volume, the system no longer has an expected magnetization direction over time and becomes, like paramagnetic atoms, superparamagnetic. That means that, in magnetic storage systems, the motivation for using materials with a large magnetocrystalline anisotropy energy (MAE) comes foremost from the drive to decrease the bit size and thus increase the storage density, and not primarily from increasing the switching speed.

As discussed in Section 4.8, the total energy of a magnetic system is determined by the strengths of its interatomic exchange interactions and magnetic anisotropies. If we consider two neighbouring regions, that is, domains, with different magnetization directions, then the ordering of the atomic moments close to the boundaries of these domains depend on the interplay by the exchange energy and the anisotropy energy of the system. If no anisotropy is present, there will be no energy barrier for the moments, and the two domains will eventually relax into a state where their magnetization locally will be ferromagnetically aligned and collinear. On the other hand, if the system does not have any finite exchange interactions between the domains, then the magnetization will abruptly change direction across the domain boundary. In real materials, both exchange and anisotropy energies are present, and the resulting magnetic structure across the interface between the two domains can be described as a smooth rotation of the magnetization: a

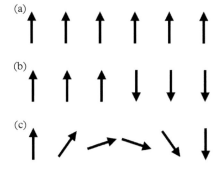

Figure 11.2 *Schematic illustration of domain wall width as a function of exchange interaction $\mathcal{J}$ and anisotropy energy K. (a) K = 0, and $\mathcal{J}$ > 0, result in a ferromagnetic configuration. (b) K ≠ 0, and $\mathcal{J}$ = 0, result in an abrupt change of the magnetization. (c) K ≠ 0, and $\mathcal{J}$ ≠ 0, give a smooth domain wall, here depicted as a Néel wall.*

domain wall. A simple illustration of these scenarios is seen in Fig. 11.2. From domain theory (Coey, 2010), the domain wall width in a magnetic system can be estimated as follows. As long as the magnetic anisotropy is much smaller than the exchange interaction strength, as is typically the case, at least for bulk materials, the width of the domain wall can be expressed as a function of the ratio between the anisotropy constant K and the micromagnetic exchange constant A, which is related to the sum of all interatomic exchange interactions $\mathcal{J}_{ij}$ as follows:

$$\delta_W = \pi\sqrt{\frac{A}{K}}. \tag{11.1}$$

Applying Eqn (11.1) to literature values for the exchange and anisotropy constants (Lilley, 1950) for bulk transition metals, the domain wall width may be estimated to be 40 nm for body-centred cubic Fe, and 15 nm for hexagonal close-packed Co. In contrast, in systems with lower dimensionality and symmetry and where the MAE is enhanced, domain walls are much thinner; for example, for a monolayer of Fe on a W(110) surface, the domain wall width has been calculated to be 1.9 nm (Chico et al., 2014). The domain wall width is important for magnetization switching because it can determine how the actual reversal process behaves. If the considered system has a domain wall width that is comparable to, or even larger than, the magnetic system, then only one domain will fit in the sample and we end up with a single-domain ferromagnet where the magnetization reversal will be governed by how the atomic moments move uniformly in the presence of an external stimulus. Since all individual atomic moments in this scenario point in the same direction as the magnetization, the system can be said to behave as a single large spin: a macrospin. If, on the other hand, the domain wall width is small compared to the system size, then several domains can form across the sample, and the reversal is determined by the movement of the domain walls. Combined, these observations mean that, more often than not, a magnetization reversal process can be said to either behave as a macrospin or be determined by domain wall motion.

11.3 Macrospin switching

Macrospin switching is typically quite easy to explain and model, although there are exceptions, including one example of the magnetization reversal of nano-islands (Etz et al., 2012). The simplicity comes from the fact that in this model, it suffices to model a single spin in an external magnetic field. First of all, the Stoner–Wolfarth (Stoner and Wohlfarth, 1948) theory is applicable to macrospin switching and, even though this theory says nothing about the dynamics of the switching, it still gives a good model for the necessary magnitude of the external field for a reversal to take place and can also describe the energetics and hysteresis of the process quite well. Regarding the dynamics of a single macrospin, it is described by the SLLG equation of motion Eqn (4.1). An important point to take note of here is that, when it comes to the expression for the effective magnetic field, the field stemming from the magnetic anisotropy, as well as the external field, is crucial for determining the dynamics, while the exchange field present in the system is not important, since it is at all times parallel to the macrospin. Another insight that can be made from the SLLG equation is that, unless the applied magnetic field is larger than the intrinsic magnetic field that comes from the MAE, for example, the anisotropy field, no reversal will take place.

As mentioned in Section 4.1, SLLG dynamics can be partitioned into two parts: a precessional part, which describes the rotation around the effective magnetic field axis, and the damping part, which describes a relaxing motion towards the axis of the magnetic field. The obvious way of reversing the direction of a macrospin is to apply a magnetic field that is larger than the anisotropy field and antiparallel to the current magnetization direction.[1] Then, the macrospin will precess around the magnetic field with a frequency proportional to the magnitude of the effective field while simultaneously relaxing towards the field axis at a rate proportional both to the effective field magnitude and to the Gilbert damping parameter α. Since, typically, $\alpha \ll 1$, the damping motion is much slower than the precessional motion, so the resulting reversal process will not be very efficient. An alternative to this damping-driven switching, potentially more efficient but with additional difficulties regarding pulse controls and timings, is what is commonly called precessional switching. Here, the driving field is not antiparallel to the starting magnetization but perpendicular. With this geometry, it is then the precessional motion of the macrospin that moves the spin towards the switched direction, while the damping motion would, in principle, drive the spin towards an unwanted state. Since the precessional motion is much faster than the damping motion, more quickly than precessional switching is carried out much damping-driven switching (Back et al., 1999; Gerrits et al., 2002; Tudosa et al., 2004). The difference between the switching speed of damping-driven switching and that of precessional switching is illustrated in Fig. 11.3, where the time evolution for a macrospin with uniaxial anisotropy is subjected to an external field. The external field is almost antiparallel to the macrospin for the damping-driven case, and perpendicular to the macrospin in the precessional case. As can be seen in Fig. 11.3, the precessional switching is more than two orders of magnitude faster

---

[1] A perfectly antiparallel magnetic field would not result in a finite torque but, in reality, there are always thermal fluctuations and/or small misalignments of the effective field that would be enough to start the motion.

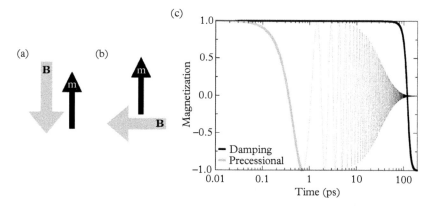

Figure 11.3 *Illustration of the direction of the applied magnetic field* **B** *(grey arrow) and magnetisation (black arrow) in (a) damping driven switching and (b) precessional switching. In (c) the z-component of the magnetization is shown, illustrating the difference in efficiency between the two different kinds of switching as simulated for a macrospin with a uniaxial anisotropy of 0.1 mRy in an applied field of 5 T (precessional switching shown in thick grey and damping switching in thick black). If the external field is present for a long-enough time in the precessional switching scenario, the z-component of the magnetization will follow the path that is indicated by the thin grey line in the figure and will relax to an unwanted state parallel to the external field. The z-direction is pointing upward in the figure.*

than the damping-driven switching. Another important advantage of precessional-driven magnetization reversal is that in precessional switching, because the field needs to be applied for a shorter time than in damping-driven switching, and the transport of energy to and from the system depends on the damping part of the motion, the energy cost and resulting heat loss of the switching are decreased significantly. As just mentioned, before, there are difficulties when performing precessional switching. One key issue is to ensure that the external field is pulsed for a well-defined time since, if the field is applied for too long, the damping process will eventually align the magnetization along the axis of the external field, which in this case is then directed between the two well-defined magnetization states (as shown by a thin grey line in Fig. 11.3 c). Since the precessional motion is very fast it can also happen that a slightly mistimed pulse can result in a whole rotation of the magnetization, that is, a return to the starting state, and not the desired half rotation that would switch the magnetization properly.

11.4 Internal-field-assisted switching

Another way to improve the efficiency of macrospin switching, in addition to optimizing the size, direction, and duration of the externally applied field and then using that as the driving stimulus, is to use intrinsic, or internal, fields. Compared to external magnetic fields, the interatomic exchange field is very large—on the order of hundreds of teslas.

So, if these internal fields can be utilized, they can be expected to assist and speed up the switching process. Since the exchange field in a ferromagnetically ordered material is parallel to the magnetization, this assisting effect can either come from anisotropy fields or in materials where interactions other than ferromagnetism are considered. One example of where the efficiency of a switching process can be drastically improved has been found from atomistic spin dynamics (ASD) simulations on artificial antiferromagnets in the form of trilayer or multilayer structures (Bergman et al., 2011). The system that was studied from this aspect was a Fe/Cr/Fe trilayer structure, but the underlying physical process is general and can, in principle, be applied to any suitable layered and shaped antiferromagnet. As a starting point, we consider a trilayer structure in which the middle layer, or spacer, is extended and connected to leads, while the ferromagnetic end layers are antiferromagnetically coupled to each other, due to the RKKY interaction mediated through the spacer, as depicted in Fig. 11.4(a). If a current is then driven through the spacer, it will give rise to an Oerstedt field in the ferromagnetic layers. This current-induced field will either be parallel to the magnetization of both ferromagnetic layers, or antiparallel to both. If the field is antiparallel, the magnetization in the ferromagnetic layers will start to rotate in the Oerstedt field. As an effect of the particular symmetry of the system, the angle between the magnetization directions in the two antiferromagnetically coupled layers change from 180° and, as a result, an internal exchange field will emerge,

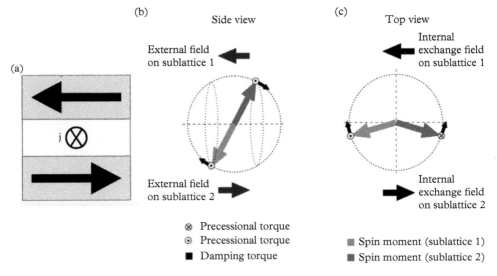

Figure 11.4 *Illustration of geometry and emerging fields in an artificial antiferromagnetic trilayer system. (a) Depiction of the system consisting of two ferromagnetic layers that are antiferromagnetically coupled across a spacer through which a current is driven. The current gives rise to Oerstedt fields in the ferromagnetic layers. (b, c) The internal and external exchange fields that occur as an effect of the precession of the ferromagnetic layers in the Oerstedt field. The results are based on atomistic spin dynamics simulations. Reprinted figure with permission from Bergman, A., Skubic, B., Hellsvik, J., Nordström, L., Delin, A., and Eriksson, O., Phys. Rev. B, 83, 224429, 2011. Copyright 2011 by the American Physical Society.*

as shown in Fig. 11.4(c). The internal exchange fields assist the magnetization switching in the two ferromagnetic layers, mostly through the precessional torque originating from the internal field, which coincides with the damping torque from the current-induced external field. As a result, simulated switching times on the order of less than 20 ps for an Oerstedt field of 100 mT was found, which is more than one order of magnitude faster than what would be the case for pure damping-driven switching in bulk Fe. While this particular geometry would prove difficult to exploit in an actual application, due in part to the large current density that would be needed to cause the needed magnitude of the Oerstedt fields, the system still works as an illustration of how internal fields can assist magnetization reversal processes.

11.5   Inertia-like switching

Internal magnetic fields can not only increase the speed of magnetization reversal processes but can also give rise to what can be described as inertia-like switching (Kimel et al., 2009), that is, a continuing switching process even in the absence of an external stimulus. Inertia in the proper sense comes from the fact that massive bodies follow Newton's second law, which can be expressed as a second-order differential equation where the acceleration is proportional to the net force. Thus, in the absence of a net force, an object moves with constant velocity. From a simple inspection of the Landau–Lifshitz–Gilbert (LLG) equation of motion for magnetic moments (Eqn (6.1)), we see that it is a first-order differential equation and that, in the absence of a finite effective field (which would be the analogue of the net force for particles), there is no motion of the moments. In the light of this observation it might seem contradictory that there have indeed been several reports of inertia-like behaviour in magnetic systems (Kimel et al., 2009; Thomas et al., 2010). In the study by Kimel et al. (2009), inertia-like dynamics was found for a canted antiferromagnet, $HoFeO_3$, where the canting occurs due to Dzyaloshinskii–Moriya interactions when the antiferromagnet is stimulated with a 100 fs, optically generated, strong magnetic field pulse. From pump-probe experiments, as will be described later in Section 11.7, it was observed that, despite the short pulse, the magnetization dynamics of the system persisted for more than 50 ps after the removal of the pulse. The behaviour was explained by introducing a formulation of the equations of motion for an antiferromagnetic ordering vector, instead of individual magnetic moments, based on a variation of the Lagrangian for the system (Bar'yakhtar et al., 1985). The Lagrangian here included the kinetic energy of the system and, as a result, the equation of motion became second order and could thus show inertial-like dynamics. It was further speculated that this dynamical behaviour could be found in materials other than for this particular rare-earth-based oxide, in particular, in artificial antiferromagnets.

These results inspired theoretical investigations of pulse-driven dynamics in artificial antiferromagnetic systems, including ASD simulations on Fe/Cr/Fe trilayers, that is, the same system for which internal field-assisted switching had been predicted (Bhattacharjee Bergman, et al., 2012). These simulations did not include

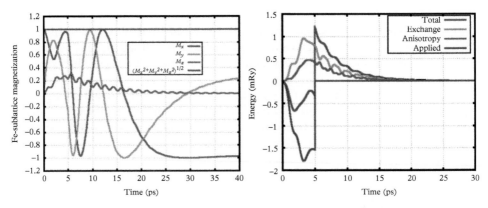

Figure 11.5 *Simulated dynamical response to a 5 ps tilted magnetic field pulse on an Fe/Cr/Fe trilayer structure. (Left) The components of the magnetization in one of the ferromagnetic Fe layers. (Right) Calculated contributions to the total energy of the system. The results are based on atomistic spin dynamics simulations. Reprinted figure from Bhattacharjee, S., Bergman, A., Taroni, A., Hellsvik, J., Sanyal, B., and Eriksson, O., Phys. Rev. X, 2, 011013, 2012. Copyright 2012 by the American Physical Society.*

Dzyaloshinskii–Moriya interactions, which were present in the experimentally studied canted antiferromagnet, but both antiferromagnetic exchange and various types of magnetocrystalline anisotropies were used in the simulations. The system was pulsed with a large magnetic field which had both an in-plane component of 3 T, and an out-of-plane component of 5 T, for a duration of 5 ps. The result of the simulations, shown in Fig. 11.5, was that there was indeed dynamics for an extended time after the removal of the magnetic field pulses. By investigating the behaviour of the internal fields and the corresponding contributions to the total energy during the simulations, the inertia-like dynamics could be explained without the need to introduce the Lagrangian for the antiferromagnetic order vector. Instead, the explanation can be simply expressed in two different but coherent ways. Using arguments focusing on the effective field that enters the LLG equation of motion, one explanation is that the strong external field pulse drives the system into a strong out-of-equilibrium state. In that state, the two antiferromagnetically coupled layers are no longer aligned antiparallel to each other, and neither layer has its magnetization aligned along an easy anisotropy axis. As a result, there will be sizeable contributions of both exchange fields and anisotropy fields to the total effective field. These internal contributions to the field are finite as long as the system is not in equilibrium, and the total effective field will thus be finite even when the external field is removed. The inertia-like dynamics is therefore nothing else than regular magnetization dynamics in an effective field, except that the external contribution is 0. The other way of explaining the post-pulse dynamics is to consider the different contributions to the total energy that are present in the system. In equilibrium, the total energy is minimized, and no motion is observed. The strong magnetic field pulse does, however, pump energy into the system, and this energy is then distributed to exchange energy and anisotropy energy as the system moves to an excited state during the pulse (Fig. 11.5). When the pulse vanishes, there is still a surplus of energy in the system and, as a result, the system evolves in time until all excess energy has dissipated from the spin system through the damping motion, as described by the second term in the LLG equation. A small Gilbert damping parameter will have a slow rate of dissipation and, as a result, the dynamics can continue for a long time, and vice versa, if the Gilbert damping is large. A similar kind of dynamics was presented in Chapter 4, Section 4.6, where the relaxation of a single spin flip in a ferromagnetic thin film was discussed.

11.6 Domain wall motion

An alternative to storing and manipulating individual and isolated macrospins is to utilize single domains in a multi domain system for information storage. In this case, we do not want to reverse the magnetization of the whole system but instead only change the magnetic configuration of selected domains. Regardless of the type of external stimulus, be it magnetic fields or spin polarized currents, the reversal processes involve the motion of domain walls. One domain-wall-driven storage solution that has attracted considerable attention is the domain wall shift register or, as it is more commonly referred to, magnetic racetrack memory (Parkin et al., 2008). Here, the idea is to create a solid state memory that does not have moving parts and where the information is stored as a sequence of domains in a magnetic, stripe-shaped material and is read by shifting the whole sequence of domains back and forth across a magnetic sensor.

The domain walls can be moved by applied magnetic fields but a more efficient way to manipulate them is to drive an electrical current through the system. Since the material is spin polarized, there will be a difference between the amount of conduction electrons with majority spin and the amount with minority spin and, as a result, the current will have a polarization. When the spin polarized current passes through a domain wall, where the magnetization direction changes, the spins of the charge carriers will align with the changed magnetization and, as a result, there will be a transfer of angular momentum between the passing current and the localized moments in the domain wall. This angular momentum transfer causes the polarization of the current to rotate towards the local magnetization direction in the domain wall, and the localized moments situated where the domain wall is rotate towards the polarization of the current. As a result, the magnetic moments present in the whole domain wall rotate and the domain wall moves. This effect is commonly referred to as spin-transfer torque (STT; Slonczewski, 1996), and more details about this important phenomenon can be found in the overview article by Ralph and Stiles (2008).

Studies of STT-driven domain wall motion have addressed a number of questions, including how pinning centers in the form of defects and impurities affect the motion, and how materials can be optimized in order to facilitate as effective a manipulation of the walls as possible. Recently, there has been an increase in the activity in this field, after the realization that thin film systems, especially ones consisting of repeated transition metal/heavy metal layer structures, show a current-induced domain wall motion that is very different from what has been found in thicker samples (Emori et al., 2013; Ryu et al., 2013, 2014). This difference has been attributed to the enhancement of spin–orbit effects such as interfacial Dzyaloshinskii–Moriya interactions and spin Hall effects that occur as a combination of artificial symmetry breaking and reduction of the thickness of the layers.

11.7   Ultrafast demagnetization: face-centred cubic Ni as an example

So far, most of the discussion in this chapter has concerned the precessional motion of localized magnetic moments on time scales down to the picosecond level but, as mentioned in Section 11.1, a very active field of research into ultrafast magnetism and all-thermal and/or all-optical control of the magnetization is currently ongoing. Following the work of Beaurepaire et al. (1996), many studies have given insight into various aspects of ultrafast magnetism, and magnetic phenomena with femtosecond time scales have been reported (e.g.in Gerrits et al., 2002; Kimel et al., 2004; Bigot et al., 2004; Tudosa et al., 2004; Stamm et al., 2005; Bigot et al., 2009; Vahaplar et al., 2012). A typical experimental set-up for these investigations is shown in Fig. 11.6. A high intensity laser pulse is used to pump the sample and, after a time delay of the order of a few femtoseconds up to typically 20 ps, a second (much weaker) laser pulse is used to probe the sample's magnetic state. As Fig. 11.6 shows, the probe step of this experiment can be done either for light reflected from the sample (Kerr mode) or for light transmitted through the sample (Faraday mode). The two measurements give the same information (Bigot et al., 2004) and rely on the fact that the magnetic signal of the probe beam reflects the magnetic state of the material via the magneto-optical effect. The probe part has subsequently also been realized with X-ray magnetic circular dichroism, giving element-specific information (Fukumoto et al., 2008).

An example of a pump-probe experiment of magnetization dynamics is the investigation of Beaurepaire et al. (1996) who studied face-centred cubic (fcc) Ni; the results from this study are show in Fig. 11.7. The magnetism is here shown as a function of the

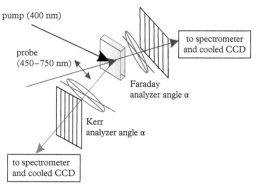

Figure 11.6 *Pump-probe experimental set-up. Figure from Bigot, J.-Y., Guidoni, L., Beaurepaire, E., and Saeta, P. N., (2004), Femtosecond spectrotemporal magneto-optics, Phys. Rev. Lett., **93**, 077401. Copyright 2012 by the American Physical Society.*

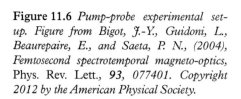

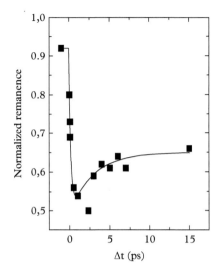

Figure 11.7 *Measured magnetization dynamics of face-centred cubic Ni. Figure from Beaurepaire, E., Merle, J.-C., Daunois, A., and Bigot, J.-Y., (1996), Ultrafast spin dynamics in ferromagnetic nickel, Phys. Rev. Lett., 76, 4250–3. Copyright 2012 by the American Physical Society.*

delay time, Δt, between the pump and the probe (as described in Fig. 11.6), and it may be seen that, over a few picoseconds, the magnetism drops to 55 % of its saturation value. At the time of this experiment, such fast magnetic phenomena were not anticipated, and so the term 'ultrafast' was coined. By now, such high speeds are more or less routinely measured, and even higher speeds down to sub-picoseconds, are observed. Experimentally, it is by now known that the pump pulse excites the electron sub-system, and the excited hot electrons become thermalized rather quickly, the order of 0.5–1.0 ps (Rhie et al., 2003) The heat of the electron subsystem is transferred to the system of atomic spins, as well as to the lattice, so that, after some 10–20 ps the three reservoirs reach thermal equilibrium. This is described in the so-called three-temperature model, which was first introduced Beaurepaire et al. (1996) and later modified by Koopmans et al. (2005); Koopmans et al. (2010). We will return to that model in Section 11.8. First, we note that, in the first 10–20 ps of pump-probe measurements, the material is not in thermal equilibrium between the different reservoirs, and the temperature changes dynamically over the interval of the probe part of the measurement. Hence, any thermal effect in the electron subsystem, the spin subsystem or the lattice subsystem may show up in the magneto-optical response. The most obvious effect is magneto-optical coupling, so that a true response of the magnetism is the dominant contribution to the data shown in Fig. 11.7.

As we shall see in Section 11.8, the three-temperature model describes the data in Fig. 11.7 rather well, but one must bear in mind that there is a conceptual difficulty in distinguishing the electron temperature from the temperature of the atomic spin bath, since it is the electrons themselves, and their pairing, that form the spin system (also discussed in Section 2.6). However, with the Born–Oppenheimer-like approximation made in the ASD approach, it becomes very meaningful to make this distinction, where slower atomic spins have a dynamical behaviour governed by one temperature, and the much faster electrons experience a different temperature. As discussed in Chapter 2, for time scales larger than some 10–100 fs, this should be a valid approximation. The electronic temperature then enters the Fermi–Dirac distribution of Kohn–Sham states, as discussed in Chapter 1, whereas the spin temperature enters the stochastic field of the stochastic Landau–Lifshitz (SLL) equation, as discussed in Chapters 4 and 5. This approach was adopted, for example, by, Chimata et al. (2012). Another conceptual difficulty that has been discussed at length is the channel for the dissipation of angular momentum in pump-probe experiments such as the one shown in Fig. 11.7. Koopmans et al. (2005); Koopmans et al. (2010) argued that Elliott–Yafet-type scattering, described by the probability that an electron flips its spin on emission or absorption of a phonon, can cause demagnetization in a process that conserves the total angular momentum. An alternative explanation was made by Battiato et al. (2010), who argued that spin currents, in a so-called superdiffusive mechanism, explain ultrafast demagnetization measurements. Spin currents were indeed reported experimentally by Eschenlohr et al. (2013), although this work was criticized by Khorsand et al. (2014). The discussion above shows that there are still unanswered questions concerning these kinds of experiments, although progress is certainly made in understanding these data better.

Different magnetic materials respond differently to pump-probe experiments, with respect to the temporal evolution of the magnetism and, in particular how the magnetism recovers after a pump has been made. This has led to the loose classification of materials in these experiments as type I and type II magnets. Among the materials classed as type I, one finds fcc Ni and, among those classed as type II, one finds, for example, hexagonal close-packed Gd. It should, of course, be remarked that a natural division of these materials into two groups is easily done via their electronic structures: the rare-earth magnets are all representatives of localized electron systems, whereas the transition metals are itinerant electron systems. The former are typically good Heisenberg systems, while the latter have a larger amount of Stoner excitations present. Anyway, it is interesting that the division of magnets into types I and II was predicted (Koopmans et al., 2010) before experiments confirmed the existence of these classes (Roth et al., 2012).

11.8   The three-temperature model

We now return to the three-temperature model. As mentioned in Section 11.7, this model was first introduced by Beaurepaire et al. (1996), and it describes how heat is communicated among the spin, electron, and lattice reservoirs. This is done in a rather intuitive approach that uses three simple equations:

$$C_e(T_e)\frac{dT_e}{dt} = -G_{el}(T_e - T_l) - G_{es}(T_e - T_s) + P(t),$$

$$C_s(T_s)\frac{dT_s}{dt} = -G_{es}(T_s - T_e) - G_{sl}(T_s - T_l), \tag{11.2}$$

$$C_l(T_l)\frac{dT_l}{dt} = -G_{el}(T_l - T_e) - G_{sl}(T_l - T_s).$$

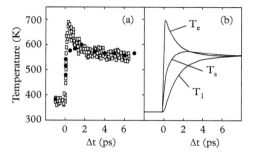

Figure 11.8 *(a) Estimated experimental spin tem-peratures (filled circles) and electron temperatures (open squares) of face-centred cubic Ni in a pump-probe experiment. (b) Temperature profile according to the three- temperature model; Δt, change in time; T_e, temperature of the electron reservoir; T_l, temperature of the lattice reservoir; T_s, tempera-ture of the spin reservoir. Figure after Roth, T., Schellekens, A. J., Alebrand, S., Schmitt, O., Steil, D., Koopmans, B., Cinchetti, M., and Aeschli-mann, M., (2012), Temperature dependence of laser-induced demagnetization in Ni: A key for identifying the underlying mechanism, Phys. Rev. X, 2, 021006. Copyright 2012 by the American Physical Society.*

In these equations, T_e, T_s, and T_l represent the temperatures of the electron reservoir, the spin reservoir and the lattice reservoir, respectively, while C_e, C_s, and C_l are the cor-responding specific heats. Coupling between the different reservoirs is provided by G_{es}, G_{el}, and G_{sl}, while $P(t)$ represents the effect provided by the laser in the pump process. The three-temperature model was found by Beaurepaire et al. (1996) to describe well the temperature of the electron subsystem and the spin subsystems, (see Fig. 11.8).

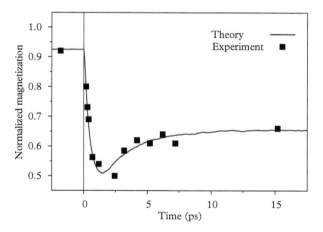

Figure 11.9 *Experimental and theoretical magnetization curves in a pump-probe experiment of face-centred cubic Ni. Figure after Evans, R. F. L., Atxitia, U., and Chantrell, R. W., (2015), Quantitative simulation of temperature-dependent magnetization dynamics and equilibrium properties of elemental ferromagnets, Phys. Rev. B, 91, 144425. Copyright 2012 by the American Physical Society.*

In principle, it is now rather straightforward to adopt this model for the temporal evolution of the electron temperature and, for each time step, solve the Kohn–Sham equation and calculate the sizes of the magnetic moment and the Heisenberg exchange interaction (in practice, this is not done at each time step; instead, interpolation between calculations at different temperature is used). In this way, the parameters of the effective spin Hamiltonian of Eqn (2.19) become time dependent. With this Hamiltonian, the effective Weiss field used in the SLL equation (Eqn (4.1)) can be updated as a function of time and, together with the stochastic field, which is determined by the spin temperature of the three-temperature model, can be used to solve the SLL equation in a way that accounts for the differences in the electron temperature and the spin temperature (Chimata et al., 2012). This approach has also been used in ASD simulations (Evans et al., 2015) to reproduce the experiments on fcc Ni. The results of these simulations (which have a quantum correction to the temperature used for the spin system (Evans et al., 2015)) are shown in Fig. 11.9, and it may be noted that theory reproduces the observations with good accuracy.

11.9   All-optical magnetization reversal

Since there is a large technological interest in achieving easier and faster control of magnetic domains, for example, when such domains are used as bits in information technology, the possibility of inducing magnetization reversal by optical means, without an applied magnetic field, is both important and scientifically interesting. Initially, it was not clear why this phenomenon would happen, but it was observed in experiments. The first experiments demonstrating this phenomenon used circularly polarized laser light on a ferrimagnetic material, such as GdFeCo alloys (Stanciu et al., 2007). In Fig. 11.10 we show schematically the details of this experiment. All compounds containing rare earths and transition metal elements have antiferromagnetic inter atomic exchange coupling for spin moments. This was explained quite some time ago by Szpunar and Kozarzewski

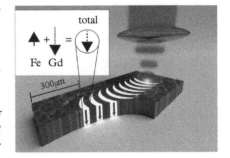

Figure 11.10 *Illustration of all-optical reversal and control of magnetism. The dark grey regions indicate magnetization pointing up, and the white regions indicate magnetization pointing down. For a GdFe alloy, each such region is composed of Fe and Gd atoms that are coupled antiparallelly but, since the Fe and Gd moments have different sizes, there is a net total moment, as shown in the inset. By changing the polarization of the circular laser light, regions of up magnetization and down magnetization can be realized. Figure redrawn after Stanciu, C. D., Hansteen, F., Kimel, A. V., Kirilyuk, A., Tsukamoto, A., Itoh, A., and Rasing, Th., (2007), All-optical magnetic recording with circularly polarized light,* Phys. Rev. Lett., **99**, 047601. *Copyright 2012 by the American Physical Society.*

(1977) and Kozarzewski as being caused by the spin-dependent hybridization between the $5d$ states of the rare-earth atom, and the $3d$ states of the transition metal, with this hybridization leading to an antiparallel coupling between the $3d$ and the $5d$ moments. The antiparallel coupling is in fact only partially induced by this spin-dependent hybridization; it is also induced by intra-atomic exchange coupling with the localized, spin polarized $4f$ shell, which couples to the $5d$ moment in a parallel fashion. In Gd this induced $5d$ moment, as discussed in Chapter 3, is of the order of 0.65 μ_B/atom. Hence, the regions of up and down magnetism in Fig. 11.10 represent in fact, ferrimagnetic regions where, in each region, the Gd and Fe (or Co) moments are antiparallel to each other (see Fig. 11.10, inset). By pumping the magnetic substrate with a laser with alternating left and right circular polarization, it was demonstrated that magnetization reversal was possible (Stanciu et al., 2007). Subsequent works reproduced this phenomenon (Mangin et al., 2014), not only for ferrimagnets but also for ferromagnetic materials (Lambert et al., 2014).

Since circularly polarized light carries spin angular momentum, the magnetization reversal illustrated in Fig. 11.10 could be explained by the spin angular moment of the pump pulse. However, surprisingly, similar phenomena were observed when linearly polarized light, which carries no angular momentum, was used (Radu et al., 2011). This pointed to other mechanisms being at play in the reversal of the magnetization of these experiments, in particular, the temperature pulse provided by the pump laser. For this reason, the phrase 'all-thermal control of magnetism' has been coined. Detailed experimental investigations of the sublattice moments also showed that, during the reversal process, the Fe and Gd moments, which initially were antiparallel, were parallel for a short period of time before they reached a stationary point where they became antiparallel again, albeit with a reversed total moment (as shown in Fig. 11.10). The antiparallel interatomic exchange coupling between Gd and Fe is rather strong, and it is surprising that just setting the magnetism of the two sublattices in motion would force their moments to become parallel, even for a short period of time. Leaving a theoretical explanation to the side for the moment, one may remark that the technological implications of these findings are large, since magnetization reversal can be achieved in these systems simply by blasting a ferrimagnet with a femtosecond laser pulse.

Different theoretical explanations of these experiments have been offered (Mentink et al., 2012; Ostler et al., 2012; Chimata et al., 2015). Although a full understanding of these experiments may not have been achieved yet, we offer here the explanation proposed by Mentink et al. (2012) and Chimata et al. (2015). Since the time scales involved in these experiments are of the order of picoseconds, the ASD approach should be valid. This is also the approach taken by Mentink et al. (2012) and Chimata et al. (2015), who used both the three-temperature model and the SLL equation (Eqn 4.1). The results of these simulations are shown in Fig. 11.11. In the lower panel, the temperature profile of the spin system is shown as a function of time, while the upper panel shows the time evolution of the Fe and Gd sublattices of an amorphous Fe-Gd alloy, for two different alloy concentrations. It can be noted that, in agreement with observations, the net moment reverses as a result of the temperature stimulus. It can also be noted that, for a short period of time, the spin moment of the Fe and Gd sublattices are parallel,

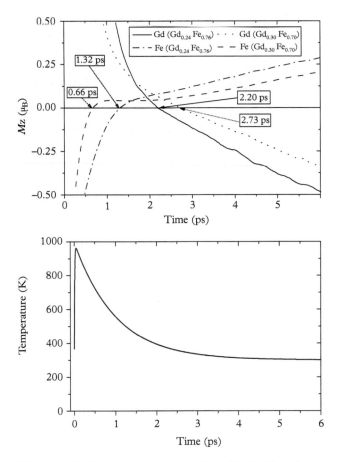

Figure 11.11 *Spin dynamics of amorphous Fe-Gd alloys. In the lower panel, the temperatures of the spin system are shown, while the upper panel shows the time evolution of the Gd and Fe sublattices, for two different alloy concentrations. Figure after Chimata, R., Isaeva, L., Kádas, K., Bergman, A., Sanyal, B., Mentink, J. H., Katsnelson, M. I., Rasing, T., Kirilyuk, A., Kimel, A., Eriksson, O., and Pereiro, M., (2015), All-thermal switching of amorphous Gd-Fe alloys: Analysis of structural properties and magnetization dynamics, Phys. Rev. B, 92, 094411. Copyright 2012 by the American Physical Society.*

again something which is in agreement with experimental findings. The different Fe-Gd alloy concentrations provide slightly different switching times, but their overall behaviours are the same, as Fig. 11.11 shows. This illustrates again the power of the atomistic SLL equation in explaining the complex magnetic phenomena that are observed in experiments. A final remark in this section is that a deeper understanding of the data in

Fig. 11.11 was provided by Mentink et al. (2012), who argued that, in these experiments, initially, the Langevin force, $\mathbf{B}_i^{\mathrm{fl}}$ in Eqn (4.1), dominates in since, at that stage, the temperature is high and the precession field, $\mathbf{B}_i$ in Eqn (4.1), is small. The latter condition comes about initially, the Fe and Gd moments are almost entirely antiparallel and are close to the ground-state configuration. In this temporal regime, angular momentum does not have to be conserved. Hence, the two sublattices evolve in time more or less independently of each other and, since the relaxation time is proportional to the sublattice moment (Mentink et al., 2012), Fe decays faster than Gd. However, when the spin system cools off and the angle between the Fe and Gd atomic moments starts to increase, the precession term in Eqn (4.1) gains in importance. In this regime, angular momentum is conserved, so that a decrease in the spin moment of one sublattice must be accompanied by an increase the spin moment of the other sublattice. The all-thermal switching in these alloys was proposed to be the result of a balance between these different regimes (Mentink et al., 2012), and this mechanism is consistent with the ASD simulations presented in Fig. 11.11.

Part 4

Conclusions and Outlook

So far, this book has described the connection between density functional theory and atomistic spin dynamics. We have given details for how this multiscale approach can be implemented, and examples of the performance of this approach have been given for various topics of magnetism. In this concluding section, we speculate about future developments of this method, and how this field may progress in the future.

12

Outlook on Magnetization Dynamics

Since its original formulation in the mid-1990s, atomistic spin dynamics has become an important tool for modelling dynamic processes in magnetic materials. So far, this book has described current methodological methods and functionalities of atomistic spin dynamics simulations. Applications of density functional theory and atomistic spin dynamics techniques to selected topics have been presented in this book, for instance, methods for the calculation of the microscopic Heisenberg and Gilbert parameters from first principles (Chapters 2 and 6), multiscale modelling of magnon spectra in bulk and thin film magnets (Chapter 9), and theoretical investigations of ultrafast switching dynamics in ferromagnets and ferrimagnets (Chapter 11) and of the exotic dynamics of topologically protected spin textures (Chapter 10). In this closing chapter, we give an outlook on recent and anticipated developments of the methodology.

12.1 Hierarchy of time scales and length scales

Methods for atomistic length-scale simulations of materials properties have developed during the last few decades to become ever more powerful and versatile. Atomistic spin dynamics is a prominent example within this family of techniques. Accompanying the progress of physical theories and approximation schemes for simulations, the continuous development of computer hardware and numerical algorithms enable larger and longer simulations, which nowadays often simultaneously span both microscopic and mesoscopic length scales. As an example, atomistic simulations based on molecular dynamics have been pivotal to understanding protein folding, thus leading to a better understanding of many diseases, such as Alzheimer's and cancer (Shirts and Pande, 2000).

In Fig. 12.1, a schematic figure of the hierarchy of existing methods for magnetization dynamics is displayed. At the bottom, we find ab initio time-dependent density functional theory (Runge and Gross, 1984; Marques et al., 2006), which, due to its computational complexity, is limited to simulations of a few (<10) spins over sub-picosecond time scales (Krieger et al., 2015; Stamenova et al., 2015; Elliott et al., 2016). As alternatives to the density functional theory, methods based on non-equilibrium Green's functions (see e.g. Fransson, 2010), many-body perturbation theory, and

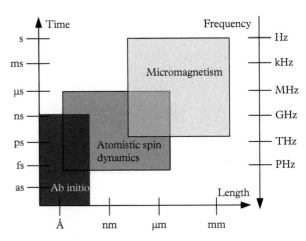

Figure 12.1 *The hierarchy of methods for magnetization dynamics in terms of length, time, and frequency scales. The atomistic spin dynamics method bridges the gap between ab initio and micromagnetic simulations. Figure adapted from Etz, C., Bergqvist, L., Bergman, A., Taroni, A., and Eriksson, O., (2015), Atomistic spin dynamics and surface magnons,* J. Phys. Condens. Matter, *27, 243202.*

non-equilibrium dynamical mean field theory (Aoki et al., 2014; Mentink and Eckstein, 2014; Mentink et al., 2015) are also relevant for modelling non-adiabatic spin dynamics.

Closely related to time-dependent density functional theory are methods for adiabatic ab-initio spin dynamics (Antropov et al., 1995, 1996; Stocks et al., 1998). The atomistic spin dynamics method, which is at the heart of this book, is a variant of adiabatic spin dynamics in which the low energy magnetic excitation spectra is parametrized to a magnetic Hamiltonian from density functional theory calculations (Chapters 1–4), or fitted from experimental data. Due to the approximations involved, most notably the adiabatic approximation and treatment of atomic magnetic moments instead of spin density, the method can take on much larger length scales and time scales than adiabatic ab-initio spin dynamics can, while nevertheless maintain materials-specific microscopic information.

At the top of the hierarchy of time and length, we find the framework for micromagnetic simulations (see e.g. Aharoni, 1996), in which the magnetization is treated as a continuous vector field coarse grained over multiples of lattice constants, wherefore more precise information on the strength and anisotropy of the typically short-range exchange interactions are replaced with an effective exchange coupling. However, micromagnetic simulations have been very successful in many fields, in particular for situations where dipolar interactions dominate. Micromagnetism can reach length scales and time scales which currently are completely beyond the scope of time-dependent density functional theory and atomistic spin dynamics methods.

In terms of the hierarchy ladder, the atomistic spin dynamics method is moving in both directions: both downwards towards time-dependent density functional theory, and upwards towards micromagnetism, as is perhaps natural, given its position in the middle. We will discuss selected recent and ongoing activity, starting off with general considerations of non-locality in space and time (Section 12.2) and exchange–correlation potentials (Section 12.3). Section 12.4 contains a brief account of time-dependent density functional theory applied to magnetization dynamics. In Section 12.5 we give more details on adiabatic ab initio atomistic spin dynamics, a method which in particular differs from standard atomistic spin dynamics in as much as it does not rely on a parametrized spin Hamiltonian to describe the magnetic interactions. A central property of the Stochastic Landau–Lifshitz–Gilbert (SLLG) equation is the restriction of the length of the magnetic moments, a property that in some situations (see e.g. Chimata et al., 2012), advantageously can be traded for a more refined description of simultaneous transversal and longitudinal fluctuations. There exist ideas and models on how to include longitudinal fluctuations in effective Hamiltonians, and these will be outlined in Section 12.6. As mentioned, the atomistic spin dynamics method is still limited in terms of length and time scales, compared to micromagnetism, and, in some particular situations, a seamless transition from atomistic spin dynamics to micromagnetism is desired (see Section 12.7). Finally, in Section 12.8 the ongoing development of methods to combine atomistic spin dynamics with molecular dynamics for lattice degrees of freedoms is outlined. This effort, initiated in seminal paper by Antropov et al. (1996), aims to create a unified theory based on ab initio density functional theory to enable simultaneous modelling of magnetic and phononic low energy bosonic excitations in an adiabatic and atomistic regime.

12.2   Non-locality in space and time

In the derivations of the atomistic spin dynamics equation of motion (see Chapter 4), the starting point was the time-dependent density functional theory and the Kohn–Sham equation. An alternative starting point is the quantum field theoretical perspective used by Bhattacharjee, Nordström, et al., (2012) to derive a generalized description of magnetization dynamics. Their starting point was an action variable defined as

$$\mathcal{S} = \mathcal{S}_{\text{int}} + \mathcal{S}_{\text{Z}} + \mathcal{S}_{\text{WZWN}}, \tag{12.1}$$

where $\mathcal{S}_{\text{int}}$ models how exchange interaction for the magnetization density is mediated by the density matrix, $\mathcal{S}_{\text{Z}}$ is the Zeeman interaction, and $\mathcal{S}_{\text{WZWN}}$ is the Wess–Zumino–Witten–Novikov term, which connects to the Berry phase accumulated by the magnetization. Minimization of the action variable with regard to fast, longitudinal magnetic fluctuations resulted in the integral equation

$$\frac{d\mathbf{m}(\mathbf{r}, t)}{dt} = \mathbf{m}(\mathbf{r}, t) \times \left[-\gamma \mathbf{B}_{\text{ext}} + \int \mathcal{D}^r(\mathbf{r}, \mathbf{r}'; t, t') \cdot \mathbf{m}(\mathbf{r}', t') \, d\mathbf{r}' dt' \right], \tag{12.2}$$

where $\mathbf{m}(\mathbf{r}, t)$ is the magnetization density, $\mathbf{B}_{\text{ext}}$ is the external magnetic field, and $\mathcal{D}^r(\mathbf{r}, \mathbf{r}'; t, t')$ is a kernel which describes how the magnetization dynamics is non-local in both space and time. Although there is no damping term in the integral equation, damping follows from the convolution over time of $\mathcal{D}^r(\mathbf{r}, \mathbf{r}'; t, t')$ and $\mathbf{m}(\mathbf{r}', t')$. This kind of time retardation is also explored in a recent paper by Thonig et al. (2015), who reported results for damped macrospin precession and the impact of time retardation on exchange-coupled systems. Likewise, convolution over space is what accounts for the non-local (in space) interaction of magnetization.

Via an expansion to second order in time $\tau = t - t'$, it is possible to connect to the usual form of the Landau–Lifshitz–Gilbert (LLG) equation for atomistic spin dynamics, Eqn (4.1), to identify exchange ($\hat{\mathbf{J}}$), damping ($\hat{\mathbf{G}}$), and inertia ($\hat{\mathbf{I}}$) tensors as

$$\hat{\mathbf{J}} = \int \mathcal{D}^r(\mathbf{r}, \mathbf{r}'; t, t')\, d\mathbf{r}'\, dt', \tag{12.3}$$

$$\hat{\mathbf{G}} = -\int \tau \mathcal{D}^r(\mathbf{r}, \mathbf{r}'; t, t')\, d\mathbf{r}'\, dt', \tag{12.4}$$

$$\hat{\mathbf{I}} = \int \frac{\tau^2}{2} \mathcal{D}^r(\mathbf{r}, \mathbf{r}'; t, t')\, d\mathbf{r}'\, dt', \tag{12.5}$$

respectively, and, after some further algebra, arrive at expressions which lend themselves to implementation in density functional theory software.

12.3 Exchange–correlation potentials

A commonly used technique, described for instance by Kübler (2009), to calculate the exchange correlation potential for non-collinear magnets is to use the local spin density approximation (LSDA) exchange–correlation potential in combination with a coordinate transformation:

$$\mathbf{B}_{\parallel}^{\mathrm{xc}}(\mathbf{r}, t) = -\frac{\delta E_{\mathrm{LSDA}}^{\mathrm{xc}}[n, \mathbf{m}]}{\delta \mathbf{m}(\mathbf{r})} = -\frac{\partial E_{\mathrm{unif}}^{\mathrm{xc}}}{\partial m}\frac{\mathbf{m}(\mathbf{r})}{m(\mathbf{r})}. \tag{12.6}$$

By construction, $\mathbf{B}_{\parallel}^{\mathrm{xc}}$ is, at each point, parallel with $\mathbf{m}(\mathbf{r})$, with the consequence that no torque is exerted on the magnetization. This property carries over also to standard forms of the generalized gradient approximation (GGA) exchange–correlation potentials. Recapitulating the compact form of the time-dependent density functional theory equation of motion as expressed in Eqn (4.9),

$$\frac{\partial n(\mathbf{r}, t)}{\partial t} = -\nabla \cdot \mathbf{J}_{\mathrm{p}}, \tag{12.7}$$

$$\frac{\partial \mathbf{s}(\mathbf{r}, t)}{\partial t} = -\nabla \cdot \mathbf{J}_{\mathrm{s}} - \gamma \hat{\mathbf{s}} \times \left[\mathbf{B}^{\mathrm{ext}}(\mathbf{r}, t) + \mathbf{B}^{\mathrm{xc}}(\mathbf{r}, t)\right], \tag{12.8}$$

but here, with the effective magnetic field written explicitly as the sum of the external magnetic field and the exchange–correlation field, we can immediately observe that, for standard LSDA and GGA potentials and no external magnetic field, the spin dynamics is entirely driven by the divergence of the spin current (Capelle et al., 2001; Katsnelson and Antropov, 2003; Krieger et al., 2015; Stamenova et al., 2015). From this perspective, it can arguably be seen as somewhat odd that, in order for the atomistic spin dynamics method to be in a form which make use of a parametrized magnetic Hamiltonian (Nowak et al., 2005; Skubic et al., 2008; Evans et al., 2014), the first term on the right hand side of Eqn (12.8) is typically omitted. That standard LSDA and GGA potentials may be insufficient when applied to antiferromagnetic and spin density wave materials has been known since long and has spurred development of new potentials (Capelle and Oliveira, 2000). With regard to magnetism, a very important starting point when constructing new potentials is the zero torque theorem for the magnetic part of the exchange–correlation potential. This theorem, which was formulated by Capelle et al. (2001) reads

$$\int d^3r \, \mathbf{m}(\mathbf{r}) \times \mathbf{B}_{xc}(\mathbf{r}) = 0 \qquad (12.9)$$

and expresses that, integrated over all space, the exchange–correlation potential cannot exert a net torque on the system. By means of a perturbative approach, Katsnelson and Antropov (2003) took into consideration the variation of the magnetization density to construct a spin angular gradient exchange correlation potential. Sharma et al. (2007) generalized the optimized effective potential method for calculations of exchange–correlation to address spin dynamics. More recently, Eich and Gross (2013), starting out from the LSDA, constructed a spin-spiral functional to take into account transverse gradients of the magnetization. The resulting local exchange–correlation field is in general not parallel to the spin magnetization finite torque, as is shown in Fig. 12.2. In a second paper partly by the same authors (Eich et al., 2013), and the starting point

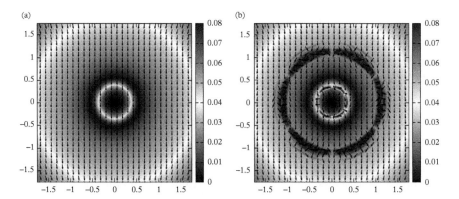

Figure 12.2 *(a) The local spin density approximation exchange–correlation field $B_{xc}(r)$ around an atom in a Cr monolayer with a 120◦ Néel antiferromagnetic state. The field is parallel to the local magnetization m(r) in all points, as expected from Eqn (12.6). Reprinted figure with permission from Eich, F. G., and Gross, E. K. U., Phys. Rev. Lett., 111, 156401, 2013. Copyright 2013 by the American Physical Society. (b) For the same magnetization density as in (a), the spin-spiral wave functional gives $B_{xc}(r)$ a more complex structure. Reprinted figure with permission from Eich, F. G., and Gross, E. K. U., Phys. Rev. Lett., 111, 156401, 2013. Copyright 2013 by the American Physical Society.*

being the GGA, a new functional to account both for transverse and longitudinal gradients was communicated. The longitudinal gradient is very important, as it gives a correction to the dynamics of the fast degrees of freedom of the density matrix, that is, the evolution in time of the charge density and of the magnitude of the magnetization. As pointed out by Eich and Gross (2013), a good strategy to follow in order to construct an explicit exchange–correlation functional for spin polarized density functional theory, in order to naturally fulfil the zero torque theorem, is to first write the exchange–correlation energy $E_{xc}[n, \mathbf{m}]$ in terms of proper scalars.

12.4   Time-dependent density functional theory, and optimal control theory

If we ignore for the moment the daunting challenge of constructing adequate exchange–correlation kernels (Marques et al., 2006), and the fact that time-dependent density functional theory simulations typically are computationally very expensive as regards CPU hours, it is undisputed that time-dependent density functional theory is a formally exact framework for simulating the dynamics contained in the many-electron Dirac equation, that is, the real-time dynamics of solids.

Recently, pioneering work on time-dependent density functional theory simulations of laser-induced demagnetization has been communicated (Krieger et al., 2015). The effect of the laser pulse was here contained in the form of a vector potential within the dipole approximation, that is, neglecting the spatial variation of the laser pulse over the extent of the sample. Unlike in atomistic spin dynamics, here the time evolution of the magnetization density can be followed, providing very direct microscopic insight into how the system responds to a perturbation. Simulations were pursued for the $3d$ metals Fe, Co, and Ni, for various time dependences for the driving field. One of the main observations was that demagnetization was seen to be a two-step process in which excitation of a fraction of the electrons was followed by spin-flip transitions of more localized electrons. For the field pulses used by Krieger et al. (2015), demagnetization could be observed on time scales as short as a few tenths of a femtosecond. The impact of the polarization of the incident electromagnetic field pulse turned out to be negligible; hence, no net demagnetization at all was observed when spin–orbit interaction was turned off in the simulations. Closely related to the pioneering exploration by Krieger et al. (2015) of time-dependent density functional theory for spin dynamics are the optimal control theory (OCT) investigations on how the temporal profile of the electromagnetic field can be tuned in order to achieve the fastest possible demagnetization (Elliott et al., 2016). Arguably, OCT is a very powerful technique which has been used to model and control charge transfer, to break chemical bonds, and for higher harmonics generation.

One main aim for the investigations on small magnetic clusters by Stamenova et al. (2015) was to learn more about the role of spin–orbit interaction in the demagnetization process. As time-dependent density functional theory simulations constitute a numerical experiment, the practitioner is at liberty to explore the net strength of all relativistic effects by means of a rescaling of the speed of light, or to focus on the spin–orbit interaction, specifically by tuning its coupling constant λ. This latter approach was taken by Stamenova et al. (2015), who communicated a λ^2 dependence for the rate of demagnetization.

12.5 Adiabatic ab initio atomistic spin dynamics

Papers by Rózsa et al. (2014) and Ma and Dudarev (2015) have reinvigorated the interest in adiabatic ab initio atomistic spin dynamics, which is an intermediate form of time-dependent density functional theory and atomistic spin dynamics, as outlined in Chapter 4.[1] Since density functional theory is a ground-state theory (see Chapter 1), the assumption is that the system, with its general magnetic configuration, would relax back to its ground state, for example, a ferromagnetic state, unless constraints are put in place (see e.g. Schwarz and Mohn, 1984; Dederichs et al., 1984).

The major advantage of adiabatic ab initio atomistic spin dynamics methods such as the constrained local moment approach by Stocks et al. (1998) is that the accuracy of the calculations are not limited by the choice of a spin Hamiltonian, and the restricted applicability that may apply for the coupling constants of such Hamiltonians for general non-equilibrium situations can, to some extent, be alleviated. Instead, at any time step, the 'optimal' effective field for a given set of moment orientations is calculated. Furthermore, longitudinal spin fluctuations are also taken into account, given that the Kohn–Sham equation is solved anew at each time step; however, unlike in time-dependent density functional theory, the time scale for the longitudinal dynamics is here approximated to be infinitely fast. The main drawback with the method is that it is computationally expensive, although less so than full-fledged time-dependent density functional theory. Being still much slower than atomistic spin dynamics, the applications will in the near future mainly be for small clusters and similar systems (Rózsa et al., 2014).

12.6 Longitudinal spin fluctuations in atomistic spin dynamics

The Heisenberg Hamiltonian, Eqn (2.19), is based on the restriction that the size of the moment does not change upon rotation, that is, rigid spin approximation. However, in real materials, this is not the case; instead, the size of the moment changes upon rotation and/or change in temperature, due to Stoner excitations, as discussed in Chapter 9. Perhaps the most straightforward way to include moment size variation, that is, longitudinal spin fluctuations (LSF), is to supplement the Heisenberg Hamiltonian with an

---

[1] Another intermediate of time-dependent density functional theory and atomistic spin dynamics is to map the electronic structure from all-electron methods to computationally more efficient tight-binding schemes for which time-dependent density functional theory can be pursued.

additional term describing how the total energy depends on the moment size, such that the total spin Hamiltonian takes the form

$$\mathcal{H}_{\text{tot}} = \mathcal{H}_{\text{S}} + \mathcal{H}_{\text{LSF}}, \tag{12.10}$$

where $\mathcal{H}_{\text{S}}$ is the Heisenberg Hamiltonian, and $\mathcal{H}_{\text{LSF}}$ the LSF Hamiltonian. Inclusion of an LSF term in the Hamiltonian was first introduced in Monte Carlo simulations by Rosengaard and Johansson (1997) but has since been employed by others (Shallcross et al., 2005; Ruban et al., 2007; Ma and Dudarev, 2012; Ležaić et al., 2013). Normally, the LSF Hamiltonian is assumed to take the form of a Landau expansion where the moment size is expanded in even powers (odd powers vanish due to time reversal symmetry) such that

$$\mathcal{H}_{\text{LSF}} = \sum_i \sum_{n=1}^{k} A_n m_i^{2n} \approx \sum_i \left(A_1 m_i^2 + A_2 m_i^4 + A_3 m_i^6 + \cdots \right), \tag{12.11}$$

where the expansion coefficient A_n can be directly obtained from the total energy of constrained density functional theory calculations in which the moment size is kept fixed. The first term in Eqn (12.11) was also derived from density functional theory (Brooks and Johansson, 1983) and is responsible for the spin-pairing energies shown in Fig. 1.1. An extension to the SLLG equation, proposed by Ma and Dudarev (2012) also includes the LSF Hamiltonian. The resulting equation, here denoted SLSF, has many similarities to the Langevin equation in molecular dynamics and takes the form

$$\frac{d\mathbf{m}_i}{dt} = -\gamma \left[\mathbf{m}_i \times \mathbf{B}_i - m_i \left(\alpha \mathbf{B}_i - \mathbf{B}_i^{\text{fl}} \right) \right], \tag{12.12}$$

where $\mathbf{B}_i$ is the effective field that appears in the SLLG equation, Eqn (4.1), and $\mathbf{B}_i^{\text{fl}}$ is a stochastic field that is similar in structure to that used in the SLLG equation. However, the main difference is that the SLSF equation no longer imposes any constraint on the size of the moment, unlike the regular SLLG equation. The choice of using the Gilbert damping parameter α of the transversal motion also for the longitudinal motion of the moment is a simplification, as discussed in Section 6.6. In principle, the longitudinal damping is expected to be different, which is an assumption that is also apparent in the first-principles treatment of damping (Chapter 6), but few studies exist to check the validity of this assumption.

12.7   A multiscale approach to atomistic spin dynamics

Micromagnetic simulations are based on a collection of atomic spins that are grouped together to create a macrospin, normally using a finite-element grid of the material of interest. On the other hand, atomistic spin dynamics treats the dynamics of each atomic moment but normally assumes that the size of the moment is constant; however, that restriction can be lifted, as outlined in Section 12.6. The Landau–Lifshitz–Bloch (LLB) equation, derived by Garanin (1997), proposes both an extension and an improvement to normal micromagnetism and can provide a basis for multiscale modelling (Kazantseva et al., 2008; Atxitia et al., 2010; Evans et al., 2012). In the LLB approach, each macrospin is allowed to change in both the transversal and longitudinal components by having two different damping parameters ($\alpha_\perp$ and $\alpha_\parallel$). Essentially, the LLB equation interpolates between the LLG equation at low temperatures, and the Ginzburg–Landau theory of phase transitions at high temperatures. Since the atomic description is lost, the LLB equation describes average properties over an ensemble of moments. In the form proposed by Evans et al. (2012), the stochastic LLB equation reads

$$\frac{d\mathbf{m}}{dt} = \gamma \mathbf{m} \times \mathbf{B} + \frac{|\gamma|\alpha_\parallel}{m^2}(\mathbf{m} \cdot \mathbf{B})\mathbf{m}$$
$$- \frac{|\gamma|\alpha_\perp}{m^2}\{\mathbf{m} \times [\mathbf{m} \times (\mathbf{B} + \xi_\perp)]\} + \xi_\parallel, \qquad (12.13)$$

where ξ is a stochastic fluctuating field. A temperature parametrization is put on the damping parameters and effective field, often taken from mean field approximation.

An alternative to the LLB approach is message passing of atomistic parameters to micromagnetic simulation and, in the opposite direction, from micromagnetic simulations to atomistic simulation. In this way, the magnetic system can be treated using micromagnetic simulations in the non-important regions of space, while an atomistic framework automatically switches on in regions of space where high accuracy is needed. This transition needs to be seamless and dynamic, and thus its implementation poses a major algorithmic and numerical challenge. The main parameters that enter micromagnetic simulations are the exchange constant A, the anisotropy energy density K, and the magnetization M. While magnetization and anisotropy are direct outputs from spin polarized density functional theory calculation (see Chapter 3), the exchange constant A requires more attention. First of all, it requires knowledge of the exchange parameters $\mathcal{J}_{ij}$ between sites i and j, obtained from Eqn (2.23). The spin wave stiffness D, which for cubic systems is the curvature of the magnon dispersion at small wavevectors(see Chapter 9), is directly related to the exchange interactions:

$$D = \frac{2}{3}\sum_{ij}\frac{\mathcal{J}_{ij}R_{ij}^2}{\sqrt{m_i m_j}}, \qquad (12.14)$$

where R_{ij} is the distance connecting the atomic sites, with indices i and j, and m_i (m_j) is the magnetic moment at site i (j). The temperature-dependent exchange constant A is in turn related to D as

$$A(T) = \frac{DM(T)}{2g\mu_B}, \qquad (12.15)$$

where g is the Landé g-factor, μ_B is the Bohr magneton, and $M(T)$ is the magnetization at temperature T.

12.8   Combined spin–lattice simulations

Atomistic spin dynamics, as described in this book, assumes a fixed, static lattice and that only the magnetic moments are fluctuating in time. However, in real materials, the atoms are also fluctuating around their equilibrium positions at elevated temperatures, giving rise to phonon excitations that, in turn, are responsible for material properties like thermal expansion, thermal conductivity, and sound wave propagation. For non-magnetic materials, simulations of these properties are commonly handled by molecular dynamics, either by using a parametrized classical force field in the same spirit as using a parametrized spin Hamiltonian in atomistic spin dynamics, or from first-principles calculations which are based on density functional theory and in which the forces are directly calculated by using the Hellmann–Feynman theorem. However, in recent years, there have been several attempts at combining spin simulations and lattice simulations, as originally proposed in the context of density functional theory, by Antropov et al. (1996). Provided that such a framework is at hand, it is then possible to study how magnetic properties depend on lattice dynamics and vice versa, a topic that is relevant for magnon–phonon coupling (Jensen and Mackintosh, 1991; Fennie and Rabe, 2006) and dynamical magneto-electric effects (Mochizuki et al., 2010a, 2011; Wang et al., 2012; Tokura et al., 2014; Mozafari et al., 2015). Due to the complexity inherent in this approach, combined spin–lattice frameworks have mainly been developed and demonstrated for Fe, but extending the method to the general case is an active field of research (Ma et al., 2008, 2012; Perera et al., 2014; Fransson et al., 2015; Perera et al., 2016).

The total Hamiltonian that describe the combined spin–lattice system has the form

$$\mathcal{H}_{tot} = \mathcal{H}_S + \mathcal{H}_L + \mathcal{H}_e, \tag{12.16}$$

where $\mathcal{H}_S$ denotes the spin Hamiltonian taken as a Heisenberg form (Eqn (2.19)), and $\mathcal{H}_L$ and $\mathcal{H}_e$ denote the lattice Hamiltonian and the electronic Hamiltonian, respectively. The lattice Hamiltonian takes the form

$$\mathcal{H}_L = \sum_i \frac{\mathbf{p}_i^2}{2m} + U(\mathbf{R}), \tag{12.17}$$

where $\mathbf{p}$ is the momentum of atom i with position vector $\mathbf{R}_i$, m is the mass, and $U(\mathbf{R})$ is the interatomic potential between all atom positions $\mathbf{R} = \{\mathbf{R}_1, \mathbf{R}_2, \ldots, \mathbf{R}_N\}$. Following the derivation by Ma et. al., no explicit form of $\mathcal{H}_e$ is specified. Instead, it is assumed that the damping and fluctuations in the lattice system result from interaction between the lattice electrons and the conduction electrons, in the Langevin equation for the atoms. The full set of equations of motion for the spin and lattice systems then takes the form

$$\frac{d\mathbf{R}_k}{dt} = \frac{\mathbf{p}_k}{m},$$ (12.18)

$$\frac{d\mathbf{p}_k}{dt} = -\frac{\partial U}{\partial \mathbf{R}_k} + \frac{1}{2}\sum_{i,j}\frac{\partial \mathcal{J}_{ij}}{\partial \mathbf{R}_k}\mathbf{m}_i \cdot \mathbf{m}_j - \frac{\alpha_{el}}{m}\mathbf{p}_k + \mathbf{f}_k,$$ (12.19)

$$\frac{d\mathbf{m}_k}{dt} = -\frac{\gamma}{1+\alpha^2}\mathbf{m}_k \times [\mathbf{B}_k + \mathbf{B}_k^{\mathrm{fl}}(t)] - \frac{\gamma\alpha}{m(1+\alpha^2)}\mathbf{m}_k \times \mathbf{m}_k \times [\mathbf{B}_k + \mathbf{B}_k^{\mathrm{fl}}(t)],$$

(12.20)

where α_{el} is an electron-spin damping parameter, and $\mathbf{f}_k$ is a stochastic field acting on atoms in the same way that $\mathbf{B}_k^{\mathrm{fl}}(t)$ is a stochastic field acting on the moment (see Chapter 5). The equation of motion for the moments is the SLLG equation, Eqn (4.1), which is repeated here for clarity. Solving Eqns (12.18), (12.19), and (12.20) by using a suitable integrator for the coupled equations result in simultaneous solution of the time propagation of the spin and lattice systems, important not least for the modelling of multiferroic and magneto-electric materials (Mochizuki et al., 2010a, 2011; Wang et al., 2012; Tokura et al., 2014). This also has relevance for theoretical studies of laser-induced demagnetization, where a strong, yet short, laser pulse can dramatically alter the magnetic state by rapidly heating up the electronic system, while the response of the lattice system is much slower (Chimata et al., 2012). Another example where magnon–phonon coupling is known to strongly influence properties is hexagonal close-packed Tb, where modification of the magnon frequencies along the *c*-axis takes place due to the presence of phonons of similar frequencies (Jensen and Mackintosh, 1991).

12.9 Conclusions

Starting from the fundamentals of quantum mechanics and electronic structure theory (Chapters 1, 2, and 3) we have, in this book, introduced a general framework for atomistic spin dynamics (Chapter 4). We have shown that, from first-principles calculations, it is possible to construct a spin model by means of a parameterized Hamiltonian, which can then be evolved in time according to the LLG equations. An important aspect of these simulations is via its effect on temperature, the equations of motion. Here, this is treated via Langevin dynamics, which results in the SLLG equation (Chapter 5). Damping from first principles, and aspects of implementation in software, are the topics of Chapters 6 and 7, respectively. After introducing the theoretical framework, we presented several applications and concepts that are highly relevant to what can be studied using atomistic spin dynamics (Chapters 8, 9, 10, and 11).

The steady increase in avilable computational power allows for more and more ambitious undertakings in the general field of computational physics. Often, this increase in capacity comes hand in hand with improved experimental capabilities, and this is indeed the case with the topic of this book: magnetization dynamics. As we have shown in this book, this has led to the very exciting situation we presently have, where it is possible to perform accurate and efficient atomistic simulations on a scale which is balancing on the edge of what is experimentally possible. In this way, atomistic spin dynamics simulations can both validate and be validated by state-of-the-art experiments, and they also can act as a predictive tool that is able to explain magnetization dynamics in experimentally inaccessible situations. Whether in the conventional form, as presented elsewhere in this book, or enhanced by methodologies introduced in this chapter, atomistic spin dynamics simulations can thus be expected to play an even larger role for future studies of magnetization dynamics. The purpose of this book has been to communicate technically relevant concepts, but a possibly even larger motivation has been to inspire interest in magnetism and magnetization dynamics, as well as in the emerging technological fields that one may foresee, for example, in magnonics, solitonics, and skyrmionics.

References

Achiezer, A. I., Barjachtar, V. G., and Peletminskij, S. V. (1968). *Spin Waves*. North-Holland, Amsterdam.

Adams, T., Mühlbauer, S., Pfleiderer, C., Jonietz, F., Bauer, A., Neubauer, A., Georgii, R., Böni, P., Keiderling, U., Everschor, K., Garst, M., and Rosch, A. (2011). Long-range crystalline nature of the skyrmion lattice in MnSi. *Phys. Rev. Lett.*, **107**, 217206.

Aeschlimann, M., Vaterlaus, A., Lutz, M., Stampanoni, M., Meier, F., Siegmann, H. C., Klahn, S., and Hansen, P. (1991). Highspeed magnetization reversal near the compensation temperature of amorphous GdTbFe. *Appl. Phys. Lett.*, **59**, 2189–91.

Aharoni, A. (1996). *Introduction to the Theory of Ferromagnetism*. Clarendon Press, Oxford.

Al Khawaja, U., and Stoof, H. (2001). Skyrmions in a ferromagnetic Bose-Einstein condensate. *Nature*, **411**, 918–20.

Andersen, O. K. (1975). Linear methods in band theory. *Phys. Rev. B*, **12**, 3060–83.

Andersen, O. K., and Jepsen, O. (1984). Explicit, first-principles tight-binding theory. *Phys. Rev. Lett.*, **53**, 2571–4.

Anderson, P. W. (1963). 'Theory of Magnetic Exchange Interactions: Exchange in Insulators and Semiconductors', in F. Seitz and D. Turnbull, eds, *Solid State Physics*, vol. 14, Academic Press, New York, NY, pp. 99–214.

Andersson, C., Sanyal, B., Eriksson, O., Nordström, L., Karis, O., Arvanitis, D., Konishi, T., Holub-Krappe, E., and Dunn, J. H. (2007). Influence of ligand states on the relationship between orbital moment and magnetocrystalline anisotropy. *Phys. Rev. Lett.*, **99**, 177207.

Anisimov, V. I., Poteryaev, A. I., Korotin, M. A., Anokhin, A. O., and Kotliar, G. (1997). First-principles calculations of the electronic structure and spectra of strongly correlated systems: Dynamical mean-field theory. *J. Phys.: Condens. Matter*, **9**, 7359–68.

Antoniak, C., Spasova, M., Trunova, A., Fauth, K., Farle, M., and Wende, H. (2009). Correlation of magnetic moments and local structure of FePt nanoparticles. *J. Phys. Conf. Ser.*, **190**, 012118.

Antropov, V. P., and Harmon, B. N. (1996). Simultaneous molecular and spin dynamics: Quasiclassical approximation and quantum effects. *J. Appl. Phys.*, **79**, 5409.

Antropov, V. P., Katsnelson, M. I., Harmon, B. N., van Schilfgaarde, M., and Kusnezov, D. (1996). Spin dynamics in magnets: Equation of motion and finite temperature effects. *Phys. Rev. B*, **54**, 1019–35.

Antropov, V. P., Katsnelson, M. I., van Schilfgaarde, M., and Harmon, B. N. (1995). Ab initio spin dynamics in magnets. *Phys. Rev. Lett.*, **75**, 729–32.

Aoki, H., Tsuji, N., Eckstein, M., Kollar, M., Oka, T., and Werner, P. (2014). Nonequilibrium dynamical mean-field theory and its applications. *Rev. Mod. Phys.*, **86**, 779–837.

Arias, R., and Mills, D. L. (1999). Extrinsic contributions to the ferromagnetic resonance response of ultrathin films. *Phys. Rev. B*, **60**, 7395–409.

Arias, R., and Mills, D. L. (2000). Extrinsic contributions to the ferromagnetic resonance response of ultrathin films. *J. Appl. Phys.*, **87**, 5455–6.

Aryasetiawan, F., and Karlsson, K. (1999). Green's function formalism for calculating spin-wave spectra. *Phys. Rev. B*, **60**, 7419–28.

Ashcroft, N. W., and Mermin, N. D. (1976). *Solid State Physics*. Saunders College Publishing, Phildelphia, PA.

Atkins, P. W., and Friedman, R. S. (2005). *Molecular Quantum Mechanics*. Oxford University Press, Oxford.

Atxitia, U., Hinzke, D., Chubykalo-Fesenko, O., Nowak, U., Kachkachi, H., Mryasov, O. N., Evans, R. F., and Chantrell, R. W. (2010). Multiscale modeling of magnetic materials: Temperature dependence of the exchange stiffness. *Phys. Rev. B*, **82**, 134440.

Babkevich, P., Poole, A., Johnson, R. D., Roessli, B., Prabhakaran, D., and Boothroyd, A. T. (2012). Electric field control of chiral magnetic domains in the high-temperature multiferroic CuO. *Phys. Rev. B*, **85**, 134428.

Back, C. H., Allenspach, R., Weber, W., Parkin, S. S. P., Weller, D., Garwin, E. L., and Siegmann, H. C. (1999). Minimum field strength in precessional magnetization reversal. *Science*, **285**, 864–7.

Baibich, M. N., Broto, J. M., Fert, A., Van Dau, F. N., Petroff, F., Etienne, P., Creuzet, G., Friederich, A., and Chazelas, J. (1988). Giant magnetoresistance of (001)Fe/(001)Cr magnetic superlattices. *Phys. Rev. Lett.*, **61**, 2472–25.

Balents, L. (2010). Spin liquids in frustrated magnets. *Nature*, **464**, 199–208.

Baňas, L. (2005). 'Numerical Methods for the Landau-Lifshitz-Gilbert Equation', in Z. Li, L. Vulkov, and J. Wåsniewski, eds, *Numerical Analysis and Its Applications: Third International Conference, NAA 2004, Rousse, Bulgaria, June 29–July 3, 2004, Revised Selected Papers*, Lecture Notes in Computer Science, vol. 3401. Springer, Berlin, Heidelberg, pp. 158–65.

Banerjee, R., Banerjee, M., Majumdar, A. K., Mookerjee, A., Sanyal, B., Hellsvik, J., Eriksson, O., and Nigam, A. K. (2011). $Fe_{3.3}Ni_{83.2}Mo_{13.5}$: A likely candidate to show spin-glass behaviour at low temperatures. *J. Phys.: Condens. Matter*, **23**, 106002.

Bar'yakhtar, V. G., Ivanov, B. A., and Chetkin, M. V. (1985). Dynamics of domain walls in weak ferromagnets. *Phys. Usp.*, **28**, 563–88.

Barth, U. von, and Hedin, L. (1972). A local exchange-correlation potential for the spin polarized case. I. *J. Phys. C: Solid State Phys.*, **5**, 1629–42.

Battiato, M., Carva, K., and Oppeneer, P. M. (2010). Superdiffusive spin transport as a mechanism of ultrafast demagnetization. *Phys. Rev. Lett.*, **105**, 027203.

Beaurepaire, E., Merle, J.-C., Daunois, A., and Bigot, J.-Y. (1996). Ultrafast spin dynamics in ferromagnetic nickel. *Phys. Rev. Lett.*, **76**, 4250–3.

Belavin, A. A., and Polyakov, A. M. (1975). Metastable states of two-dimensional isotropic ferromagnets. *JETP Lett.*, **22**, 245–8.

Belhadji, B., Bergqvist, L., Zeller, R., Dederichs, P. H., Sato, K., and Katayama-Yoshida, H. (2007). Trends of exchange interactions in dilute magnetic semiconductors. *J. Phys. Condens. Matter*, **19**, 436227.

Belletti, F., Cotallo, M., Cruz, A., Fernandez, L., Gordillo-Guerrero, A., Guidetti, M., Maiorano, A., Mantovani, F., Marinari, E., Martin-Mayor, V., Sudupe, A., Navarro, D., Parisi, G., Perez-Gaviro, S., Ruiz-Lorenzo, J., Schifano, S., Sciretti, D., Tarancon, A., Tripiccione, R., Velasco, J., and Yllanes, D. (2008). Nonequilibrium spin-glass dynamics from picoseconds to a tenth of a second. *Phys. Rev. Lett.*, **101**, 157201.

Berger, L. (1996). Emission of spin waves by a magnetic multilayer traversed by a current. *Phys. Rev. B*, **54**, 9353–8.

Bergman, A., Skubic, B., Hellsvik, J., Nordström, L., Delin, A., and Eriksson, O. (2011). Ultrafast switching in a synthetic antiferromagnetic magnetic random-access memory device. *Phys. Rev. B*, **83**, 224429.

Bergman, A., Taroni, A, Bergqvist, L., Hellsvik, J., Hjörvarsson, B., and Eriksson, O. (2010). Magnon softening in a ferromagnetic monolayer: A first-principles spin dynamics study. *Phys. Rev. B*, **81**, 144416.

Bergqvist, L., Taroni, A., Bergman, A., Etz, C., and Eriksson, O. (2013). Atomistic spin dynamics of low-dimensional magnets. *Phys. Rev. B*, **87**, 144401.

Berkov, D. V., Gorn, N. L., Schmitz, R., and Stock, D. (2006). Langevin dynamic simulations of fast remagnetization processes in ferrofluids with internal magnetic degrees of freedom. *J. Phys. Condens. Matter*, **18**, S2595–S2622.

Berthier, L., and Young, A. (2004). Aging dynamics of the Heisenberg spin glass. *Phys. Rev. B*, **69**, 184423.

Bertotti, G., Mayergoyz, I. D., and Serpico, C. (2009). *Nonlinear Magnetization Dynamics in Nanosystems*. Elsevier, Oxford.

Bhagat, S. M., and Lubitz, P. (1974). Temperature variation of ferromagnetic relaxation in the 3d transition metals. *Phys. Rev. B*, **10**, 179–85.

Bhattacharjee, S., Bergman, A., Taroni, A., Hellsvik, J., Sanyal, B., and Eriksson, O. (2012). Theoretical analysis of inertia-like switching in magnets: Applications to a synthetic antiferromagnet. *Phys. Rev. X*, **2**, 011013.

Bhattacharjee, S., Nordström, L., and Fransson, J. (2012). Atomistic spin dynamic method with both damping and moment of inertia effects included from first principles. *Phys. Rev. Lett.*, **108**, 057204.

Bigot, J.-Y., Guidoni, L., Beaurepaire, E., and Saeta, P. N. (2004). Femtosecond spectrotemporal magneto-optics. *Phys. Rev. Lett.*, **93**, 077401.

Bigot, J.-Y., Vomir, M., and Beaurepaire, E. (2009). Coherent ultrafast magnetism induced by femtosecond laser pulses. *Nat. Phys.*, **5**, 515–20.

Binasch, G., Grünberg, P., Saurenbach, F., and Zinn, W. (1989). Enhanced magnetoresistance in layered magnetic structures with antiferromagnetic interlayer ex- change. *Phys. Rev. B*, **39**, 4828–30.

Binder, K. (1981). Finite size scaling analysis of Ising model block distribution functions. *Zeitschrift für Physik B Condensed Matter*, **43**, 119–40.

Bode, M., Heide, M., von Bergmann, K., Ferriani, P., Heinze, S., Bihlmayer, G., Kubetzka, A., Pietzsch, O., Blügel, S., and Wiesendanger, R. (2007). Chiral magnetic order at surfaces driven by inversion asymmetry. *Nature*, **447**, 190–3.

Bogdanov, A., and Hubert, A. (1994). Thermodynamically stable magnetic vortex states in magnetic crystals. *J. Magn. Magn. Mater.*, **138**, 255–69.

Bogdanov, A., and Yablonskii, D. A. (1989). Thermodynamically stable "vortices" in magnetically ordered crystals. The mixed state of magnets. *Zh. Eksp. Teor. Fiz.* **95**,178–82.

Böttcher, D., Ernst, A., and Henk, J. (2012). Temperature-dependent Heisenberg exchange coupling constants from linking electronic-structure calculations and Monte Carlo simulations. *J. Magn. Magn. Mater.*, **324**, 610–15.

Box, G. E. P., and Muller, M. E. (1958). A note on the generation of random normal deviates. *Ann. Math. Statist.*, **29**, 610–11.

Bradley, C. J., and Cracknell, A. P. (1972). *The Mathematical Theory of Symmetry in Solids. Representation Theory for Point Groups and Space Groups*. Clarendon Press, Oxford.

Bramwell, S. T., and Gingras, M. J. P. (2001). Spin ice state in frustrated magnetic pyrochlore materials. *Science*, **294**, 1495–501.

Brataas, A., Tserkovnyak, Y., and Bauer, G. E. W. (2008). Scattering theory of Gilbert damping. *Phys. Rev. Lett.*, **101**, 037207.

Brooks, H. (1940). Ferromagnetic anisotropy and the itinerant electron model. *Phys. Rev.*, **58**, 909–18.

Brooks, M. S. S., Auluck, S., Gasche, T., Trygg, J., Nordström, L., Severin, L., and Johansson, B. (1992). Theory of the Curie temperatures of the rare earth metals. *J. Magn. Magn. Mater.*, **104**, 1496–8.

Brooks, M. S. S., Eriksson, O., Johansson, B., Franse, J. J. M., and Frings, P. H. (1988). Chemical bonding and magnetism in 3d-5f intermetallics. *J. Phys. F: Met. Phys.*, **18**, L33–L40.

Brooks, M. S. S., and Johansson, B. (1983). Exchange integral matrices and cohesive energies of transition metal atoms. *J. Phys. F: Met. Phys.*, **13**, L197–L202.

Brown, G., Novotny, M. A., and Rikvold, P. A. (2001). Langevin simulation of thermally activated magnetization reversal in nanoscale pillars. *Phys. Rev. B*, **64**, 134422.

Brown, R. H., Nicholson, D. M. C., Wang, X., and Schultthess, T. C. (1999). First principles theory of spin waves in Fe, Co, and Ni. *J. Appl. Phys.*, **85**, 4830–2.

Brown, W. F., Jr. (1963). Thermal fluctuations of a single-domain particle. *J. Appl. Phys.*, **34**, 1319–20.

Bruno, P. (1989). Tight-binding approach to the orbital magnetic moment and magnetocrystalline anisotropy of transition metal monolayers. *Phys. Rev. B*, **39**, 865–8.

Buczek, P., Ernst, A., Bruno, P., and Sandratskii, L. (2009). Energies and lifetimes of magnons in complex ferromagnets: A first-principle study of Heusler alloys. *Phys. Rev. Lett.*, **102**, 247206.

Buczek, P., Ernst, A., and Sandratskii, L. M. (2011). Different dimensionality trends in the Landau damping of magnons in iron, cobalt, and nickel: Time-dependent density functional study. *Phys. Rev. B*, **84**, 174418.

Burkert, T., Eriksson, O., Simak, S. I., Ruban, A. V., Sanyal, B., Nordström, L., and Wills, J. M. (2005). Magnetic anisotropy of $L10$ FePt and Fe1-xMnxPt. *Phys. Rev. B*, **71**, 134411.

Butler, W. H. (1985). Theory of electronic transport in random alloys: Korringa-Kohn-Rostoker coherent-potential approximation. *Phys. Rev. B*, **31**, 3260–77.

Callaway, J. (1968). Spin waves in ferromagnetic metals. *Phys. Lett. A*, **27**, 215–16.

Callen, H. (1951). Irreversibility and generalized noise. *Phys. Rev.*, **83**, 34–40.

Capelle, K., and Oliveira, L. (2000). Density-functional theory for spin-density waves and antiferromagnetic systems. *Phys. Rev. B*, **61**, 15228–40.

Capelle, K., Vignale, G., and Györffy, B. (2001). Spin currents and spin dynamics in time-dependent density-functional theory. *Phys. Rev. Lett.*, **87**, 206403.

Cardy, J. L. (1996). *Scaling and Renormalization in Statistical Physics*. Cambridge University Press, Cambridge.

Carva, K., Turek, I., Kudrnovský, J., and Bengone, O. (2006). Disordered magnetic multilayers: Electron transport within the coherent potential approximation. *Phys. Rev. B*, **73**, 144421.

Ceperley, D. M., and Alder, B. J. (1980). Ground state of the electron gas by a stochastic method. *Phys. Rev. Lett.*, **45**, 566–9.

Chadov, S., Minar, J., Katsnelson, M. I., Ebert, H., Ködderitzsch, D., and Lichtenstein, A. I. (2008). Orbital magnetism in transition metal systems: The role of local correlation effects. *EPL*, **82**, 37001.

Cheetham, A. K., and Hope, D. A. O. (1983). Magnetic ordering and exchange effects in the antiferromagnetic solid solutions MnxNi1-xO. *Phys. Rev. B*, **27**, 6964–7.

Chen, K., and Landau, D. P. (1994). Spin-dynamics study of the dynamic critical behavior of the three-dimensional classical Heisenberg ferromagnet. *Phys. Rev. B*, **49**, 3266–74.

Chen, Z., Yi, M., Chen, M., Li, S., Zhou, S., and Lai, T. (2012). Spin waves and small intrinsic damping in an in-plane magnetized FePt film. *Appl. Phys. Lett.*, **101**, 224402.

Chico, J., Etz, C., Bergqvist, L., Fransson, J., Delin, A., Eriksson, O., and Bergman, A. (2014). Thermally driven domain-wall motion in Fe on W(110). *Phys. Rev. B*, **90**, 014434.

Chikazumi, S. (1997). *Physics of Ferromagnetism* (2nd edn). Clarendon Press, Oxford.

Chimata, R., Bergman, A., Bergqvist, L., Sanyal, B., and Eriksson, O. (2012). Microscopic model for ultrafast remagnetization dynamics. *Phys. Rev. Lett.*, **109**, 157201.

Chimata, R., Isaeva, L., Kádas, K., Bergman, A., Sanyal, B., Mentink, J. H., Katsnelson, M. I., Rasing, T., Kirilyuk, A., Kimel, A., Eriksson, O., and Pereiro, M. (2015). All-thermal switching of amorphous Gd-Fe alloys: Analysis of structural properties and magnetization dynamics. *Phys. Rev. B*, **92**, 094411.

Christensen, N. E. (1984). Relativistic band structure calculations. *Int. J. Quantum Chem.*, **25**, 233–61.

Coey, J. M. D. (2010). *Magnetism and Magnetic Materials*. Cambridge University Press, Cambridge.

Colarieti-Tosti, M., Simak, S. I., Ahuja, R., Nordström, L., Eriksson, O., Åberg, D., Edvardsson, S., and Brooks, M. S. S. (2003). Origin of magnetic anisotropy of Gd metal. *Phys. Rev. Lett.*, **91**, 157201.

Cooke, J. F., Lynn, J. W., and Davis, H. L. (1980). Calculations of the dynamic susceptibility of nickel and iron. *Phys. Rev. B*, **21**, 4118–31.

Costa, A. T., Jr, Muniz, R. B., and Mills, D. L. (2004a). Theory of large-wave-vector spin waves in ultrathin ferromagnetic films: Sensitivity to electronic structure. *Phys. Rev. B*, **70**, 054406.

Costa, A. T., Jr, Muniz, R. B., and Mills, D. L. (2004b). Theory of spin waves in ultrathin ferromagnetic films: The case of Co on Cu(100). *Phys. Rev. B*, **69**, 064413.

Craig, P. P., Nagle, D. E., Steyert, W. A., and Taylor, R. D. (1962). Paramagnetism of Fe impurities in transition metals. *Phys. Rev. Lett.*, **9**, 12–14.

Daalderop, G. H. O., Kelly, P. J., and Schuurmans, M. F. H. (1991). Magnetocrystalline anisotropy and orbital moments in transition-metal compounds. *Phys. Rev. B*, **44**, 12054–7.

D'Aquino, M., Serpico, C., Coppola, G., Mayergoyz, I. D., and Bertotti, G. (2006). Midpoint numerical technique for stochastic Landau-Lifshitz-Gilbert dynamics. *J. Appl. Phys.*, **99**, 08B905.

D'Aquino, M., Serpico, C., and Miano, G. (2005). Geometrical integration of Landau–Lifshitz–Gilbert equation based on the mid-point rule. *J. Comput. Phys.*, **209**, 730–53.

Deák, A., Szunyogh, L., and Ujfalussy, B. (2011). Thickness-dependent magnetic structure of ultrathin Fe/Ir(001) films: From spin-spiral states toward ferromagnetic order. *Phys. Rev. B*, **84**, 224413–22.

Dederichs, P. H., Blügel, S., Zeller, R., and Akai, H. (1984). Ground states of constrained systems: Application to cerium impurities. *Phys. Rev. Lett.*, **53**, 2512–15.

Delamotte, B. (2004). A hint of renormalization. *Am. J. Phys.*, **72**, 170–84.

Depondt, P., and Mertens, F. G. (2009). Spin dynamics simulations of two-dimensional clusters with Heisenberg and dipole–dipole interactions. *J. Phys. Condens. Matter*, **21**, 336005.

Dyakonov, M. I., and Perel, V. I. (1971). Current-induced spin orientation of electrons in semiconductors. *Physics Letters A*, **35**, 459–60.

Dzyaloshinskii, I. E. (1957). Thermodynamic theory of weak ferromagnetism in antiferromagnetic substances. *Sov. Phys. JETP*, **5**, 1259–62.

Eastman, D. E., Himpsel, F. J., and Knapp, J. A. (1980). Experimental exchange-split energy-band dispersions for Fe, Co, and Ni. *Phys. Rev. Lett.*, **44**, 95–8.

Ebert, H., Ködderitzsch, D., and Minár, J. (2011). Calculating condensed matter properties using the KKR-Green's function method: Recent developments and applications. *Reports Prog. Phys.*, **74**, 096501.

Ebert, H., Mankovsky, S., Chadova, K., Polesya, S., Minár, J., and Ködderitzsch, D. (2015). Calculating linear-response functions for finite temperatures on the basis of the alloy analogy model. *Phys. Rev. B*, **91**, 165132.

Economou, E. N. (1979). *Green's Functions in Quantum Physics*, Springer Series in Solid-State Sciences, vol. 7. Springer, Berlin, Heidelberg.

Edwards, D. M. (1962). Spin waves in ferromagnetic metals. *Proc. R. Soc. Lond. A*, **269**, 338–51.

Edwards, D. M. (1967). Spin wave energies in the band theory of ferromagnetism. *Proc. R. Soc. Lond. A*, **300**, 373–90.

Edwards, D. M., and Muniz, R. B. (1985). Spin waves in ferromagnetic transition metals. i. general formalism and application to nickel and its alloys. *J. Phys. F: Met. Phys.*, **15**, 2339–56.

Edwards, D. M., and Rahman, M. A. (1978). The transverse dynamical susceptibility of ferromagnets and antiferromagnets within the local exchange approximation. *J. Phys. F: Met. Phys.*, **8**, 1501–12.

Eich, F., Pittalis, S., and Vignale, G. (2013). Transverse and longitudinal gradients of the spin magnetization in spin-density-functional theory. *Phys. Rev. B*, **88**, 245102.

Eich, F. G., and Gross, E. K. U. (2013). Transverse spin-gradient functional for noncollinear spin-density-functional theory. *Phys. Rev. Lett.*, **111**, 156401.

Einstein, A. (1906). The theory of the Brownian motion. *Ann. Phys.*, **19**, 371–81.

Elliott, P., Krieger, K., Dewhurst, J. K., Sharma, S., and Gross, E. K. U. (2016). Optimal control of laser-induced spin–orbit mediated ultrafast demagnetization. *New J. Phys.*, **18**, 013014.

Elmers, H. J. (1995). Ferromagnetic monolayers. *Int. J. Mod. Phys. B*, **9**, 3115–80.

Elmers, H. J., Hauschild, J., and Gradmann, U. (1996). Critical behavior of the uniaxial ferromagnetic monolayer Fe(110) on W(110). *Phys. Rev. B*, **54**, 15224–33.

Emori, S., Bauer, U., Ahn, S.-M., Martinez, E., and Beach, G., S. D. (2013). Current-driven dynamics of chiral ferromagnetic domain walls. *Nat. Mater.*, **12**, 611–16.

Eriksson, O. (1989). 'Electronic Structure, Magnetic and Cohesive Properties of Actinide, Lanthanide and Transition Metal Systems', PhD thesis, Uppsala University, Uppsala.

Eriksson, O., Brooks, M. S. S., and Johansson, B. (1989). Relativistic Stoner theory applied to PuSn$_3$. *Phys Rev. B*, **39**, 13115–19.

Eriksson, O., Nordström, L., Pohl, A., Severin, L., Boring, A. M., and Johansson, B. (1990). Spin and orbital magnetism in 3d systems. *Phys. Rev. B*, **41**, 11807–12.

Eriksson, O., Sjöström, J., Johansson, B., Häggström, L., and Skriver, H. L. (1988). Itinerant ferromagnetism in Fe$_2$P. *J. Magn. Magn. Mater.*, **74**, 347–58.

Eriksson, T., Bergqvist, L., Burkert, T., Felton, S., Tellgren, R., Nordblad, P., Eriksson, O., and Andersson, Y. (2005). Cycloidal magnetic order in the compound IrMnSi. *Phys. Rev. B*, **71**, 174420.

Eschenlohr, A., Battiato, M., Maldonado, P., Pontius, N., Kachel, T., Holldack, K., Mitzner, R., Föhlisch, A., Oppeneer, P. M., and Stamm, C. (2013). Ultrafast spin transport as key to femtosecond demagnetization. *Nat. Mater.*, **12**, 332–6.

Etz, C., Bergqvist, L., Bergman, A., Taroni, A., and Eriksson, O. (2014). Atomistic spin dynamics and surface magnons. *Psi-k Highlight*, **123**, 1–34.

Etz, C., Bergqvist, L., Bergman, A., Taroni, A., and Eriksson, O. (2015). Atomistic spin dynamics and surface magnons. *J. Phys.: Condens. Matter*, **27**, 243202.

Etz, C., Costa, M., Eriksson, O., and Bergman, A. (2012). Accelerating the switching of magnetic nanoclusters by anisotropy-driven magnetization dynamics. *Phys. Rev. B*, **86**, 224401.

Etzkorn, M., Anil-Kumar, P. S., Vollmer, R., Ibach, H., and Kirschner, J. (2004). Spin waves in ultrathin Co-films measured by spin polarized electron energy loss spectroscopy. *Surf. Sci.*, **566**, 241–5.

Evans, R. F. L., Atxitia, U., and Chantrell, R. W. (2015). Quantitative simulation of temperature-dependent magnetization dynamics and equilibrium properties of elemental ferromagnets. *Phys. Rev. B*, **91**, 144425.

Evans, R. F. L., Fan, W. J., Chureemart, P., Ostler, T. A., Ellis, M. O. A., and Chantrell, R. W. (2014). Atomistic spin model simulations of magnetic nanomaterials. *J. Phys.: Condens. Matter*, **26**, 103202.

Evans, R. F. L., Hinzke, D., Atxitia, U., Nowak, U., Chantrell, R. W., and Chubykalo- Fesenko, O. (2012). Stochastic form of the Landau-Lifshitz-Bloch equation. *Phys. Rev. B*, **85**, 014433.

Everschor, K., Garst, M., Duine, R. A., and Rosch, A. (2011). Current- induced rotational torques in the skyrmion lattice phase of chiral magnets. *Phys. Rev. B*, **84**, 064401.

Ewald, P. P. (1921). Die berechnung optischer und elektrostatischer Gitterpotentiale. *Ann. Phys.*, **369**, 253–87.

Farle, M., Silva, T., and Woltersdorf, G. (2013). 'Spin Dynamics in the Time and Frequency Domain', in H. Zabel and M. Farle, eds, *Magnetic Nanostructures: Spin Dynamics and Spin Transport*, Springer Tracts in Modern Physics, vol. 246. Springer, Berlin, Heidelberg, pp. 37–83.

Fawcett, E. (1988). Spin-density-wave antiferromagnetism in chromium. *Rev. Mod. Phys.*, **60**, 209–83.

Fawcett, E., Alberts, H. L., Galkin, V. Y., Noakes, D. R., and Yakhmi, J. V. (1994). Spin-density-wave antiferromagnetism in chromium alloys. *Rev. Mod. Phys.*, **66**, 25–127.

Fähnle, M., Drautz, R., Singer, R., Steiauf, D., and Berkov, D. V. (2005). A fast ab initio approach to the simulation of spin dynamics. *Comput. Mater. Sci.*, **32**, 118–22.

Fähnle, M., Singer, R., Steiauf, D., and Antropov, V. (2006). Role of nonequilibrium conduction electrons on the magnetization dynamics of ferromagnets in the *s-d* model. *Phys. Rev. B*, **73**, 172408.

Fedorova, N. S., Ederer, C., Spaldin, N. A., and Scaramucci, A. (2015). Biquadratic and ring exchange interactions in orthorhombic perovskite manganites. *Phys. Rev. B*, **91**, 165122.

Fennie, C. J., and Rabe, K. M. (2006). Magnetically induced phonon anisotropy in $ZnCr_2O_4$ from first principles. *Phys. Rev. Lett.*, **96**, 205505.

Fernandez, V., Vettier, C., de Bergevin, F., Giles, C., and Neubeck, W. (1998). Observation of orbital moment in NiO. *Phys. Rev. B*, **57**, 7870–6.

Ferrenberg, A. M., Landau, D. P., and Wong, Y. J. (1992). Monte Carlo simulations: Hidden errors from "good" random number generators. *Phys. Rev. Lett.*, **69**, 3382–4.

Fert, A., Cros, V., and Sampaio, J. (2013). Skyrmions on the track. *Nat. Nanotechnol.*, **8**, 152–6.

Fisher, M. E. (1964). Magnetism in one-dimensional systems: The Heisenberg model for infinite spin. *Am. J. Phys.*, **32**, 343–6.

Fisher, M. E. (1967). The theory of equilibrium critical phenomena. *Reports Prog. Phys.*, **30**, 615–731.

Fisher, M. E. (1974). The renormalization group in the theory of critical behavior. *Rev. Mod. Phys.*, **46**, 597–616.

Fletcher, G. C. (1954). Calculations of the 1st ferromagnetic anisotropy coefficient, gyromagnetic ratio and spectroscopic splitting for nickel. *Proc. Phys. Soc. A*, **67A**, 505–19.

Foldy, L. L., and Wouthuysen, S. A. (1950). On the Dirac theory of spin 1/2 particles and its non-relativistic limit. *Phys. Rev.*, **78**, 29–36.

Frank, J., Huang, W., and Leimkuhler, B. (1997). Geometric integrators for classical spin systems. *J. Comput. Phys.*, **133**, 160–72.

Fransson, J. (2010). *Non-Equilibrium Nano-Physics: A Many-Body Approach*, Lecture Notes in Physics, vol. 809. Springer Netherlands, Dordrecht.

Fransson, J., Hellsvik, J., and Nordström, L. (2015). Microscopic theory for coupled magnetization and lattice dynamics. arXiv: 1505:08005.

Frigo, M., and Johnson, S. G. (2005). The design and implementation of FFTW3. *Proc. IEEE*, **93**, 216–31.

Frota-Pessôa, S., Muniz, R. B., and Kudrnovský, J. (2000). Exchange coupling in transition-metal ferromagnets. *Phys. Rev. B*, **62**, 5293–6.

Fujii, H., Komura, S., Takeda, T., Okamoto, T., Ito, Y., and Akimitsu, J. (1979). Polarized neutron diffraction study of Fe_2P single crystal. *J. Phys. Soc. Jpn*, **46**, 1616–21.

Fukumoto, K., Matsushita, T., Osawa, H., Nakamura, T., Muro, T., Arai, K., Kimura, T., Otani, Y., and Kinoshita, T. (2008). Construction and development of a time-resolved x-ray magnetic circular dichroism: Photoelectron emission microscopy system using femtosecond laser pulses at BL25SU SPring-8. *Rev. Sci. Instrum.*, **79**, 063903.

Garanin, D. A. (1997). Fokker-Planck and Landau-Lifshitz-Bloch equations for classical ferromagnets. *Phys. Rev. B*, **55**, 3050–7.

García-Palacios, J., and Lázaro, F. (1998). Langevin-dynamics study of the dynamical properties of small magnetic particles. *Phys. Rev. B*, **58**, 14937–58.

Gerrits, T., Van Den Berg, H. A. M., Hohlfeld, J., Bär, L., and Rasing, T. (2002). Ultrafast precessional magnetization reversal by picosecond magnetic field pulse shaping. *Nature*, **418**, 509–12.

Gilbert, T. L. (2004). A phenomenological theory of damping in ferromagnetic materials. *IEEE Trans. Magn.*, **40**, 3443–9.

Gilmore, K., Idzerda, Y. U., and Stiles, M. D. (2007). Identification of the dominant precession-damping mechanism in Fe, Co, and Ni by first-principles calculations. *Phys. Rev. Lett.*, **99**, 027204.

Giovannetti, G., Kumar, S., Stroppa, A., van den Brink, J., Picozzi, S., and Lorenzana, J. (2011). High-*Tc* ferroelectricity emerging from magnetic degeneracy in cupric oxide. *Phys. Rev. Lett.*, **106**, 026401.

Giulinani, G. F., and Vignale, G. (2005). *Quantum Theory of the Electron Liquid*. Butterworth-Heinemann, London.

Glatzmaier, G. A., and Roberts, P. H. (1996). Rotation and magnetism of earth's inner core. *Science*, **274**, 1887–91.

Goedecker, S., and Hoisie, A. (2001). *Performance Optimization of Numerically Intensive Codes*. Society for Industrial and Applied Mathematics, Philadelphila, PA.

Gokhale, M. P., and Mills, D. L. (1994). Spin excitations of a model itinerant fer- romagnetic film: Spin waves, Stoner excitations, and spin-polarized electron-energy-loss spectroscopy. *Phys. Rev. B*, **49**, 3880–93.

Goldstein, H., Poole, C. P., and Safko, J. (2002). *Classical Mechanics* (3rd edn). Pearson Education, Upper Saddle River, NJ.

Gradmann, U. (1993). 'Magnetism in Ultrathin Transition Metal Films' in K. H. J. Buschow, ed., *Handbook of Magnetic Materials*, vol. 7. Elsevier, Amsterdam, pp. 1–96.

Grånäs, O., Di Marco, I., Eriksson, O., Nordström, L., and Etz, C. (2014). Electronic structure, cohesive properties, and magnetism of $SrRuO_3$. *Phys. Rev. B*, **90**, 165130.

Grånäs, O., Di Marco, I., Thunström, P., Nordström, L., Eriksson, O., Björkman, T., and Wills, J. M. (2012). Charge self-consistent dynamical mean-field theory based on the full-potential linear muffin-tin orbital method: Methodology and applications. *Comput. Mater. Sci.*, **55**, 295–302.

Greengard, L., and Rokhlin, V. (1987). A fast algorithm for particle simulations. *J. Comput. Phys.*, **73**, 325–48.

Griffiths, J. H. E. (1946). Anomalous high-frequency resistance of ferromagnetic metals. *Nature*, **158**, 670–1.

Grotheer, O., Ederer, C., and Fähnle, M. (2001). Fast ab initio methods for the calculation of adiabatic spin wave spectra in complex systems. *Phys. Rev. B*, **63**, 100401.

Gunnarsson, O. (1976). Band model for magnetism of transition metals in the spin-density-functional formalism. *J. Phys. F: Met. Phys.*, **6**, 587–606.

Gunnarsson, O., and Lundqvist, B. I. (1976). Exchange and correlation in atoms, molecules, and solids by the spin-density-functional formalism. *Phys. Rev. B*, **13**, 4274–98.

Gurevihvc, A. G., and Melkov, G. A. (1996). *Magnetization Oscillations and Waves*. CRC, Boca Raton, FL.

Gyorffy, B. L., Pindor, A. J., Staunton, J., Stocks, G. M., and Winter, H. (1985). A first-principles theory of ferromagnetic phase transitions in metals. *J. Phys. F: Met. Phys.*, **15**, 1337–86.

Hairer, E., Lubich, C., and Wanner, G. (2006). *Geometric Numerical Integration: Structure-Preserving Algorithms for Ordinary Differential Equations* (2nd edn), Springer Series in Computational Mathematics, vol. 31. Springer, Berlin, Heidelberg.

Halilov, S., Eschrig, H., Perlov, A., and Oppeneer, P. (1998). Adiabatic spin dynamics from spin-density-functional theory: Application to Fe, Co, and Ni. *Phys. Rev. B*, **58**, 293–302.

Harrison, W. A. (1989). *Electronic Structure and the Properties of Solids: The Physics of the Chemical Bond*. Dover Publications, Mineola, NY.

Hayden, L. X., Kaplan, T. A., and Mahanti, S. D. (2010). Frustrated classical Heisenberg and XY models in two dimensions with nearest-neighbor biquadratic exchange: Exact solution for the ground-state phase diagram. *Phys. Rev. Lett.*, **105**, 047203.

Hedin, L. (1965). New method for calculating the one-particle Green's function with application to the electron-gas problem. *Phys. Rev.*, **139**, A796–A823.

Hedin, L., and Lundqvist, B. I. (1971). Explicit local exchange-correlation potentials. *J. Phys. C: Solid State Phys.*, **4**, 2064–83.

Heinrich, B., and Frait, Z. (1966). Temperature dependence of the FMR linewidth of iron single-crystal platelets. *Phys. Status Solidi (b)*, **16**, K11–K14.

Heinze, S., Bergmann, K. von, Menzel, M., Brede, J., Kubetzka, A., Wiesendanger, R., Bihl-mayer, G., and Blügel, S. (2011). Spontaneous atomic-scale magnetic skyrmion lattice in two dimensions. *Nat. Phys.*, **7**, 713–18.

Hellsvik, J., Balestieri, M., Usui, T., Stroppa, A., Bergman, A., Bergqvist, L., Prabhakaran, D., Eriksson, O., Picozzi, S., Kimura, T., and Lorenzana, J. (2014). Tuning order-by-disorder multiferroicity in CuO by doping. *Phys. Rev. B*, **90**, 014437.

Hellsvik, J., Skubic, B., Nordström, L., Sanyal, B., Eriksson, O., Nordblad, P., and Sved-lindh, P. (2008). Dynamics of diluted magnetic semiconductors from atomistic spin-dynamics simulations: Mn-doped GaAs. *Phys. Rev. B*, **78**, 144419.

Herring, C. (1966). *Magnetism*. Academic Press, New York, NY.

Hjortstam, O., Trygg, J., Wills, J. M., Johansson, B., and Eriksson, O. (1996). Calculated spin and orbital moments in the surfaces of the $3d$ metals Fe, Co, and Ni and their overlayers on Cu(001). *Phys. Rev. B*, **53**, 9204–13.

Hohenberg, P., and Kohn, W. (1964). Inhomogeneous electron gas. *Phys. Rev.*, **136**, B864–B871.

Hohenberg, P. C., and Halperin, B. I. (1977). Theory of dynamic critical phenomena. *Rev. Mod. Phys.*, **49**, 435–79.

Holstein, T., and Primakoff, H. (1940). Field dependence of the intrinsic domain magnetization of a ferromagnet. *Phys. Rev.*, **58**, 1098–113.

Ibach, H., Balden, M., and Lehwald, S. (1996). Recent advances in electron energy loss spectroscopy of surface vibrations. *J. Chem. Soc., Faraday Trans.*, **92**, 4771–4.

Ishida, S., Asano, S., and Ishida, J. (1987). Electronic structures and magnetic properties of T_2P (T=Mn, Fe, Ni). *J. Phys. F: Met. Phys.*, 17, 475–82.

Ishikawa, Y., and Arai, M. (1984). Magnetic phase diagram of MnSi near critical temperature studied by neutron small angle scattering. *J. Phys. Soc. Jpn*, 53, 2726–733.

Iwasaki, J., Beekman, A. J., and Nagaosa, N. (2014). Theory of magnon-skyrmion scattering in chiral magnets. *Phys. Rev. B*, 89, 064412.

Iwasaki, J., Mochizuki, M., and Nagaosa, N. (2012). Universal current-velocity relation of skyrmion motion in chiral magnets. *Nat. Commun.*, 4, 1463.

Iwasaki, J., Mochizuki, M., and Nagaosa, N. (2013). Current-induced skyrmion dynamics in constricted geometries. *Nat. Nanotechnol.*, 8, 742–7.

James, P., Eriksson, O., Johansson, B., and Abrikosov, I. A. (1999). Calculated magnetic properties of binary alloys between Fe, Co, Ni, and Cu. *Phys. Rev. B*, 59, 419–30.

Janthon, P., Luo, S. A., Kozlov, S. M., Vives, F., Limtrakul, J., Truhlar, D. G., and Illas, F. (2014). Bulk properties of transition metals: A challenge for the design of universal density functionals. *J. Chem. Theory Comput.*, 10, 3832–9.

Jensen, J., and Mackintosh, A. R. (1991). *Rare Earth Magnetism*. Oxford University Press, Oxford.

Jiang, W., Upadhyaya, P., Zhang, W., Yu, G., Jungfleisch, M. B., Fradin, F. Y., Pearson, J. E., Tserkovnyak, Y., Wang, K. L., Heinonen, O., te Velthuis, S. G. E., and Hoffmann, A. (2015). Blowing magnetic skyrmion bubbles. *Science*, 349, 283–6.

Jorgensen, C. K. (1962). Electron transfer spectra of lanthanide complexes. *Mol. Phys.*, 5, 271–7.

Kamberský, V. (1970). On the Landau-Lifshitz relaxation in ferromagnetic metals. *Can. J. Phys.*, 48, 2906–11.

Kamberský, V. (1976). On ferromagnetic resonance damping in metals. *Czech. J. Phys. B*, 26, 1366–83.

Kamberský, V. (2007). Spin-orbital Gilbert damping in common magnetic metals. *Phys. Rev. B*, 76, 134416.

Katsnelson, M., and Antropov, V. (2003). Spin angular gradient approximation in the density functional theory. *Phys. Rev. B*, 67, 140406.

Katsura, H., Nagaosa, N., and Balatsky, A. V. (2005). Spin current and magnetoelectric effect in noncollinear magnets. *Phys. Rev. Lett.*, 95, 057205.

Kawamura, H. (1992). Chiral ordering in Heisenberg spin glasses in two and three dimensions. *Phys. Rev. Lett.*, 68, 3785–8.

Kazantseva, N., Hinzke, D., Nowak, U., Chantrell, R., Atxitia, U., and Chubykalo- Fesenko, O. (2008). Towards multiscale modeling of magnetic materials: Simulations of FePt. *Phys. Rev. B*, 77, 184428.

Keffer, F., and Kittel, C. (1952). Theory of antiferromagnetic resonance. *Phys. Rev.*, 85, 329–37.

Khorsand, A. R., Savoini, M., Kirilyuk, A., and Rasing, T. (2014). Optical excitation of thin magnetic layers in multilayer structures. *Nat. Mater.*, 13, 101.

Kimel, A. V., Ivanov, B. A., Pisarev, R. V., Usachev, P. A., Kirilyuk, A., and Rasing, T. (2009). Inertia-driven spin switching in antiferromagnets. *Nat. Phys.*, 5, 727–31.

Kimel, A. V., Kirilyuk, A., Tsvetkov, A., Pisarev, R. V., and Rasing, T. (2004). Laser- induced ultrafast spin reorientation in the antiferromagnet $TmFeO_3$. *Nature*, 429, 850–3.

Kittel, C. (1948). On the theory of ferromagnetic resonance absorption. *Phys. Rev.*, 73, 155–61.

Kittel, C, and Fong, C.-Y. (1987). *Quantum Theory of Solids* (2nd rev. edn). Wiley, New York, NY.

Kloeden, P. E., and Platen, E. (1992). *Numerical Solution of Stochastic Differential Equations*, Applications of Mathematics, vol 23. Springer, Berlin, Heidelberg.

Koelling, D. D., and Harmon, B. N. (1977). A technique for relativistic spin-polarised calculations. *J. Phys. C: Solid State Phys.*, **10**, 3107–14.

Kohn, K. (1977). A new ferrimagnet Cu_2SeO_4. *J. Phys. Soc. Jpn*, **42**, 2065–6.

Kohn, W., and Sham, L. J. (1965). Self-consistent equations including exchange and correlation effects. *Phys. Rev.*, **140**, A1133–A1138.

Kondorsky, E. I. (1974). Review of theory of magnetic anisotropy in Ni. *IEEE Transactions on Magnetics*, **MA10**, 132–6.

Kondorsky, E. I., and Straube, E. (1972). Magnetic anisotropy of nickel. *Zh. Eksp. Teor. Fiz.*, **63**, 356.

Koopmans, B., Kicken, H. H. J. E., van Kampen, M., and de Jonge, W. J. M. (2005). Microscopic model for femtosecond magnetization dynamics. *J. Magn. Magn. Mater.*, **286**, 271–5.

Koopmans, B., Malinowski, G., Dalla Longa, F., Steiauf, D., Fähnle, M., Roth, T., Cinchetti, M., and Aeschlimann, M. (2010). Explaining the paradoxical diversity of ultrafast laser-induced demagnetization. *Nat. Mater.*, **9**, 259–65.

Korzhavyi, P. A., Abrikosov, I. A., Smirnova, E. A., Bergqvist, L., Mohn, P., Mathieu, R., Svedlindh, P., Sadowski, J., Isaev, E. I., Vekilov, Yu. Kh., and Eriksson, O. (2002). Defect-induced magnetic structure in $(Ga_{1-x}Mn_x)As$. *Phys. Rev. Lett.*, **88**, 187202.

Kotliar, G., Savrasov, S. Y., Haule, K., Oudovenko, V. S., Parcollet, O., and Marianetti, C. A. (2006). Electronic structure calculations with dynamical mean-field theory. *Rev. Mod. Phys.*, **78**, 865–951.

Krieger, K., Dewhurst, J. K., Elliott, P., Sharma, S., and Gross, E. K. U. (2015). Laser-induced demagnetization at ultrashort time scales: Predictions of TDDFT. *J. Chem. Theory Comput.*, **11**, 4870.

Kubetzka, A., Ferriani, P., Bode, M., Heinze, S., Bihlmayer, G., von Bergmann, K., Pietzsch, O., Blügel, S., and Wiesendanger, R. (2005). Revealing antiferromagnetic order of the Fe monolayer on W(001): Spin-polarized scanning tunneling microscopy and first-principles calculations. *Phys. Rev. Lett.*, **94**, 087204.

Kübler, J., Hock, K.-H., Sticht, J., and Williams, A. R. (1988). Density functional theory of non-collinear magnetism. *J. Phys. F: Met. Phys.*, **18**, 469–84.

Kübler, J. (2000). *Theory of Itinerant Electron Magnetism*, International Series of Monographs on Physics, vol. 106. Clarendon Press, Oxford.

Kübler, J. (2009). *Theory of Itinerant Electron Magnetism* (rev. edn), International Series of Monographs on Physics, vol. 106. Oxford University Press, Oxford.

Kubo, R. (1966). The fluctuation-dissipation theorem. *Reports Prog. Phys.*, **29**, 255–84.

Kubo, R. and Hashitsu, N. (1970). Brownian motion of spins. *Prog. Theor. Phys. Suppl.*, **46**, 210–20.

Kudrnovský, J., Máca, F., Turek, I., and Redinger, J. (2009). Substrate-induced antiferromagnetism of a Fe monolayer on the Ir(001) surface. *Phys. Rev. B*, **80**, 064405.

Kudrnovský, J., Turek, I., Drchal, V., Máca, F., Weinberger, P., and Bruno, P. (2004). Exchange interactions in III-V and group-IV diluted magnetic semiconductors. *Phys. Rev. B*, **69**, 115208.

Kuneš, J., and Kamberský, V. (2002). First-principles investigation of the damping of fast magnetization precession in ferromagnetic 3*d* metals. *Phys. Rev. B*, **65**, 212411.

Kurz, P., Bihlmayer, G., Hirai, K., and Blügel, S. (2001). Three-dimensional spin structure on a two-dimensional lattice: Mn/Cu(111). *Phys. Rev. Lett.*, **86**, 1106–9.

Kvashnin, Y. O., Grånäs, O., Di Marco, I., Katsnelson, M. I., Lichtenstein, A. I., and Eriksson, O. (2015). Exchange parameters of strongly correlated materials: Extraction from spin-polarized density functional theory plus dynamical mean-field theory. *Phys. Rev. B*, **91**, 1–10.

Lambert, C.-H., Mangin, S., Varaprasad, B. S. D. C. S., Takahashi, Y. K., Hehn, M., Cinchetti, M., Malinowski, G., Hono, K., Fainman, Y., Aeschlimann, M., and Fullerton, E. E. (2014). All-optical control of ferromagnetic thin films and nanostructures. *Science*, 345, 1337–40.

Landau, D. P., and Binder, K. (2005). *A Guide to Monte Carlo Simulations in Statistical Physics.* Cambridge University Press, New York, NY.

Landau, L., and Lifshitz, E. (1935). On the theory of the dispersion of magnetic permeability in ferromagnetic bodies. *Phyz. Z. Sowjetunion*, 8, 153–69.

Landau, L. D. (1946). On the vibrations of the electronic plasma. *J. Phys. USSR*, 10, 25–34.

Landau, L. D., and Lifshitz, E. M. (1980). *Statistical Physics* (3rd edn). Butterworth-Heinemann, Oxford.

Lax, B., and Button, K. J. (1962). *Microwave Ferrites and Ferrimagnetics.* McGraw-Hill, New York, NY.

Lee, M., Kang, W., Onose, Y., Tokura, Y., and Ong, N. P. (2009). Unusual Hall effect anomaly in MnSi under pressure. *Phys. Rev. Lett.*, 102, 186601.

Ležaić, M., Mavropoulos, P., Bihlmayer, G., and Blügel, S. (2013). Exchange interactions and local-moment fluctuation corrections in ferromagnets at finite temperatures based on noncollinear density-functional calculations. *Phys. Rev. B*, 88, 134403.

Lenz, K., Wende, H., Kuch, W., Baberschke, K., Nagy, K., and Jánossy, A. (2006). Two-magnon scattering and viscous Gilbert damping in ultrathin ferromagnets. *Phys. Rev. B*, 73, 144424.

Li, C., Freeman, A. J., and Fu, C. L. (1990). Electronic structure and magnetism of surfaces and interfaces: Selected examples. *J. Magn. Magn. Mater.*, 83, 51–6.

Liechtenstein, A. I., Katsnelson, M. I., Antropov, V. P., and Gubanov, V. A. (1987). Local spin density functional approach to the theory of exchange interactions in ferromagnetic metals and alloys. *J. Magn. Magn. Mater.*, 67, 65–74.

Liechtenstein, A. I., Katsnelson, M. I., and Gubanov, V. A. (1984). Exchange interactions and spin-wave stiffness in ferromagnetic metals. *J. Phys. F: Met. Phys.*, 14, L125– L128.

Lilley, B. A. (1950). LXXI. Energies and widths of domain boundaries in ferromagnetics. *Philos. Mag.*, 41, 792–813.

Liu, Y., Starikov, A. A., Yuan, Z., and Kelly, P. J. (2011). First-principles calculations of magnetization relaxation in pure Fe, Co, and Ni with frozen thermal lattice disorder. *Phys. Rev. B*, 84, 014412.

Lizarraga, R., Nordström, L., Bergqvist, L., Bergman, A., Sjöstedt, E., Mohn, P., and Eriksson, O. (2004). Conditions for noncollinear instabilities of ferromagnetic materials. *Phys. Rev. Lett.*, 93, 107205.

Lorenz, R., and Hafner, J. (1996). Magnetic structure and anisotropy of thin Fe films on Cu(001) substrates. *Phys. Rev. B*, 54, 15937–49.

Lounis, S., Costa, A. T., Muniz, R. B., and Mills, D. L. (2010). Dynamical magnetic excitations of nanostructures from first principles. *Phys. Rev. Lett.*, 105, 187205.

Lundin, U., and Eriksson, O. (2001). Novel method of self-interaction corrections in density functional calculations. *Int. J. Quant. Chem.*, 81, 247–52.

Lynn, J. W. (1975). Temperature dependence of the magnetic excitations in iron. *Phys. Rev. B*, 11, 2624–37.

Ma, P.-W., and Dudarev, S. L. (2011). Langevin spin dynamics. *Phys. Rev. B*, 83, 134418.

Ma, P.-W., and Dudarev, S. L. (2012). Longitudinal magnetic fluctuations in Langevin spin dynamics. *Phys. Rev. B*, 86, 054416.

Ma, P.-W., and Dudarev, S. L. (2015). Constrained density functional for non-collinear magnetism. *Phys. Rev. B*, 91, 054420.

Ma, P.-W., Dudarev, S. L., Semenov, A. A., and Woo, C. H. (2010). Temperature for a dynamic spin ensemble. *Phys. Rev. E*, **82**, 031111.

Ma, P.-W., Dudarev, S. L., and Woo, C. H. (2012). Spin-lattice-electron dynamics simulations of magnetic materials. *Phys. Rev. B*, **85**, 184301.

Ma, P.-W., Woo, C., and Dudarev, S. (2008). Large-scale simulation of the spin-lattice dynamics in ferromagnetic iron. *Phys. Rev. B*, **78**, 024434.

MacDonald, A. H., Girvin, S. M., and Yoshioka, D. (1988). $\frac{t}{u}$ expansion for the Hubbard model. *Phys. Rev. B*, **37**, 9753–6.

Mackintosh, A. K., and Andersen, O. K. (1975). 'The Electronic Structure of Transition Metals', in M. Springford, ed., *Electrons at the Fermi Surface*. Cambridge University Press, Cambridge, pp. 149–224.

Mangin, S., Gottwald, M., Lambert, C.-H., Steil, D., Uhlir, V., Pang, L., Hehn, M., Alebrand, S., Cinchetti, M., Malinowski, G., Fainman, Y., Aeschlimann, M., and Fullerton, E. E. (2014). Engineered materials for all-optical helicity-dependent magnetic switching. *Nat. Mater.*, **13**, 286–92.

Mankovsky, S., Ködderitzsch, D., Woltersdorf, G., and Ebert, H. (2013). First-principles calculation of the Gilbert damping parameter via the linear response formalism with application to magnetic transition metals and alloys. *Phys. Rev. B*, **87**, 014430.

Marder, P. (2010). *Condensed Matter Physics*. Wiley, Hoboken, NJ.

Marques, M. A. L., Ullrich, C. A., Nogueira, F., Rubio, A., Burke, K., and Gross, E. K. U. (2006). *Time-Dependent Density Functional Theory*, Lecture Notes in Physics, vol. 706. Springer, Berlin, Heidelberg.

Marsaglia, G., and Tsang, W. W. (2000). The ziggurat method for generating random variables. *J. Stat. Softw.*, **5**, 1–7.

Marshall, W., and Lovesey, S. W. (1971). *Theory of Thermal Neutron Scattering*. Oxford University Press, Oxford.

Martin, V., Meyer, W., Giovanardi, C., Hammer, L., Heinz, K., Tian, Z., Sander, D., and Kirschner, J. (2007). Pseudomorphic growth of Fe monolayers on Ir(001)-(1 × 1): From a fct precursor to a bct film. *Phys. Rev. B*, **76**, 205418.

Matsumoto, M., and Nishimura, T. (1998). Mersenne Twister: A 623-dimensionally equidistributed uniform pseudo-random number generator. *ACM Trans. Model. Comput. Simul.*, **8**, 3–30.

McLachlan, R. I., Modin, K., and Verdier, O. (2014). Symplectic integrators for spin systems. *Phys. Rev. E*, **89**, 061301(R).

McLachlan, Robert I., Modin, Klas, and Verdier, Olivier (2016). Geometry of discrete-time spin systems. *J. Nonlinear Sci.* 26, 1507.

McShane, E. J. (1974). *Stochastic Calculus and Stochastic Models*. Academic Press, New York, NY.

Mentink, J. H., Balzer, K., and Eckstein, M. (2015). Ultrafast and reversible control of the exchange interaction in Mott insulators. *Nat. Commun.*, **6**, 6708.

Mentink, J. H., and Eckstein, M. (2014). Ultrafast quenching of the exchange interaction in a Mott insulator. *Phys. Rev. Lett.*, **113**, 057201.

Mentink, J. H., Hellsvik, J., Afanasiev, D. V., Ivanov, B. A., Kirilyuk, A., Kimel, A. V., Eriksson, O., Katsnelson, M. I., and Rasing, T. (2012). Ultrafast spin dynamics in multisublattice magnets. *Phys. Rev. Lett.*, **108**, 057202.

Mentink, J. H., Tretyakov, M. V., Fasolino, A., Katsnelson, M. I., and Rasing, T. (2010). Stable and fast semi-implicit integration of the stochastic Landau-Lifshitz equation. *J. Phys.: Condens. Matter*, **22**, 176001.

Mermin, N. D., and Wagner, H. (1966). Absence of ferromagnetism or antiferromagnetism in one- or two-dimensional isotropic Heisenberg models. *Phys. Rev. Lett.*, 17, 1133–6.

Metzner, W., and Vollhardt, D. (1989). Correlated lattice fermions in $d = 1$dimensions. *Phys. Rev. Lett.*, 62, 324–7.

Mills, D. L., and Arias, R. (2006). The damping of spin motions in ultrathin films: Is the Landau-Lifschitz-Gilbert phenomenology applicable? *Physica B: Cond. Matter.*, 384, 147–51.

Milshtein, G. N., and Tret'yakov, M. V. (1994). Numerical solution of differential equations with colored noise. *J. Stat. Phys.*, 77, 691–715.

Milstein, G. N., Repin, Yu. M., and Tretyakov, M. V. (2002). Numerical methods for stochastic systems preserving symplectic structure. *SIAM J. Numer. Anal.*, 40, 1583–604.

Milstein, G. N., and Tretyakov, M. V. (2004). *Stochastic Numerics for Mathematical Physics*, Scientific Computation. Springer, Berlin, Heidelberg.

Mochizuki, M., Furukawa, N., and Nagaosa, N. (2010a). Spin model of magnetostrictions in multiferroic Mn perovskites. *Phys. Rev. Lett.*, 105, 137205.

Mochizuki, M., Furukawa, N., and Nagaosa, N. (2010b). Theory of electromagnons in the multiferroic Mn perovskites: The vital role of higher harmonic components of the spiral spin order. *Phys. Rev. Lett.*, 104, 177206.

Mochizuki, M., Furukawa, N., and Nagaosa, N. (2011). Theory of spin-phonon coupling in multiferroic manganese perovskites RMnO {3}. *Phys. Rev. B*, 84, 144409.

Mochizuki, M., Yu, X. Z., Seki, S., Kanazawa, N., Koshibae, W., Zang, J., Mostovoy, M., Tokura, Y., and Nagaosa, N. (2014). Thermally driven ratchet motion of a skyrmion microcrystal and topological magnon Hall effect. *Nat. Mater.*, 13, 241–6.

Modin, K., and Söderlind, G. (2011). Geometric integration of Hamiltonian systems perturbed by Rayleigh damping. *BIT Numer. Math.*, 51, 977–1007.

Mohn, P. (2003). *Magnetism in the Solid State: An Introduction*, Solid-State Sciences, vol 134. Springer, Berlin, Heidelberg.

Mohn, P. (2006). *Magnetism in the Solid State: An Introduction* (corrected 2nd printing), Solid-State Sciences, vol 134. Springer, Berlin, Heidelberg.

Moodera, J. S., Kinder, L. R., Wong, T. M., and Meservey, R. (1995). Large magnetoresistance at room temperature in ferromagnetic thin film tunnel junctions. *Phys. Rev. Lett.*, 74, 3273–6.

Moon, R. M., Koehler, W. C., Cable, J. W., and Child, H. R. (1972). Distribution of magnetic moment in metallic gadolinium. *Phys. Rev. B*, 5, 997–1016.

Moriya, T. (1960). Anisotropic superexchange interaction and weak ferromagnetism. *Phys. Rev.*, 120, 91–8.

Mozafari, E., Alling, B., Steneteg, P., and Abrikosov, I. A. (2015). Role of N defects in paramagnetic CrN at finite temperatures from first principles. *Phys. Rev. B*, 91, 094101.

Muniz, R. B., and Mills, D. L. (2002). Theory of spin excitations in Fe(110) monolayers. *Phys. Rev. B*, 66, 174417.

Munkres, J. (2000). *Topology*. Prentice Hall, Upper Saddle River, NJ.

Mühlbauer, S., Binz, B., Jonietz, F., Pfleiderer, C., Rosch, A., Neubauer, A., Georgii, R., and Böni, P. (2009). Skyrmion lattice in a chiral magnet. *Science*, 323, 915–19.

Miyazaki, T., and Tezuka, N. (1995). Giant magnetic tunneling effect in Fe/Al_2O_3/Fe junction. *J. Magn. Magn. Mater.*, 139, L231–L234.

Neubauer, A., Pfleiderer, C., Binz, B., Rosch, A., Ritz, R., Niklowitz, P. G., and Böni, P. (2009). Topological Hall effect in the *A*phase of MnSi. *Phys. Rev. Lett.*, 102, 186602.

Newman, M. E. J., and Barkema, G. T. (1999). *Monte Carlo Methods in Statistical Physics*. Oxford University Press, Oxford.

Nickolls, J., Buck, I., Garland, M., and Skadron, K. (2008). Scalable parallel programming with CUDA. *Queue*, **6**, 40–53.

Niu, Q., Wang, X., Kleinman, L., Liu, W.-M., Nicholson, D. M. C., and Stocks, G. M. (1999). Adiabatic dynamics of local spin moments in itinerant magnets. *Phys. Rev. Lett.*, **83**, 207–10.

Nordström, L., and Singh, D. (1996). Noncollinear intra-atomic magnetism. *Phys. Rev. Lett.*, **76**, 4420–3.

Norman, M. R. (1991). Crystal-field polarization and the insulating gap in FeO, CoO, NiO, and La$_2$CuO$_4$. *Phys. Rev. B*, **44**, 1364–7.

Nowak, U., Mryasov, O., Wieser, R., Guslienko, K., and Chantrell, R. (2005). Spin dynamics of magnetic nanoparticles: Beyond Brown's theory. *Phys. Rev. B*, **72**, 172410.

Nugent, L. J. (1970). Theory of the tetrad effect in the lanthanide(iii) and actinide(iii) series. *J. Inorg. Nucl. Chem.*, **32**, 3485–91.

Omelyan, I. P., Mryglod, I. M., and Folk, R. (2001). Algorithm for molecular dynamics simulations of spin liquids. *Phys. Rev. Lett.*, **86**, 898–901.

Oogane, M., Wakitani, T., Yakata, S., Yilgin, R., Ando, Y., Sakuma, A., and Miyazaki, T. (2006). Magnetic damping in ferromagnetic thin films. *Jpn J. Appl. Phys.*, **45**, 3889–91.

Ostler, T. A., Barker, J., Evans, R. F. L., Chantrell, R. W., Atxitia, U., Chubykalo-Fesenko, O., El Moussaoui, S., Le Guyader, L., Mengotti, E., Heyderman, L. J., Nolting, F., Tsukamoto, A., Itoh, A., Afanasiev, D., Ivanov, B. A., Kalashnikova, A. M., Vahaplar, K., Mentink, J., Kirilyuk, A., Rasing, T., and Kimel, A. V. (2012). Ultrafast heating as a sufficient stimulus for magnetization reversal in a ferrimagnet. *Nat. Commun.*, **3**, 666.

Pajda, M., Kudrnovský, J., Turek, I., Drchal, V., and Bruno, P. (2000). Oscillatory Curie temperature of two-dimensional ferromagnets. *Phys. Rev. Lett.*, **85**, 5424–7.

Pajda, M., Kudrnovský, J., Turek, I., Drchal, V., and Bruno, P. (2001). Ab initio calculations of exchange interactions, spin-wave stiffness constants, and Curie temperatures of Fe, Co, and Ni. *Phys. Rev. B*, **64**, 174402.

Pal, P., Banerjee, R., Banerjee, R., Mookerjee, A., Kaphle, G. C., Sanyal, B., Hellsvik, J., Eriksson, O., Mitra, P., Majumdar, A. K., and Nigam, A. K. (2012). Magnetic ordering in Ni-rich NiMn alloys around the multicritical point: Experiment and theory. *Phys. Rev. B*, **85**, 174405.

Parkin, S. S. P., Hayashi, M., and Thomas, L. (2008). Magnetic domain-wall racetrack memory. *Science*, **320**, 190–4.

Pasrija, K., and Kumar, S. (2013). High-temperature noncollinear magnetism in a classical bilinear-biquadratic Heisenberg model. *Phys. Rev. B*, **88**, 144418.

Pauthenet, R. (1982a). Experimental verification of spin-wave theory in high fields. *J. Appl. Phys.*, **53**, 8187–92.

Pauthenet, R. (1982b). Spin-waves in nickel, iron, and yttrium-iron garnet. *J. Appl. Phys.*, **53**, 2029–31.

Perdew, J. P., and Zunger, A. (1981). Self-interaction correction to density-functional approximations for many-electron systems. *Phys. Rev. B*, **23**, 5048–79.

Perera, D., Eisenbach, M., Nicholson, D. M., Stocks, G. M., and Landau, D. P. (2016). Reinventing atomistic magnetic simulations with spin- orbit coupling. *Phys. Rev. B*, **93**, 060402.

Perera, D., Landau, D. P., Nicholson, D. M., Stocks, G. M., Eisenbach, M., Yin, J., and Brown, G. (2014). Combined molecular dynamics-spin dynamics simulation of bcc iron. *J. Phys: Conf. Ser.*, **487**, 012007.

Plischke, M., and Bergersen, B. (1994). *Equilibrium Statistical Physics* (2nd edn). World Scientific, Singapore.

Popescu, V., Ebert, H., Nonas, B., and Dederichs, P. H. (2001). Spin and orbital magnetic moments of 3d and 4d impurities in and on the (001) surface of bcc Fe. *Phys. Rev. B*, **64**, 184407.

Prokop, J., Tang, W., Zhang, Y., Tudosa, I., Peixoto, T., Zakeri, Kh., and Kirschner, J. (2009). Magnons in a ferromagnetic monolayer. *Phys. Rev. Lett.*, **102**, 177206.

Qian, X., and Hübner, W. (2001). *Ab initio* magnetocrystalline anisotropy calculations for Fe/W(110) and Fe/Mo(110). *Phys. Rev. B*, **64**, 092402.

Qian, Z., and Vignale, G. (2002). Spin dynamics from time-dependent spin-density-functional theory. *Phys. Rev. Lett.*, **88**, 056404.

Radu, I., Vahaplar, K., Stamm, C., Kachel, T., Pontius, N., Dürr, H. A., Ostler, T. A., Barker, J., Evans, R. F. L., Chantrell, R. W., Tsukamoto, A., Itoh, A., Kirilyuk, A., Rasing, T., and Kimel, A. V. (2011). Transient ferromagnetic-like state mediating ultrafast reversal of antiferromagnetically coupled spins. *Nature*, **472**, 205–8.

Rajeswari, J., Ibach, H., and Schneider, C. M. (2014). Standing spin waves in ultra-thin magnetic films: A method to test for layer-dependent exchange coupling. *Phys. Rev. Lett.*, **112**, 127202.

Ralph, D. C., and Stiles, M. D. (2008). Spin transfer torques. *J. Magn. Magn. Mater.*, **320**, 1190–216.

Rhie, H.-S., Dürr, H. A., and Eberhardt, W. (2003). Femtosecond electron and spin dynamics in Ni/W(110) films. *Phys. Rev. Lett.*, **90**, 247201.

Risken, H. (1989). *The Fokker-Planck Equation: Methods of Solution and Applications* (2nd edn), Springer Series in Synergetics, vol 18. Springer, Berlin, Heidelberg.

Rodrigues, O. (1840). Des lois géométriques qui régissent les déplacements d'un système solide dans l'espace, et de la variation des cordonnées provenant de ces déplacements considérés indpendamment des causes qui peuvent les produire. *J. Math. Pures Appl.*, **5**, 380–440.

Roeland, L. W., Cock, G. J., Muller, F. A., Moleman, A. C., McEwen, K. A., Jordan, R. G., and Jones, D. W. (1975). Conduction electron polarization of gadolinium metal. *J. Phys. F: Met. Phys.*, **5**, L233–L237.

Rohart, S., and Thiaville, A. (2013). Skyrmion confinement in ultrathin film nanos- tructures in the presence of Dzyaloshinskii-Moriya interaction. *Phys. Rev. B*, **88**, 184422.

Romming, N., Hanneken, C., Menzel, M., Bickel, J. E., Wolter, B., von Bergmann, K., Kubetzka, A., and Wiesendanger, R. (2013). Writing and deleting single magnetic skyrmions. *Science*, **341**, 636–9.

Romming, N., Kubetzka, A., Hanneken, C., Bergmann, K. von, and Wiesendanger, R. (2015). Field-dependent size and shape of single magnetic skyrmions. *Phys. Rev. Lett.*, **114**, 177203.

Rosengaard, N. M., and Johansson, B. (1997). Finite-temperature study of itinerant ferromagnetism in Fe, Co, and Ni. *Phys. Rev. B*, **55**, 14975–86.

Roth, T., Schellekens, A. J., Alebrand, S., Schmitt, O., Steil, D., Koopmans, B., Cinchetti, M., and Aeschlimann, M. (2012). Temperature dependence of laser-induced demagnetization in Ni: A key for identifying the underlying mechanism. *Phys. Rev. X*, **2**, 021006.

Rößler, U. K., Bogdanov, A. N., and Pfleiderer, C. (2006). Spontaneous skyrmion ground states in magnetic metals. *Nature*, **442**, 797–801.

Rózsa, L., Udvardi, L., and Szunyogh, L. (2014). Langevin spin dynamics based on ab initio calculations: Numerical schemes and applications. *J. Phys.: Condens. Matter*, **26**, 216003.

Ruban, A. V., Khmelevskyi, S., Mohn, P., and Johansson, B. (2007). Temperature-induced longitudinal spin fluctuations in Fe and Ni. *Phys. Rev. B*, **75**, 054402.

Runge, E., and Gross, E. K. U. (1984). Density-functional theory for time-dependent systems. *Phys. Rev. Lett.*, **52**, 997–1000.

Rümelin, W. (1982). Numerical treatment of stochastic differential equations. *SIAM J. Numer. Anal.*, **19**, 604–13.

Ryu, K.-S., Thomas, L., Yang, S.-H., and Parkin, S. (2013). Chiral spin torque at magnetic domain walls. *Nat. Nanotechnol.*, **8**, 527–33.

Ryu, K.-S., Yang, S.-H., Thomas, L., and Parkin, S. S. P. (2014). Chiral spin torque arising from proximity-induced magnetization. *Nat. Commun.*, **5**, 3910.

Safonov, V., and Neal Bertram, H. (2001). Spin-wave dynamic magnetization reversal in a quasi-single-domain magnetic grain. *Phys. Rev. B*, **63**, 094419.

Sampaio, J., Cros, V., Rohart, S., Thiaville, A., and Fert, A. (2013). Nucleation, stability and current-induced motion of isolated magnetic skyrmions in nanostructures. *Nat. Nanotechnol.*, **8**, 839–44.

Sandratskii, L. M. (1991a). Symmetry analysis of electronic states for crystals with spiral magnetic order. I. General properties. *J. Phys.: Condens. Matter*, **3**, 8565–86.

Sandratskii, L. M. (1991b). Symmetry analysis of electronic states for crystals with spiral magnetic order. II. Connection with limiting cases. *J. Phys.: Condens. Matter*, **3**, 8587–96.

Sandratskii, L. M. (1998). Noncollinear magnetism in itinerant-electron systems: Theory and applications. *Adv. Phys.*, **47**, 91–160.

Sandratskii, L. M. (2010). Stable and variable features of the magnetic structure of fcc Fe/Cu(001) films. *Phys. Rev. B*, **81**, 064417.

Sandratskii, L. M., and Bruno, P. (2006). Exchange interactions in ZnMeO (Me = Mn, Fe, Co, Ni): Calculations using the frozen-magnon technique. *Phys. Rev. B*, **73**, 045203.

Sandratskii, L. M., Singer, R., and Şaşıoğlu, E. (2007). Heisenberg Hamiltonian de- scription of multiple-sublattice itinerant-electron systems: General considerations and applications to NiMnSb and MnAs. *Phys. Rev. B*, **76**, 184406.

Sato, K., Bergqvist, L., Kudrnovský, J., Dederichs, P. H., Eriksson, O., Turek, I., Sanyal, B., Bouzerar, G., Katayama-Yoshida, H., Dinh, V. A., Fukushima, T., Kizaki, H., and Zeller, R. (2010). First-principles theory of dilute magnetic semiconductors. *Rev. Mod. Phys.*, **82**, 1633–90.

Savrasov, S. Y. (1998). Linear response calculations of spin fluctuations. *Phys. Rev. Lett.*, **81**, 2570–3

Saxena, S. K., Shen, G., and Lazor, P. (1994). Temperatures in Earth's core based on melting and phase transformation experiments on iron. *Science*, **264**, 405–7.

Schadler, G., Weinberger, P., Boring, A. M., and Albers, R. C. (1986). Relativistic spin-polarized electronic structure of Ce and Pu. *Phys. Rev. B*, **34**, 713–22.

Schieback, C., Kläu, M., Nowak, U., Rüdiger, U., and Nielaba, P. (2007). Numerical investigation of spin-torque using the Heisenberg model. *Eur. Phys. J. B*, **59**, 429433.

Schmitz, D., Charton, C., Scroll, A., Carbone, C., and Eberhardt, W. (1999). Magnetic moments of fcc Fe overlayers on Cu(100) and Co(100). *Phys. Rev. B*, **59**, 4327–33.

Schütte, C., and Garst, M. (2014). Magnon-skyrmion scattering in chiral magnets. *Phys. Rev. B*, **90**, 094423.

Schwarz, K., and Mohn, P. (1984). Itinerant metamagnetism in YCo_2. *J. Phys. F: Met. Phys.*, **14**, L129–L134.

Secchi, A., Brener, S., Lichtenstein, A. I., and Katsnelson, M. I. (2013). Non-equilibrium magnetic interactions in strongly correlated systems. *Ann. Phys.*, **333**, 221–71.

Secchi, A., Lichtenstein, A. I., and Katsnelson, M. I. (2015). Spin and orbital exchange interactions from Dynamical Mean Field Theory. *J. Magn. Magn. Mater.*, **400**, 112–16.

Seki, S., Yu, X. Z., Ishiwata, S., and Tokura, Y. (2012). Observation of skyrmions in a multiferroic material. *Science*, **336**, 198–201.

Sergienko, I. A., and Dagotto, E. (2006). Role of the Dzyaloshinskii-Moriya interaction in multiferroic perovskites. *Phys. Rev. B*, 73, 094434.

Serpico, C., Mayergoyz, I. D., and Bertotti, G. (2001). Numerical technique for integration of the Landau-Lifshitz equation. *J. Appl. Phys.*, 89, 6991.

Shallcross, S., Kissavos, A., Meded, V., and Ruban, A. (2005). An ab initio effective Hamiltonian for magnetism including longitudinal spin fluctuations. *Phys. Rev. B*, 72, 104437.

Sharma, S., Pittalis, S., Kurth, S., Shallcross, S., Dewhurst, J., and Gross, E. (2007). Comparison of exact-exchange calculations for solids in current-spin-density- and spin-density-functional theory. *Phys. Rev. B*, 76, 100401.

Shastry, B. S. (1984). Spin dynamics of paramagnetic iron. *Phys. Rev. Lett.*, 53, 1104–7.

Shibata, K., Yu, X. Z., Hara, T., Morikawa, D., Kanazawa, N., Kimoto, K., Ishiwata, S., Matsui, Y., and Tokura, Y. (2013). Towards control of the size and helicity of skyrmions in helimagnetic alloys by spin-orbit coupling. *Nat. Nanotechnol.*, 8, 723–8.

Shirts, M., and Pande, V. S. (2000). Screen savers of the world unite! *Science*, 290, 1903–4.

Shubin, S., and Zolotukhin, M. (1936). [No title]. *Zh. Eksp. Teor. Fiz.*, 6, 105.

Sinclair, R. N., and Brockhouse, B. N. (1960). Dispersion relation for spin waves in a fcc cobalt alloy. *Phys. Rev.*, 120, 1638–40.

Singh, D. J., and Ashkenazi, J. (1992). Magnetism with generalized-gradient- approximation density functionals. *Phys. Rev. B*, 46, 11570–7.

Skriver, H. L. (1985). Crystal structure from one-electron theory. *Phys. Rev. B*, 31, 1909–23.

Skubic, B., Hellsvik, J., Nordström, L., and Eriksson, O. (2008). A method for atomistic spin dynamics simulations: Implementation and examples. *J. Phys.: Condens. Matter*, 20, 315203.

Skubic, B., Peil, O., Hellsvik, J., Nordblad, P., Nordström, L., and Eriksson, O. (2009). Atomistic spin dynamics of the Cu-Mn spin-glass alloy. *Phys. Rev. B*, 79, 024411.

Skyrme, T. H. R. (1961). A non-linear field theory. *Proc. R. Soc. A*, 260, 127–38.

Slater, J. C. (1937). The theory of ferromagnetism: Lowest energy levels. *Phys. Rev.*, 52, 198–214.

Sliwko, V., Mohn, P., and Schwarz, K. -H. (1994). The electronic and magnetic structures of alpha- and beta-manganese. *J. Phys.: Condens. Matter*, 6, 6557–64.

Slonczewski, J. C. (1996). Current-driven excitation of magnetic multilayers. *J. Magn. Magn. Mater.*, 159, L1–L7.

Slonczewski, J. C. (1962). Band theory of anisotropy. *J. Phys. Soc. Jpn Suppl. B1*, 17, 34.

Smit, J., and Beljers, H. C. (1955). Ferromagnetic resonance absorption in $BaFe_{12}O_{19}$, a highly anisotropic crystal. *Philips Res. Rep.*, 10, 1113–30.

Solovyev, I. V. (2005). Orbital polarization in itinerant magnets. *Phys. Rev. Lett.*, 95, 267205.

Soven, P. (1967). Coherent-potential model of substitutional disordered alloys. *Phys. Rev.*, 156, 809–13.

Söderlind, P., Ahuja, R., Eriksson, O., Wills, J. M., and Johansson, B. (1994). Crystal structure and elastic-constant anomalies in the magnetic 3d transition metals. *Phys. Rev. B*, 50, 5918–27.

Squires, G. L. (1997). *Introduction to the Theory of Thermal Neutron Scattering*. Dover Publications, Mineola, NY.

Squires, G. L. (2012). *Introduction to the Theory of Thermal Neutron Scattering* (3rd edn). Cambridge University Press, Cambridge.

Stamenova, M., Simoni, J., and Sanvito, S. (2016). The role of spin-orbit interaction in the ultrafast demagnetization of small iron clusters. *Phys. Rev. B*, 94, 014423.

Stamm, C., Tudosa, I., Siegmann, H. C., Stöhr, J., Dobin, A. Yu., Woltersdorf, G., Heinrich, B., and Vaterlaus, A. (2005). Dissipation of spin angular momentum in magnetic switching. *Phys. Rev. Lett.*, 94, 197603.

Stanciu, C. D., Hansteen, F., Kimel, A. V., Kirilyuk, A., Tsukamoto, A., Itoh, A., and Rasing, T. (2007). All-optical magnetic recording with circularly polarized light. *Phys. Rev. Lett.*, **99**, 047601.

Stanley, H. E. (1999). Scaling, universality, and renormalization: Three pillars of modern critical phenomena. *Rev Mod Phys*, **71**, S358–S366.

Starikov, A., Kelly, P., Brataas, A., Tserkovnyak, Y., and Bauer, G. E. W. (2010). Unified first-principles study of Gilbert damping, spin-flip diffusion, and resistivity in transition metal alloys. *Phys. Rev. Lett.*, **105**, 236601.

Staunton, J., Gyorffy, B. L., Pindor, A. J., Stocks, G. M., and Winter, H. (1984). The "disordered local moment" picture of itinerant magnetism at finite temperatures. *J. Magn. Magn. Mater.*, **45**, 15–22.

Staunton, J. B., Ostanin, S., Razee, S. S. A., Gyorffy, B. L., Szunyogh, L., Ginatempo, B., and Bruno, E. (2004). Temperature dependent magnetic anisotropy in metallic magnets from an *ab initio* electronic structure theory: $L1_0$-ordered FePt. *Phys. Rev. Lett.*, **93**, 257204–8.

Stearns, M. B. (1984). 'Magnetostriction coefficients', in H. P. J. Wijn, ed., *3d, 4d and 5d Elements, Alloys and Compounds*, Landolt-Börnstein: Group III Condensed Matter, vol 19a, Springer, Berlin, Heidelberg, pp. 48–52.

Steiauf, D., and Fähnle, M. (2005). Damping of spin dynamics in nanostructures: An ab initio study. *Phys. Rev. B*, **72**, 064450.

Steinigeweg, R., and Schmidt, H. J. (2006). Symplectic integrators for classical spin systems. *Comput. Phys. Commun.*, **174**, 853–861.

Stocks, G. M., Ujfalussy, B., Wang, X., Nicholson, D. M. C., Shelton, W. A., Wang, Y., Canning, A., and Györffy, B. L. (1998). Towards a constrained local moment model for first principles spin dynamics. *Philos. Mag. B*, **78**, 665–73.

Stone, J. E., Gohara, D., and Shi, G. (2010). OpenCL: A parallel programming standard for heterogeneous computing systems. *IEEE Des. Test*, **12**, 66–73.

Stoner, E. C. (1936). Collective electron specific heat and spin paramagnetism in metals. *Proc. R. Soc. A*, **154**, 656–78.

Stoner, E. C. (1938). Collective electron ferromagnetism. *Proc. R. Soc. A*, **165**, 372–414.

Stoner, E. C., and Wohlfarth, E. P. (1948). A mechanism of magnetic hysteresis in heterogeneous alloys. *Proc. R. Soc. A*, **240**, 599–642.

Suhl, H. (1955). Ferromagnetic resonance in nickel ferrite between one and two kilomegacycles. *Phys. Rev.*, **97**, 555–7.

Sun, C. J., Zhou, Y. Z., Chen, J. S., Chow, G. M., Fecher, G. H., Lin, H. J., and Hwu, Y. K. (2006). Field dependence of spin and orbital moments of Fe in $L1_0$ FePt magnetic thin films. *J. Magn. Magn. Mater.*, **303**, e247–e250.

Sun, Q., and Xie, X. C. (2005). Definition of the spin current: The angular spin current and its physical consequences. *Phys. Rev. B*, **72**, 245305.

Szilva, A., Costa, M., Bergman, A., Szunyogh, L., Nordström, L., and Eriksson, O. (2013). Interatomic exchange interactions for finite-temperature magnetism and nonequilibrium spin dynamics. *Phys. Rev. Lett.*, **111**, 127204.

Szpunar, B., and Kozarzewski, B. (1977). The application of CPA to calculations of the mean magnetic moment in the Gd_{1-x}-, Ni_x, Gd_{1-x}-, Fe_x, Gd_{1-x}-, Co_x, and Y_{1-x}-, Co_x intermetallic compounds. *Phys. Status Solidi (b)*, **82**, 205–11.

Şaşıoğlu, E., Schindlmayr, A., Friedrich, C., Freimuth, F., and Blügel, S. (2010). Wannier-function approach to spin excitations in solids. *Phys. Rev. B*, **81**, 054434.

Stöhr, J., and Siegmann, H. C. (2006). *Magnetism: From Fundamentals to Nanoscale Dynamics*. Springer, Berlin, Heidelberg.

Tang, G., and Nolting, W. (2007). Effects of dilution and disorder on magnetism in diluted spin systems. *Phys. Status Solidi (b)*, **244**, 735–47.

Tang, H., Pihal, M., and Mills, D. (1998). Theory of the spin dynamics of bulk Fe and ultrathin Fe(100) films. *J. Magn. Magn. Mater.*, **187**, 23–46.

Tang, W. X., Zhang, Y., Tudosa, I., Prokop, J., Etzkorn, M., and Kirschner, J. (2007). Large wave vector spin waves and dispersion in two monolayer Fe on W(110). *Phys. Rev. Lett.*, **99**, 087202.

Tao, X., Landau, D. P., Schulthess, T. C., and Stocks, G. M. (2005). Spin waves in paramagnetic bcc iron: Spin dynamics simulations. *Phys. Rev. Lett.*, **95**, 087207.

Taroni, A., Bergman, A., Bergqvist, L., Hellsvik, J., and Eriksson, O. (2011). Suppression of standing spin waves in low-dimensional ferromagnets. *Phys. Rev. Lett.*, **107**, 037202.

Thiele, A. A. (1973). Steady-state motion of magnetic domains. *Phys. Rev. Lett.*, **30**, 230–3.

Thomas, L., Moriya, R., Rettner, C., and Parkin, S. S. P. (2010). Dynamics of magnetic domain walls under their own inertia. *Science*, **330**, 1810–13.

Thonig, D. (2013). 'Magnetization Dynamics and Magnetic Ground State Properties from First Principles', PhD thesis, Martin Luther University, Halle.

Thonig, D., Henk, J., and Eriksson, O. (2015). Gilbert-like damping caused by time retardation in atomistic magnetization dynamics. *Phys. Rev. B*, **92**, 104403.

Tokura, Y., Seki, S., and Nagaosa, N. (2014). Multiferroics of spin origin. *Rep. Prog. Phys.*, 77, 076501.

Tsai, S.-H., Lee, H. K., and Landau, D. P. (2005). Molecular and spin dynamics simulations using modern integration methods. *Am. J. Phys.*, **73**, 615.

Tudosa, I., Stamm, C., Kashuba, A. B., King, F., Siegmann, H. C., Stöhr, J., Ju, G., Lu, B., and Weller, D. (2004). The ultimate speed of magnetic switching in granular recording media. *Nature*, **428**, 831–3.

Turek, I., Drchal, V., Kudrnovský, J., Šob, M., and Weinberger, P. (1997). *Electronic Structure of Disordered Alloys, Surfaces and Interfaces*. Springer US, Boston, MA.

Turek, I., Kudrnovský, J., Drchal, V., and Bruno, P. (2006). Exchange interactions, spin waves, and transition temperatures in itinerant magnets. *Philos. Mag.*, **86**, 1713–52.

Turek, I., Kudrnovský, J., Drchal, V., Bruno, P., and Blügel, S. (2003). Ab initio theory of exchange interactions in itinerant magnets. *Phys. Status Solidi (b)*, **236**, 318–24.

Udvardi, L., and Szunyogh, L. (2009). Chiral asymmetry of the spin-wave spectra in ultrathin magnetic films. *Phys. Rev. Lett.*, **102**, 207204.

Udvardi, L., Szunyogh, L., Palotás, K., and Weinberger, P. (2003). First-principles relativistic study of spin waves in thin magnetic films. *Phys. Rev. B*, **68**, 104436.

Újfalussy, B., Lazarovits, B., Szunyogh, L., Stocks, G., and Weinberger, P. (2004). Ab initio spin dynamics applied to nanoparticles: Canted magnetism of a finite Co chain along a Pt(111) surface step edge. *Phys. Rev. B*, **70**, 100404–8.

Újfalussy, B., Wang, X.-D., Nicholson, D. M. C., Shelton, W. A., Stocks, G. M., Wang, Y., and Gyorffy, B. L. (1999). Constrained density functional theory for first principles spin dynamics. *J. Appl. Phys.*, **85**, 4824–6.

Vahaplar, K., Kalashnikova, A. M., Kimel, A. V., Gerlach, S., Hinzke, D., Nowak, U., Chantrell, R., Tsukamoto, A., Itoh, A., Kirilyuk, A., and Rasing, T. (2012). All-optical magnetization reversal by circularly polarized laser pulses: Experiment and multiscale modeling. *Phys. Rev. B*, **85**, 104402.

van Kampen, N. G. (2007). *Stochastic Processes in Physics and Chemistry* (3rd edn). Elsevier, Amsterdam.

van Kranendonk, J., and Van Vleck, J. H. (1958). Spin waves. *Rev. Mod. Phys.*, **30**, 1–23.

van Schilfgaarde, M., Abrikosov, I. A., and Johansson, B. (1999). Origin of the Invar effect in iron-nickel alloys. *Nature*, **400**, 46–9.

Vaz, C. A. F., Bland, J. A. C., and Lauhoff, G. (2008). Magnetism in ultrathin film structures. *Rep. Prog. Phys.*, **71**, 056501.

Visscher, P. B., and Feng, X. (2002). Quaternion-based algorithm for micromagnetics. *Phys. Rev. B*, **65**, 104412.

Vollmer, R., Etzkorn, M., Anil Kumar, P. S., Ibach, H., and Kirschner, J. (2003). Spin-polarized electron energy loss spectroscopy of high energy, large wave vector spin waves in ultrathin fcc Co films on Cu(001). *Phys. Rev. Lett.*, **91**, 147201.

Vollmer, R., Etzkorn, M., Anil-Kumar, P. S., Ibach, H., and Kirschner, J. (2004). Spin-wave excitation in ultrathin Co and Fe films on Cu(001) by spin-polarized electron energy loss spectroscopy. *J. Appl. Phys.*, **95**, 7435.

Vonsovskii, S. V. (1966). *Ferromagnetic Resonance*. Pergamon Press, Oxford.

Wang, D., Weerasinghe, J., and Bellaiche, L. (2012). Atomistic molecular dynamic simulations of multiferroics. *Phys. Rev. Lett.*, **109**, 067203.

Wäppling, R., Häggstrom, L., Ericsson, T., Devanarayanan, S., Karlsson, E., Carlsson, B., and Rundquist, S. (1975). First order magnetic transition, magnetic structure, and vacancy distribution in Fe_2P. *J. Solid State Chem.*, **13**, 258–71.

Wieser, R. (2015). Description of a dissipative quantum spin dynamics with a Landau-Lifshitz/Gilbert like damping and complete derivation of the classical Landau-Lifshitz equation. *Eur. Phys. J. B*, **88**, 77.

Woo, C. H., Wen, H., Semenov, A. A., Dudarev, S. L., and Ma, P. W. (2015). Quantum heat bath for spin-lattice dynamics. *Phys. Rev. B*, **91**, 104306.

Wright, D. C., and Mermin, N. D. (1989). Crystalline liquids: The blue phases. *Rev. Mod. Phys.*, **61**, 385–432.

Wu, R., and Freeman, A. J. (1992). Magnetic properties of Fe overlayers on W(001) and the effects of oxygen adsorption. *Phys. Rev. B*, **45**, 7532–5.

Yablonskii, D. A. (1990). Mean-field theory of magnetic order in cupric oxide. *Phys. C Supercond.*, **171**, 454–6.

Young, A. P. (1998). *Spin Glasses and Random Fields*. World Scientific, Singapore.

Young, D. A. (1991). *Phase Diagrams of the Elements*. University of California Press, Oakland, CA.

Yu, X. Z., Kanazawa, N., Zhang, W. Z., Nagai, T., Hara, T., Kimoto, K., Matsui, Y., Onose, Y., and Tokura, Y. (2012). Skyrmion flow near room temperature in an ultralow current density. *Nat. Commun.*, **3**, 998.

Yu, X. Z., Onose, Y., Kanazawa, N., Park, J. H., Han, J. H., Matsui, Y., Nagaosa, N., and Tokura, Y. (2010). Real-space observation of a two-dimensional skyrmion crystal. *Nature*, **465**, 901–4.

Zakeri, K., and Kirschner, J. (2013). 'Probing Magnons by Spin-Polarized Electrons' in S. O. Demokritov and A. N. Slavin, eds, *Magnonics: From Fundamentals to Applications,* Topics in Applied Physics, vol. 125. Springer, Berlin, Heidelberg, pp. 83–99.

Zakeri, Kh. (2014). Elementary spin excitations in ultrathin itinerant magnets. *Phys. Rep.*, **545**, 47–93.

Zakeri, Kh., Chuang, T.-H., Ernst, A., Sandratskii, L. M., Buczek, P., Qin, H. J., Zhang, Y., and Kirschner, J. (2013). Direct probing of the exchange interaction at buried interfaces. *Nat. Nanotechnol.*, **8**, 853–8.

Zakeri, Kh., Lindner, J., Barsukov, I., Meckenstock, R., Farle, M., von Hörsten, U., Wende, H., Keune, W., Rocker, J., Kalarickal, S., Lenz, K., Kuch, W., Baberschke, K., and Frait, Z. (2007). Spin dynamics in ferromagnets: Gilbert damping and two-magnon scattering. *Phys. Rev. B*, **76**, 104416.

Zakeri, Kh., Zhang, Y., Chuang, T.-H., and Kirschner, J. (2012). Magnon lifetimes on the Fe(110) surface: The role of spin-orbit coupling. *Phys. Rev. Lett.*, **108**, 197205.

Zakeri, Kh., Zhang, Y., and Kirschner, J. (2013). Surface magnons probed by spin- polarized electron energy loss spectroscopy. *J. Electron Spectrosc. Relat. Ph.*, **189**, 157–63.

Zakeri, Kh., Zhang, Y., Prokop, J., Chuang, T.-H., Sakr, N., Tang, W. X., and Kirschner, J. (2010). Asymmetric spin-wave dispersion on Fe(110): Direct evidence of the Dzyaloshinskii-Moriya interaction. *Phys. Rev. Lett.*, **104**, 137203.

Zhang, S., and Li, Z. (2004). Roles of nonequilibrium conduction electrons on the magnetization dynamics of ferromagnets. *Phys. Rev. Lett.*, **93**, 127204.

Zhang, Y., Chuang, T.-H., Zakeri, Kh., and Kirschner, J. (2012). Relaxation time of terahertz magnons excited at ferromagnetic surfaces. *Phys. Rev. Lett.*, **109**, 087203.

Zimmermann, B., Heide, M., Bihlmayer, G., and Blügel, S. (2014). First-principles analysis of a homochiral cycloidal magnetic structure in a monolayer Cr on W(110). *Phys. Rev. B*, **90**, 115427.

Index